T0406755

Lakhmi C. Jain and Ngoc Thanh Nguyen (Eds.)

Knowledge Processing and Decision Making in Agent-Based Systems

Studies in Computational Intelligence, Volume 170

Editor-in-Chief
Prof. Janusz Kacprzyk
Systems Research Institute
Polish Academy of Sciences
ul. Newelska 6
01-447 Warsaw
Poland
E-mail: kacprzyk@ibspan.waw.pl

Further volumes of this series can be found on our homepage: springer.com

Vol. 149. Roger Lee (Ed.)
Software Engineering, Artificial Intelligence, Networking and Parallel/Distributed Computing, 2008
ISBN 978-3-540-70559-8

Vol. 150. Roger Lee (Ed.)
Software Engineering Research, Management and Applications, 2008
ISBN 978-3-540-70774-5

Vol. 151. Tomasz G. Smolinski, Mariofanna G. Milanova and Aboul-Ella Hassanien (Eds.)
Computational Intelligence in Biomedicine and Bioinformatics, 2008
ISBN 978-3-540-70776-9

Vol. 152. Jarosław Stepaniuk
Rough – Granular Computing in Knowledge Discovery and Data Mining, 2008
ISBN 978-3-540-70800-1

Vol. 153. Carlos Cotta and Jano van Hemert (Eds.)
Recent Advances in Evolutionary Computation for Combinatorial Optimization, 2008
ISBN 978-3-540-70806-3

Vol. 154. Oscar Castillo, Patricia Melin, Janusz Kacprzyk and Witold Pedrycz (Eds.)
Soft Computing for Hybrid Intelligent Systems, 2008
ISBN 978-3-540-70811-7

Vol. 155. Hamid R. Tizhoosh and M. Ventresca (Eds.)
Oppositional Concepts in Computational Intelligence, 2008
ISBN 978-3-540-70826-1

Vol. 156. Dawn E. Holmes and Lakhmi C. Jain (Eds.)
Innovations in Bayesian Networks, 2008
ISBN 978-3-540-85065-6

Vol. 157. Ying-ping Chen and Meng-Hiot Lim (Eds.)
Linkage in Evolutionary Computation, 2008
ISBN 978-3-540-85067-0

Vol. 158. Marina Gavrilova (Ed.)
Generalized Voronoi Diagram: A Geometry-Based Approach to Computational Intelligence, 2009
ISBN 978-3-540-85125-7

Vol. 159. Dimitri Plemenos and Georgios Miaoulis (Eds.)
Artificial Intelligence Techniques for Computer Graphics, 2009
ISBN 978-3-540-85127-1

Vol. 160. P. Rajasekaran and Vasantha Kalyani David
Pattern Recognition using Neural and Functional Networks, 2009
ISBN 978-3-540-85129-5

Vol. 161. Francisco Baptista Pereira and Jorge Tavares (Eds.)
Bio-inspired Algorithms for the Vehicle Routing Problem, 2009
ISBN 978-3-540-85151-6

Vol. 162. Costin Badica, Giuseppe Mangioni, Vincenza Carchiolo and Dumitru Dan Burdescu (Eds.)
Intelligent Distributed Computing, Systems and Applications, 2008
ISBN 978-3-540-85256-8

Vol. 163. Pawel Delimata, Mikhail Ju. Moshkov, Andrzej Skowron and Zbigniew Suraj
Inhibitory Rules in Data Analysis, 2009
ISBN 978-3-540-85637-5

Vol. 164. Nadia Nedjah, Luiza de Macedo Mourelle, Janusz Kacprzyk, Felipe M.G. França and Alberto Ferreira de Souza (Eds.)
Intelligent Text Categorization and Clustering, 2009
ISBN 978-3-540-85643-6

Vol. 165. Djamel A. Zighed, Shusaku Tsumoto, Zbigniew W. Ras and Hakim Hacid (Eds.)
Mining Complex Data, 2009
ISBN 978-3-540-88066-0

Vol. 166. Constantinos Koutsojannis and Spiros Sirmakessis (Eds.)
Tools and Applications with Artificial Intelligence, 2009
ISBN 978-3-540-88068-4

Vol. 167. Ngoc Thanh Nguyen and Lakhmi C. Jain (Eds.)
Intelligent Agents in the Evolution of Web and Applications, 2009
ISBN 978-3-540-88070-7

Vol. 168. Andreas Tolk and Lakhmi C. Jain (Eds.)
Complex Systems in Knowledge-based Environments: Theory, Models and Applications, 2009
ISBN 978-3-540-88074-5

Vol. 169. Nadia Nedjah, Luiza de Macedo Mourelle and Janusz Kacprzyk (Eds.)
Innovative Applications in Data Mining, 2009
ISBN 978-3-540-88044-8

Vol. 170. Lakhmi C. Jain and Ngoc Thanh Nguyen (Eds.)
Knowledge Processing and Decision Making in Agent-Based Systems, 2009
ISBN 978-3-540-88048-6

Lakhmi C. Jain
Ngoc Thanh Nguyen
(Eds.)

Knowledge Processing and Decision Making in Agent-Based Systems

Prof. Lakhmi C. Jain
University of South Australia
Adelaide City
Mawson Lakes Campus
South Australia SA 5095
Australia
Email: Lakhmi.jain@unisa.edu.au

Prof. Ngoc Thanh Nguyen
Institute of Informatics
Wroclaw University of Technology
Str. Janiszewskiego 11/17
50-370 Wroclaw
Poland
Email: Ngoc-Thanh.Nguyen@pwr.wroc.pl

ISBN 978-3-540-88048-6 e-ISBN 978-3-540-88049-3

DOI 10.1007/978-3-540-88049-3

Studies in Computational Intelligence ISSN 1860949X

Library of Congress Control Number: 2008938351

Typeset & Cover Design: Scientific Publishing Services Pvt. Ltd., Chennai, India.

Printed in acid-free paper

9 8 7 6 5 4 3 2 1

springer.com

Preface

Agent technology has been successfully applied in the last decade to numerous business applications ranging from entertainment and education to electronic commerce and reliable intelligent manufacturing [1]. Today, agent technology has touched virtually every field in the world, some with a limited success. A number of researchers are engaged in the design, development and innovative applications of intelligent agents due to their characteristics such as autonomous behaviour and their ability to work cooperatively in teams. Knowledge processing plays an important part in the implementation of intelligent machines which can mimic the human intelligence in a limited way. The characteristics such as knowledge extraction from data, learning and teaming are important to realize intelligent decision support systems. Decision support systems can be benefitted by the better models of knowledge processing and intelligent agent for delivering the best decisions to the user.

This book presents a sample of research results in the knowledge processing and decision making in agent-based systems. The chapters range from theoretical foundations to the practical applications.

We wish to express our sincere gratitude to the authors and reviewers for their contributions. Thanks are due to the Springer-Verlag for their editorial assistance. The editorial assistance provided by the SCI Data Processing Team of Scientific Publishing Services Private Limited is acknowledged.

Lakhmi C. Jain
Ngoc Thanh Nguyen

Reference

[1] Nishida, T.: Foreword in a book on Intelligent Agents and Their Applications. In: Jain, L.C., et al. (eds.) Springer, Heidelberg (2002)

Contents

1 Innovations in Knowledge Processing and Decision Making in Agent-Based Systems

Lakhmi C. Jain[1], Chee Peng Lim[2], and Ngoc Thanh Nguyen[3]

[1] School of Electrical & Information Engineering
University of South Australia, Australia
[2] School of Electrical & Electronic Engineering
University of Science Malaysia, Malaysia
[3] Institute of Information Science and Engineering
Wroclaw University of Technology, Poland

Abstract. This chapter introduces knowledge processing and decision making using agent-based technologies. The importance of creating effective and efficient computerized systems for extracting information and processing knowledge as well as for supporting decision making activities is highlighted. Then, an overview covering agent-based software tools and development methodologies, and usability and challenges of agent-based systems in industrial applications is presented. The contribution of each chapter included in this book is also described.

1.1 Introduction

Most real life problems are complex and multi-facet, and involve many criteria in the decision making process. As a result, researchers have investigated and proposed a variety of methodologies and techniques to design and develop computerised systems for decision support applications. In general, a Decision Support System (DSS) is a computerized information system that supports decision-making activities in various domains such as business, finance, management, manufacturing, and biomedicine. A useful DSS is able to compile and extract meaningful information from raw data and to suggest potential solutions for users to make informed decisions.

A useful conceptual framework for classifying and describing DSSs is proposed in [1]. Five generic DSS types are identified and defined based upon the dominant technology component including communications-driven, data-driven, document-driven, knowledge-driven, and model-driven DSSs. A DSS can be developed for specific or general-purpose applications, and can be used by individuals or groups. The enabling technology of the DSS can be a mainframe computer, a client/server LAN, a spreadsheet, or a web-based architecture [1].

Many methodologies have been proposed to help build and understand DSSs. One of the approaches for developing DSS is agent-based technologies. From the literature, a lot of agent-based DSSs can be found, and a lot of successful applications of agent-based DSSs are reported. In this book, a small fraction of DSSs that utilize agent-based technologies for knowledge processing and decision making are presented. The main aim is to share and disseminate information pertaining to recent

L.C. Jain, N.T. Nguyen (Eds.): Knowl. Proc. & Dec. Mak. in Agent-Based Sys., SCI 170, pp. 1–12.
springerlink.com

advancements in theoretical and practical aspects of agent-based DSSs for tackling real-life problems in various domains.

1.2 Knowledge Processing and Decision Making

Humans make decisions, either consciously or sub-consciously in daily activities. The process of decision making is so essential that it has become an integral part of life, and automated tools and systems for decision support are always in demand. The demand is exacerbated by the rapid development and wide-spread usage of the internet as a resource for information and knowledge sharing and reuse.

The world-wide-web contains many heterogeneous data sources ranging from text documents to multimedia images; from audio files to video streams. However, it is difficult for users to extract relevant materials from such a complex environment, due to information overload [2, 3]. In general, data are raw, numeric records. Information is concerned with data that have been processed and analysed, and knowledge covers actionable information that is comprehensible for humans to reason and infer decisions. Therefore, the heterogeneous data sources from the internet, or from other sources, need to be processed and analysed sensibly, and automatically, so that meaningful knowledge that are relevant and useful to decision makers can be retrieved.

For a computerized DSS to be useful for practical implementation, several crucial properties that enable the DSS to combine different types of data and information from various sources in a seamlessly manner and without much user intervention should be established. These properties are related to knowledge processing and decision making activities such as knowledge representation, knowledge management and reuse, reasoning and inference techniques, as well as risk analysis. In this field, one of the emerging technologies to facilitate and manipulate knowledge processing and decision making in DSS is agent-based systems, as described in [4, 5, 6].

1.3 Agent-Based Systems

The rapid advancement of agent-based technologies has opened up the way for the development of a new and exciting paradigm for the establishment of intelligent software systems operating in dynamic and complex environments. There are a lot of areas in agent-based systems that have attracted attention of researchers. These include formal frameworks for collaboration and cooperation between agents, methodologies for development of multi-agent systems, as well as model and techniques for managing inter-agent relationships (e.g., belief, trust, and reputation).

Agent-based technologies have emerged from the field of distributed artificial intelligence [7]. Before embarking on the overview of agent-based technologies, a number of terminologies of agents that are useful for understanding their problem-solving and decision making characteristics, as defined in [8], are provided. An (intelligent) agent is an autonomous, problem-solving computational entity capable of operating in dynamic and open environments. Agent properties refer to the fundamental characteristics of agents, which include autonomous decision making and response, with the ability to communicate, negotiate, and cooperate with other agents. Agent-based solutions to a decision-making problem explore agents as autonomous decision-making units and their interactions to achieve global goals. Other agent-related terminologies and definitions can be found in [8].

In [9], three main domains of research and development in agent-based technologies are discussed: (i) development of tools (environments, languages, specific components); (ii) development of methodologies; and (iii) search for and development of reusable solutions and components. In this aspect, some available software tools for development and deployment of agent-based systems include MASDK [10], JACK [11], JADE [12], and AgentBuilder [13]. There are a number of development methodologies for agent-based system. They include Gaia [14], MaSE [15], Tropos [16], Prometheus [17], and MESSAGE [18]. On the other hand, some useful web resources pertaining to software tools as well as publications of agent-based systems are easily accessible, refer [19-21].

While agent properties are useful for solving complex, real-world industrial problems, there are some challenges faced by the practical implementation of agent-based systems. The challenges arising from application of agent-based systems include ontology management (e.g. negotiation [22], perspectives [23]), reusable agent repository, and spectral analysis automation [24]. Besides, although agent-based systems are capable of addressing much richer and broader range of possible tasks, e.g. for qualification and deployment of flight maintenance systems, the process of verification and validation of agent-based software systems becomes considerably harder [25].

In industrial environments, agent-based solutions are investigated for tackling problems in three main domains [8]. They are real-time manufacturing control problems, complex operation management problems, and virtual enterprise problems. As pointed out in [8], five key industrial application areas that are suitable for deploying agent-based solutions are real-time control of high-volume, high variety, discrete manufacturing operations; monitoring and control of physically highly distributed systems; production management of frequently disrupted operations; coordination of organizations with conflicting goals; and frequently reconfigured, automated environments. On one hand, the benefits of adopting agent-based solutions in industrial environments include feasibility, robustness and flexibility, reconfigurablity, and redeployability. On the other hand, the barriers include cost, guarantees for operational performance, and scalability [8].

1.4 Chapters Included in This Book

The book includes thirteen chapters. Chapter one introduces knowledge processing and decision making in agent-based systems. It also outlines the contributions made in the book. Chapter two presents a novel real-world HTN planning agent. It is demonstrated by the authors that the proposed approach is efficient than other replanning methods. Chapter three presents a mobile agent-based system for distributed software maintenance. The authors have considered scenarios such as maintenance process, software modification in their case studies. Chapter four presents a combination of artificial intelligence mechanisms and computational economics concepts for tackling problems in telecommunication services. The mechanisms and concepts are implemented in agent-based electronic markets, and novel discovery and negotiation models to improve the efficiency of the created multi-agent system are described.

Chapter five presents a framework for multi-agent reasoning based on paraconsistent annotated system. It is demonstrated that the proposed framework is ideal for dealing with imprecise, inconsistent knowledge, conflicting beliefs and awareness. Chapter six presents an agent-based negotiation platform for a collaborative decision making in construction supply chain. It is shown that the platform improves the collaborative working environment. Chapter seven presents a novel method of using meta-agents for searching services using event driven algorithms. The novelty of the method is demonstrated.

Chapter eight presents a generic mobile agent towards ambient intelligence. It is shown that the proposed approach offers advantages such as more responsive interface between the users and the computing infrastructure. Chapter nine presents a systematic approach of developing actionable trading strategies. The methods are validated using market data and the merit is demonstrated. Chapter ten illustrates the agent-based uncertainty theory and its connection with quantum mechanics where agents are interpreted in terms of particles.

Chapter eleven presents the development of an agent transportation layer adaptation system model. The conceptual model is based on a blackboard architecture where each element represents a block of functionality required to automate a process in order to complete a specific task. Chapter twelve proposes to use software agent-based negotiation as a means to compose services to meet the requirement of consumers. The final chapter presents the intelligent decision making process required to realise an autonomous decentralised flexible manufacturing system using automated guided vehicles and matching centres. The development of a real-time reasoning model and its improvement are presented.

1.5 Summary

This chapter has presented an overview of knowledge processing and decision making using agent-based systems. It has been explained that information overload has become an issue in today's digital era, and DSSs are needed to help extract and elicit meaningful and actionable information and knowledge so that decision makers can reach informed decisions. In this aspect, agent-based technologies offer a suitable platform for developing DSSs for applications in various domains. In addition, the available software tools and development methodologies of agent-based systems have been described. Issues related to the practicality and challenges faced by industrial adoption of agent-based solutions have also been discussed.

References

[1] Power, D.J.: Specifying an expanded framework for classifying and describing decision support systems. Communications of the Association for Information Systems 13, 158–166 (2004)

[2] Maes, P.: Agents that reduce work and information overload. Communications of the ACM 37, 31–40 (1994)

[3] Shaw, N.G., Mian, A., Yadav, S.B.: A comprehensive agent-based architecture for intelligent information retrieval in a distributed heterogeneous environment. Decision Support Systems 32, 401–415 (2002)
[4] Hess, T.J., Rees, L.P., Rakes, T.R.: Using autonomous software agents to create the next generation of decision support systems. Decision Sciences 31, 1–31 (2000)
[5] Heintz, F., Doherty, P.: DyKnow: An Approach to Middleware for Knowledge Processing. Journal of Intelligent and Fuzzy Systems 15, 3–13 (2004)
[6] Heintz, F., Doherty, P.: A knowledge processing middleware framework and its relation to the JDL data fusion model. Journal of Intelligent and Fuzzy Systems 17, 335–351 (2006)
[7] Jennings, N.R., Sycara, K., Wooldridge, M.: A roadmap of agent research and development. International Journal of Autonomous Agents and Multi-Agent Systems 1, 7–38 (1998)
[8] Marik, V., McFarlane, D.: Industrial adoption of agent-based technologies. IEEE Intelligent Systems 20, 27–35 (2005)
[9] Karsaev, O.: Technology of agent-based decision making system development. In: Gorodetsky, V., Liu, J., Skormin, V.A. (eds.) AIS-ADM 2005. LNCS, vol. 3505, pp. 108–122. Springer, Heidelberg (2005)
[10] Gorodetski, V., Karsaev, O., Samoylov, V., Konushy, V., Mankov, E., Malyshev, A.: Multi agent system development kit: MAS software tool implementing Gaia methodology. In: Shi, Z., He, O. (eds.) IIP 2004, pp. 69–78. Springer, Heidelberg (2004)
[11] `http://www.agent-software.com/shared/products/index.html` (accessed on July 25, 2008)
[12] `http://jade.tilab.com/` (accessed on July 25, 2008)
[13] `http://www.agentbuilder.com/` (accessed on July 25, 2008)
[14] Wooldridge, M., Jennings, N.R., Kinny, D.: The Gaia methodology for agent-oriented analysis and design. Journal of Autonomous Agents and Multi-Agent Systems 3, 285–312 (2000)
[15] DeLoach, S.A., Wood, M.F., Sparkman, C.H.: Multi-agent Systems Engineering. International Journal of Software Engineering and Knowledge Engineering 11, 231–258 (2001)
[16] Bresciani, P., Giorgini, P., Giunchiglia, F., Mylopolous, J., Perini, A.: Tropos: an agent-oriented software development methodology. Journal of Autonomous Multi-Agent Systems 8, 203–236 (2004)
[17] Padgham, L., Winikoff, M.: Prometheus: a methodology for developing intelligent agents. In: Giunchiglia, F., Odell, J.J., Weiss, G. (eds.) AOSE 2002. LNCS, vol. 2585, pp. 174–185. Springer, Heidelberg (2003)
[18] Evans, R., Kearny, P., Stark, J., Caire, G., Garijo, F., Gomez-Sanz, J.J., Leal, F., Chainho, P., Massonet, P.: MESSAGE: methodology for engineering systems of software agents, Technical Report P907, EURESCOM (2001)
[19] `http://agents.umbc.edu/` (accessed on July 25, 2008)
[20] `http://www.agentlink.org/index.php` (accessed on July 25, 2008)
[21] `http://www.sics.se/isl/abc/survey.html` (accessed on July 25, 2008)
[22] Bailin, S., Truszkowski, W.: Ontology Negotiation between intelligent information agents. The Knowledge Engineering Review 17, 7–19 (2002)
[23] Bailin, S., Truszkowski, W.: Perspectives: an analysis of multiple viewpoints in agent-based systems. In: van Elst, L., Dignum, V., Abecker, A. (eds.) AMKM 2003. LNCS, vol. 2926, pp. 368–387. Springer, Heidelberg (2004)

[24] Truszkowski, W.: Challenges arising from applications of agent-based system. In: Hinchey, M.G., Rash, J.L., Truszkowski, W.F., Rouff, C.A., Gordon-Spears, D.F. (eds.) FAABS 2002. LNCS (LNAI), vol. 2699, pp. 269–273. Springer, Heidelberg (2003)
[25] Pecheur, C.: Challenges arising from applications. In: Hinchey, M.G., Rash, J.L., Truszkowski, W.F., Rouff, C.A., Gordon-Spears, D.F. (eds.) FAABS 2002. LNCS (LNAI), vol. 2699, pp. 236–238. Springer, Heidelberg (2003)

1.6 Resources

Following is a sample of resources on knowledge processing, intelligent decision making and multiagents systems.

1.6.1 Journals

- IEEE Intelligent Systems, IEEE Press, USA
  `www.computer.org/intelligent/`
- International Journal of Knowledge-Based Intelligent Engineering Systems, IOS Press, The Netherlands.
  `http://www.kesinternational.org/journal/`
- International Journal of Hybrid Intelligent Systems, IOS Press, The Netherlands.
  `http://www.iospress.nl/html/14485869.html`
- Intelligent Decision Technologies: An International Journal, IOS Press, The Netherlands.
  `http://www.iospress.nl/html/18724981.html`

1.6.2 Special Issue of Journals

- Jain, L.C., Lim, C.P. and Nguyen, N.T. (Guest Editors), Recent Advances in Intelligent Paradigms Fusion and Their Applications, International Journal of Hybrid Intelligent Systems, Volume 5, Issue 3, 2008.
- Lim, C.P., Jain, L.C., Nguyen, N.T. and Balas, V. (Guest Editors), Advances in Computational Intelligence Paradigms and Applications, An International Journal on Fuzzy Optimization and Decision Making, Kluwer Academic Publisher, in press.
- Nguyen, N.T., Lim, C.P., Jain, L.C. and Balas, V.E. (Guest Editors), Theoretical Advances and Applications of Intelligent Paradigms, Journal of Intelligent and Fuzzy Systems, IOS Press, Volume 19, Issue 6, 2008, in press.
- Abraham, A., Jarvis, D., Jarvis, J. and Jain, L.C. (Guest Editors), Special issue on Innovations in agents: An International Journal on Multiagent and Grid Systems, IOS Press, Volume 4, Issue 4, 2008, in press.
- Palade, V. and Jain, L.C. (Guest Editors), Practical Applications of Neural Networks, Journal of "Neural Computing and Applications", Springer-Verlag, Germany, Volume 14, No. 2, 2005.
- Abraham, A. and Jain, L.C. (Guest Editors), Computational Intelligence on the Internet, Journal of Network and Computer Applications, Elsevier Publishers, Volume 28, Number 2, 2005.

- Abraham, A. and Jain, L.C. (Guest Editors), Special issue on Optimal Knowledge Mining, Journal of Fuzzy Optimization and Decision Making, Kluwer Academic Publishers, Volume 3, Number 2, 2004.
- Palade, V., Ghaoui, C. and Jain, L.C. (Guest Editors), Special issue on Intelligent Instructional Environments, Journal of Interactive Technology and Smart Education,Troubador Publishing Ltd, UK, Volume 1, Issue 3, August 2004.
- Palade, V., Ghaoui, C. and Jain, L.C. (Guest Editors), Special issue on Engineering Applications of Computational Intelligence, Journal of Intelligent and Fuzzy systems, IOS Press, Volume 15, Number 3, 2004.
- Alahakoon, D., Abraham, A. and Jain, L.C. (Guest Editors), Special issue on Neural Networks for Enhanced Intelligence, Neural Computing and Applications, Springer-Verlag, UK, Volume 13, No. 2, June 2004.
- Abraham, A., Jonkar, I., Barakova, E., Jain, R. and Jain, L.C. (Guest Editors), Special issue on Hybrid Neurocomputing, Neurocomputing, Elsevier, The Netherlands, Volume 13, No. 2, June 2004.
- Abraham, A. and Jain, L.C., (Guest Editors), Special issue on Knowledge Engineering, Journal of Intelligent and Fuzzy Systems, The IOS Press, The Netherlands, Volume 14, Number 3, 2003
- Jain, L.C. (Guest Editor), Special issue on Fusion of Neural Nets, Fuzzy Systems and Genetic Algorithms in Industrial Applications, IEEE Transactions on Industrial Electronics, USA, December 1999.
- De Silva, C. and Jain, L.C. (Guest Editors), Special Issue on Intelligent Electronic Systems, Engineering Applications of Artificial Intelligence, an international journal, USA, January 1998.
- Jain, L.C. (Guest Editor), Special issue on Intelligent Systems: Design and Applications, Journal of Network and Computer Applications (An International Journal published by Academic Press, England). Vol. 2, April 1996.
- Jain, L.C. (Guest Editor), Special issue on Intelligent Systems: Design and Applications, Journal of Network and Computer Applications (An International Journal published by Academic Press, England). Vol. 1, January, 1996.

1.6.3 Conferences

- AAAI Conference on Artificial Intelligence
  `www.aaai.org/aaai08.php`
- KES International Conference Series
  `www.kesinternational.org/`

1.6.4 Conference Proceedings

- Lovrek, I., Howlett, R.J. and Jain, L.C. (Editors), Knowledge-Based Intelligent Information and Engineering Systems, Lecture Notes in Artificial Intelligence, **Volume 1, LNAI**, KES 2008, Springer-Verlag, Germany, 2008.
- Lovrek, I., Howlett, R.J. and Jain, L.C. (Editors), Knowledge-Based Intelligent Information and Engineering Systems, Lecture Notes in Artificial Intelligence, **Volume 2, LNAI**, KES 2008, Springer-Verlag, Germany, 2008.

- Lovrek, I., Howlett, R.J. and Jain, L.C. (Editors), Knowledge-Based Intelligent Information and Engineering Systems, Lecture Notes in Artificial Intelligence, **Volume 3, LNAI**, KES 2008, Springer-Verlag, Germany, 2008.
- Nguyen, N.T., Jo, G.S., Howlett, R.J. and Jain, L.C. (Editors), Agents and Multi-Agents Systems: Technologies and Applications, Lecture Notes in Artificial Intelligence, Springer-Verlag, Germany, 2008.
- Apolloni, B., Howlett, R.J. and Jain, L.C. (Editors), Knowledge-Based Intelligent Information and Engineering Systems, Lecture Notes in Artificial Intelligence, **Volume 1, LNAI 4692**, KES 2007, Springer-Verlag, Germany, 2007.
- Apolloni, B.,Howlett, R.J.and Jain, L.C. (Editors), Knowledge-Based Intelligent Information and Engineering Systems, Lecture Notes in Artificial Intelligence, **Volume 2, LNAI 4693,** , KES 2007, Springer-Verlag, Germany, 2007.
- Apolloni, B.,Howlett, R.J.and Jain, L.C. (Editors), Knowledge-Based Intelligent Information and Engineering Systems, Lecture Notes in Artificial Intelligence, **Volume 3, LNAI 4694,** KES 2007, Springer-Verlag, Germany, 2007.
- Nguyen, N.T., Grzech, A., Howlett, R.J. and Jain, L.C., Agents and Multi-Agents Systems: Technologies and Applications, Lecture Notes in artificial Intelligence, **LNAI 4696,** Springer-Verlag, Germany, 2007.
- Howlett, R.P., Gabrys, B. and Jain, L.C. (Editors), Knowledge-Based Intelligent Information and Engineering Systems, Lecture Notes in Artificial Intelligence, KES 2006, Springer-Verlag, Germany, Vol. **4251**, 2006.
- Howlett, R.P., Gabrys, B. and Jain, L.C. (Editors), Knowledge-Based Intelligent Information and Engineering Systems, Lecture Notes in Artificial Intelligence, KES 2006, Springer-Verlag, Germany, Vol. **4252**, 2006.
- Howlett, R.P., Gabrys, B. and Jain, L.C. (Editors), Knowledge-Based Intelligent Information and Engineering Systems, Lecture Notes in Artificial Intelligence, KES 2006, Springer-Verlag, Germany, Vol. **4253**, 2006.
- Liao, B.-H., Pan, J.-S., Jain, L.C., Liao, M., Noda, H. and Ho, A.T.S., Intelligent Information Hiding and Multimedia Signal Processing, IEEE Computer Society Press, USA, 2007. ISBN: 0-7695-2994-1.
- Khosla, R., Howlett, R.P., and Jain, L.C. (Editors), Knowledge-Based Intelligent Information and Engineering Systems, Lecture Notes in Artificial Intelligence, KES 2005, Springer-Verlag, Germany, Vol. **3682**, 2005.
- Skowron, A., Barthes, P., Jain, L.C., Sun, R.,Mahoudeaux, P., Liu, J. and Zhong, N.(Editors), Proceedings of the 2005 IEEE/WIC/ACM International Conference on Intelligent Agent Technology, Compiegne, France, IEEE Computer Society Press, USA, 2005.
- Khosla, R., Howlett, R.P., and Jain, L.C. (Editors), Knowledge-Based Intelligent Information and Engineering Systems, Lecture Notes in Artificial Intelligence, KES 2005, Springer-Verlag, Germany, Vol. **3683**, 2005.
- Khosla, R., Howlett, R.P., and Jain, L.C. (Editors), Knowledge-Based Intelligent Information and Engineering Systems, Lecture Notes in Artificial Intelligence, KES 2005, Springer-Verlag, Germany, Vol. **3684**, 2005.
- Khosla, R., Howlett, R.P., and Jain, L.C. (Editors), Knowledge-Based Intelligent Information and Engineering Systems, Lecture Notes in Artificial Intelligence, KES 2005, Springer-Verlag, Germany, Vol. **3685**, 2005.

- Negoita, M., Howlett, R.P., and Jain, L.C. (Editors), Knowledge-Based Intelligent Engineering Systems, KES 2004, Lecture Notes in Artificial Intelligence, Vol. **3213**, Springer, 2004.
- Negoita, M., Howlett, R.P., and Jain, L.C. (Editors), Knowledge-Based Intelligent Engineering Systems, KES 2004, Lecture Notes in Artificial Intelligence, Vol. **3214**, Springer, 2004
- Negoita, M., Howlett, R.P., and Jain, L.C. (Editors), Knowledge-Based Intelligent Engineering Systems, KES 2004, Lecture Notes in Artificial Intelligence, Vol. **3215**, Springer, 2004
- Murase, K., Jain, L.C., Sekiyama, K. and Asakura, T. (Editors), Proceedings of the Fourth International Symposium on Human and Artificial Intelligence Systems, University of Fukui, Japan, 2004.
- Palade, V., Howlett, R.P., and Jain, L.C. (Editors), Knowledge-Based Intelligent Engineering Systems, Lecture Notes in Artificial Intelligence, Vol. **2773**, Springer, 2003
- Palade, V., Howlett, R.P., and Jain, L.C. (Editors), Knowledge-Based Intelligent Engineering Systems, Lecture Notes in Artificial Intelligence, Vol. **2774**, Springer, 2003
- Damiani, E., Howlett, R.P., Jain, L.C. and Ichalkaranje, N. (Editors), Proceedings of the Fifth International Conference on Knowledge-Based Intelligent Engineering Systems, **Volume 1**, IOS Press, The Netherlands, 2002.
- Damiani, E., Howlett, R.P., Jain, L.C. and Ichalkaranje, N. (Editors), Proceedings of the Fifth International Conference on Knowledge-Based Intelligent Engineering Systems, **Volume 2**, IOS Press, The Netherlands, 2002.
- Baba, N., Jain, L.C. and Howlett, R.P. (Editors), Proceedings of the Fifth International Conference on Knowledge-Based Intelligent Engineering Systems (KES'2001), **Volume 1,** IOS Press, The Netherlands, 2001.
- Baba, N., Jain, L.C. and Howlett, R.P. (Editors), Proceedings of the Fifth International Conference on Knowledge-Based Intelligent Engineering Systems (KES'2001), **Volume 2,** IOS Press, The Netherlands, 2001.
- Howlett, R.P. and Jain, L.C.(Editors), Proceedings of the Fourth International Conference on Knowledge-Based Intelligent Engineering Systems, IEEE Press, USA, 2000. **Volume 1.**
- Howlett, R.P. and Jain, L.C.(Editors), Proceedings of the Fourth International Conference on Knowledge-Based Intelligent Engineering Systems, IEEE Press, USA, 2000. **Volume 2**.
- Jain, L.C.(Editor), Proceedings of the Third International Conference on Knowledge-Based Intelligent Engineering Systems, IEEE Press, USA, 1999.
- Jain, L.C. and Jain, R.K. (Editors), Proceedings of the Second International Conference on Knowledge-Based Intelligent Engineering Systems, **Volume 1**, IEEE Press, USA, 1998.
- Jain, L.C. and Jain, R.K. (Editors), Proceedings of the Second International Conference on Knowledge-Based Intelligent Engineering Systems, **Volume 2**, IEEE Press, USA, 1998.

- Jain, L.C. and Jain, R.K. (Editors), Proceedings of the Second International Conference on Knowledge-Based Intelligent Engineering Systems, **Volume 3**, IEEE Press, USA, 1998.
- Jain, L.C. (Editor), Proceedings of the First International Conference on Knowledge-Based Intelligent Engineering Systems, **Volume 1**, IEEE Press, USA, 1997.
- Jain, L.C. (Editor), Proceedings of the First International Conference on Knowledge-Based Intelligent Engineering Systems, **Volume 2**, IEEE Press, USA, 1997.
- Narasimhan, V.L., and Jain, L.C. (Editors), The Proceedings of the Australian and New Zealand Conference on Intelligent Information Systems, IEEE Press, USA, 1996.

1.6.5 Book Series

- Advanced Intelligence and Knowledge Processing, Springer-Verlag, Germany `www.springer.com/series/4738`
- Computational Intelligence and its Applications Series, Idea group Publishing, USA `http://www.igi-pub.com/bookseries/details.asp?id=5`
- The CRC Press International Series on Computational Intelligence, The CRC Press, USA
- Advanced Information Processing, Springer-Verlag, Germany.
- Knowledge-Based Intelligent Engineering Systems Series, IOS Press, The Netherlands. `http://www.kesinternational.org/bookseries.php`
- International series on Natural and artificial Intelligence, AKI. `http://www.innoknowledge.com`
- World Scientific Book Series on Innovative Intelligence `http://www.worldscibooks.com/series/sii_series.shtml`

1.6.6 Books

- Phillips-Wren, G., Ichalkaranje, N. and Jain, L.C. (Editors), Intelligent Decision Making-An AI-Based Approach, Springer-Verlag, 2008.
- Fulcher, J. and Jain, L.C., Computational Intelligence: A Compendium, Springer-Verlag, 2008.
- Phillips-Wren, G. and Jain, L.C. (Editors), Intelligent Decision Support Systems in Agent-Mediated Environments, IOS Press, The Netherlands, 2005.
- Khosla, R., Ichalkaranje, N. and Jain, L.C. (Editors), Design of Intelligent Multi-Agent Systems, Springer-Verlag, Germany, 2005.
- Fulcher, J. and Jain, L.C. (Editors), Applied Intelligent Systems, Springer-Verlag, Germany, 2004.
- Resconi, G. and Jain, L.C., Intelligent Agents: Theory and Applications, Springer-Verlag, Germany, 2004.
- Howlett, R., Ichalkaranje, N., Jain, L.C. and Tonfoni, G. (Editors), Internet-Based Intelligent Information Processing, World Scientific Publishing Company Singapore, 2002.

- Jain, L.C., et al. (Editors), Intelligent Agents and Their Applications, Springer-Verlag, Germany, 2002.
- Jain, L.C. (Editor), Soft Computing Techniques in Knowledge-Based Intelligent Engineering Systems, Springer-Verlag, Germany, 1997.
- Mentzas, G., et al., Knowledge Asset Management, Springer-Verlag, London, 2003.
- Nguyen, N.T., Advanced Methods for Inconsistent Knowledge Management, Springer-Verlag, London, 2008.
- Meisels, A., Distributed Search by Constrained Agents, Springer-Verlag, London, 2008.
- Maloof, M.A. (Editor), Machine Learning and Data Mining for Computer Security, Springer-Verlag, London, 2006.
- Loia, V. (Editor), Soft Computing Agents, IOS Press, The Netherlands.
- Namatame, A., et al. (Editors), Agent-Based Approaches in Economic and Social Complex Systems, IOS Press, The Netherlands.

1.6.7 Book Chapters

- Pedrycz, W., Ichalkaranje, N., Phillips-Wren, G., and Jain, L.C., Introduction to Computational Intelligence for Decision Making, Springer-Verlag, 2008, pp. 75-93, Chapter 3.
- Tweedale, J., Ichalkaranje, N., Sioutis, C., Urlings, P. and Jain, L.C., Future Directions: Building a Decision Making Framework using Agent Teams, Springer-Verlag, 2008, pp. 381-402, Chapter 14.
- Virvou, M. and Jain, L.C., Intelligent Interactive Systems in Knowledge-Based Environments: An Introduction, Springer-Verlag, 2008, pp. 1-8, Chapter 1.
- Tran, C., Abraham, A. and Jain, L., Soft Computing Paradigms and Regression Trees in Decision Support Systems, in Advances in Applied Artificial Intelligence, Idea Group Publishing, 2006, pp. 1-28, Chapter 1.
- Jarvis, B., Jarvis, D. and Jain, L., Teams in Multi-Agent Systems, in IFIP International Federation for Information Processing, Vol. **228**, Intelligent Information Processing III, Springer, 2006, pp. 1-10, Chapter 1.

1.6.8 Research Papers

- Minz, S. and Jain, R., Refining Decision Tree Classifiers using Rough Set Tools, International Journal of Hybrid Intelligent Systems, IOS Press, Volume 2, Number 2, 2005, pp. 133-148.
- Ren, X., Thompson, H.A., and Fleming, P.J., An Agent-based System for Distributed Fault Diagnosis, International Journal of Knowledge-Based and Intelligent Engineering Systems, IOS Press, The Netherlands, Volume 10, Number 5, 2005.
- Tweedale, J., Ichalkaranje, N., Sioutis, C., Urlings, P. and Jain, L.C., Building a Decision Making Framework using Agent Teams, International Journal of Intelligent Decision Technologies, IOS Press, The Netherlands, Volume 1, Number 4, 2007, pp. 175-181.

- Zaki, Y. and Pierre, S., Mobile Agents in Distributed Meeting Scheduling: A Case Study for Distributed Applications, International Journal of Intelligent Decision Technologies, IOS Press, The Netherlands, Volume 1, Numbers 1-2, 2007, pp. 71-82.
- Kasap, Z. and Magnenat-Thalmann, Intelligent Virtual Humans with Autonomy and Personality: State-of-the-Art, International Journal of Intelligent Decision Technologies, IOS Press, The Netherlands, Volume 1, Numbers 1-2, 2007, pp. 3-15.
- Lambert, D. and Scholz, J., Ubiquitous Command and Control, International Journal of Intelligent Decision Technologies, IOS Press, The Netherlands, Volume 1, Numbers 3, 2007, pp. 157-173.
- Balbo, F. and Pinson, S., A Transportation Decision Support System in Agent-Based Environment, International Journal of Intelligent Decision Technologies, IOS Press, The Netherlands, Volume 1, Numbers 3, 2007, pp. 97-115.
- Loia, V. and Senatore, S., A Synergistic Approach to Efficient web Searching, International Journal of Intelligent Decision Technologies, IOS Press, The Netherlands, Volume 1, Numbers 1-2, 2007, pp. 83-98.
- Caragea, D., Silvescu, A. and Honavar, V., A Framework for Learning from Distributed Data using Sufficient Statistics and its Application to Learning Decision Trees, International Journal of Hybrid Intelligent Systems, IOS Press, Volume 1, Number 2, 2004, pp. 80-89.
- Tolk, A., Turnitsa, C. and Diallo, S., Implied Ontological Representation within the Levels of Conceptual Interoperability Model, International Journal of Intelligent Decision Technologies, IOS Press, The Netherlands, Volume 2, Number 1, 2008, pp. 3-19.
- Lehmann, T. and Karcher, A., Ontology Enabled Decision Support and Situational Awareness, International Journal of Intelligent Decision Technologies, IOS Press, The Netherlands, Volume 2, Number 1, 2008, pp. 21-31.
- Hasebrook, J. and Saha, A., Infoviz for Strategic Decision Making, International Journal of Intelligent Decision Technologies, IOS Press, The Netherlands, Volume 2, Number 2, 2008, pp. 89-102.

2

Towards Real-World HTN Planning Agents

Hisashi Hayashi, Seiji Tokura, and Fumio Ozaki

Corporate Research and Development Center, Toshiba Corporation
1 Komukai Toshiba-cho, Saiwai-ku, Kawasaki, 212-8582 Japan
{hisashi3.hayashi,seiji.tokura,fumio.ozaki}@toshiba.co.jp

Abstract. HTN planning, especially forward-chaining HTN planning, is becoming important in the areas of agents and robotics, which have to deal with the dynamically changing world. Therefore, replanning in "forward-chaining" HTN planning has become an important subject. This chapter first presents an online planning agent algorithm that integrates forward-chaining HTN planning, execution, belief updates, and plan modifications. In stratified multi-agent planning, the parent planning agent and its child planning agents work together to achieve a goal. The parent planning agent executes a rough plan for a goal, and the child planning agents execute detailed plans for subgoals. Although this stratified multi-agent planning is efficient, it is difficult for the parent planning agent to change the plan while a child planning agent is working. This chapter also shows how to realize online planning in stratified multi-agent planning.

2.1 Introduction

Agents working in the real world need to adapt to the dynamically changing world. In dynamic environments, planning is an effective way to achieve goals based on the current situation. However, this is insufficient. If the world changes unexpectedly while a plan is being executed, the plan might become invalid and the agent might not achieve the goal after the plan execution. Therefore, online planning agents that interleave planning, action execution, belief updates, and plan modifications are required.

HTN planning [10, 26, 29, 32, 33] is used in many dynamic application domains such as robotics [4], RoboCup simulation [27], mobile network agents [22], web service composition [31], and interactive storytelling [7]. HTN planners make plans by decomposing an abstract task in a plan to subtasks. HTN planning is efficient because of the expressive power of the domain control knowledge as explained in [19, 34]. Also, by postponing some task decompositions until the tasks need to be executed, it is possible to adapt to the situation at the time of the task execution. Therefore, HTN planning is often used in many dynamic domains.

The first aim of this chapter is to show how to make online HTN planning agents. The planning agent makes plans via task decompositions. While executing a plan, it keeps and incrementally modifies some alternative plans. When the

L.C. Jain, N.T. Nguyen (Eds.): Knowl. Proc. & Dec. Mak. in Agent-Based Sys., SCI 170, pp. 13–41.
springerlink.com

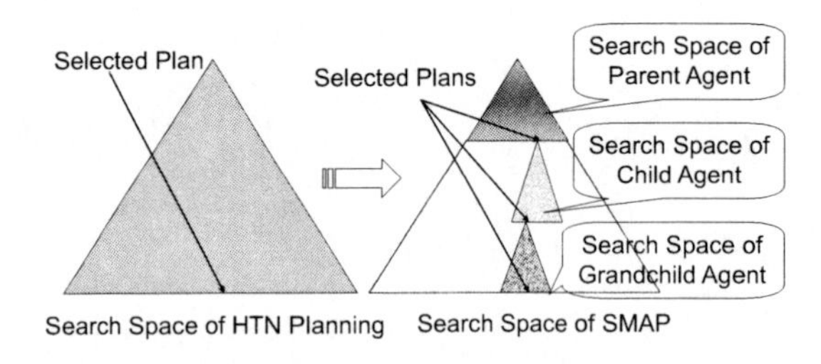

Fig. 2.1. Search Space of HTN Planning and SMAP

plan becomes invalid or an alternative plan becomes more attractive in terms of costs, the planning agent switches the current plan to an alternative plan. This is how the online HTN planning agents of Dynagent [24] work. Furthermore, this chapter extends the Dynagent algorithm so that the agents can suspend unnecessary action execution after replanning. This is important in order to adapt to the dynamic world as soon as possible.

The second aim of this chapter is to show how to make stratified multi-agent systems using online HTN planning agents where the parent planning agent makes and executes a rough plan and its child planning agent makes and executes more detailed plans. In order to find the best plan with regard to the cost in HTN planning, the planner has to compare a large number of alternative plans. One way to solve this problem is to stratify HTN planning agents. As shown in Figure 2.1, by stratifying HTN planning agents, it is possible to reduce the search space of HTN planning because once the parent planning agent makes plans and selects a plan from them, the other alternative plans of the parent planning agent are not taken into consideration unless the child planning agent fails to make a more detailed (sub)plan.

It seems that the stratification of online HTN planning agents is promising. However, if the planning agent cannot stop action execution, this is a problem. Normally, the planning agents do not stop the action execution once it starts it. They replan after the current action execution and before the next action execution. However, in stratified multi-agent systems, in order to execute an action of the parent planning agent, the child planning agent makes and executes a subplan. The action execution time of the parent planning agent is generally long and changes the world greatly. Therefore, when the parent planning agent changes the plan, it is important to suspend unnecessary plan execution of the child planning agent. This chapter also shows how to tackle this problem.

This chapter is organized as follows. Section 2.2 briefly explains online forward-chaining HTN planning of Dynagent. Section 2.3 introduces the museum guide scenario as an example. Section 2.4 defines belief and planning knowledge. Section 2.5 shows how to describe the museum guide domain using belief and planning knowledge. Section 2.6 defines the terminology that will be used for the online planning algorithm. Section 2.7 defines invalid plans and new valid plans after belief updates. Section 2.8 defines the planning algorithm. Section 2.9

intuitively shows how to start replanning quickly in the stratified multi-agent planning system. Section 2.10 integrates planning, execution, belief updates, and plan modifications in the stratified multi-agent planning system. Section 2.11 shows the experimental result of the museum guide scenario. Section 2.12 analyzes the efficiency of replanning in the stratified multi-agent planning system. Section 2.13 discusses related work. Section 2.14 is the conclusion.

2.2 Online Forward-Chaining HTN Planning of Dynagent

This section briefly explain online forward-chaining HTN planning of Dynagent.

HTN planning [10, 26, 29, 32, 33] is different from standard planning which just connects the "preconditions" and "effects" of actions. It makes plans, instead, by recursively decomposing abstract tasks into more concrete subtasks or subplans, which is similar to Prolog[1] that decomposes goals into subgoals.

Forward-chaining HTN planners such as SHOP [26] and Dynagent [24] decompose tasks in plans in the same order that they will be executed. Therefore, when SHOP and Dynagent decompose a task in a plan, they know what actions (= primitive tasks) to execute before the task, and they can easily check the precondition of the task. Because there are several ways to decompose a task, the search space will be an or-tree of alternative plans.

Dynagent is an online forward-chaining HTN planner that continuously modifies plans while executing a plan. As shown in Figure 2.2, Dynagent keeps several alternative plans and incrementally modifies the alternative plans while executing a plan. It is unnecessary to decompose all the abstract tasks in the alternative plans in advance. When comparing the costs of abstract plans that have abstract tasks, A*-like cost functions are used. When the current plan becomes invalid or another plan becomes more attractive in terms of costs, Dynagent changes the plan. When changing to an alternative plan, the agent decomposes abstract tasks in the alternative plan further and starts executing it.

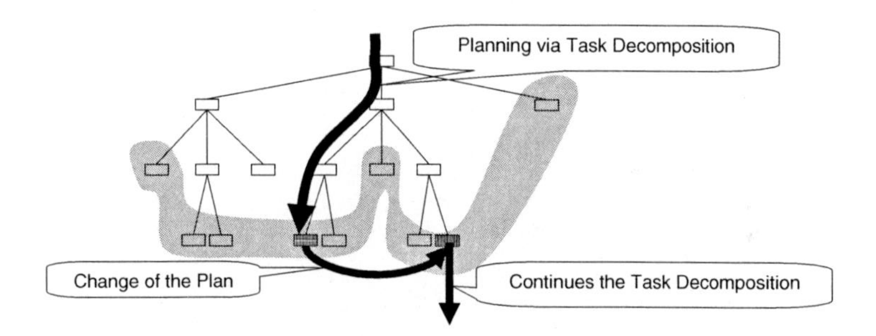

Fig. 2.2. Planning Search Tree

[1] Prolog is the best-known logic programming language which is often used in the area of artificial intelligence. For details, see a textbook such as [9]. There are many other textbooks available. Note that Prolog is not a planner.

2.3 Museum Guide Scenario

This section introduces a museum guide scenario as an example to illustrate stratified multi-agent replanning. Subsequently, this scenario will also be used for experimental evaluation. Figure 2.3 shows the map of a museum where the robot moves. Nodes are places where the robot localizes itself relative to the map with the help of, for example, markers which can be recognized through image processing. In particular, nodes are set at intersections of paths or at points of interests. The robot moves from one node to the next node along an arc. When the user specifies the destination (node), the robot takes the person there.

The museum is divided into areas. Some areas are connected by a door. Given the destination, considering the doors, the parent planning agent first searches a rough route that connects only areas. Then the child planning agent searches a detailed route in the first area that connects nodes. For example, when moving from $n1$ ($area1$) to $n40$ ($area8$), the parent planning agent first makes a rough plan: $area1 \rightarrow area3 \rightarrow area5 \rightarrow area7 \rightarrow area8$. The child planning agent then thinks about how it should move in the first area: $n1 \rightarrow n6 \rightarrow n10$.

The robot can detect if a door is open or not. If the robot notices that a door on the route is closed, the parent planning agent should change the route. If the robot notices that a door is open, the parent planning agent might be able to find a better plan. Therefore, the robot must be able to replan.

In general, actions are primitive tasks that agents execute quickly, and the time for action execution is not taken into account. Also, the action execution cannot be stopped normally. However, in this scenario, one action of the parent planning agent is an area movement. Therefore, the parent planning agent should be able to suspend the current action execution and change the plan.

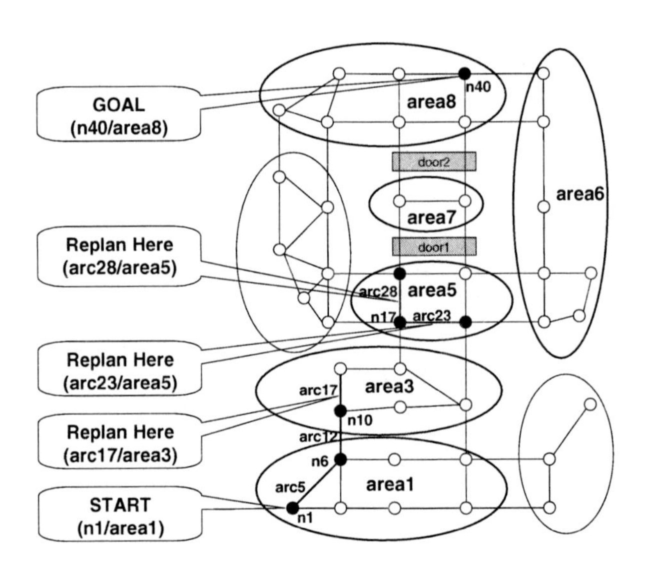

Fig. 2.3. The Map

For example, suppose that the robot is on *arc*28 in *area*5 and moving towards *area*8 via *area*7 when it finds *door*2 is closed. If the parent planning agent cannot replan immediately and stop the current action execution, then the robot continues to move to *area*7 and the parent planning agent would change the plan there. The new route would be to go to *area*8 via *area*5 and *area*6, and the robot would go back to *area*5. This is completely correct replanning. However, the robot should change the route in *area*5 without going to *area*7. This movement is meaningless and time-consuming.

2.4 Belief and Planning Knowledge

This section defines the belief and planning knowledge. Belief expresses the state of the domain. Planning knowledge expresses the effects of actions and rules for task decompositions.

As with Prolog, fluents (corresponding to positive literals in Prolog or predicates in classical logic) and clauses (corresponding to Horn clauses in Prolog or implications in classical logic) are defined as follows using *constants*, *variables*, *functions*, and *predicates*.

Definition 2.1. *A* **term** *is a constant, a variable, or a complex term where a* **complex term** *is of the form:* $\mathsf{F}(\mathsf{T}_1, \cdots, \mathsf{T}_n)$ *where* $n \geq 0$*,* F *is an n-ary function, and each* T_i *(*$1 \leq i \leq n$*) is a term.*

Definition 2.2. *A* **fluent** *is of the form:* $\mathsf{P}(\mathsf{T}_1, \cdots, \mathsf{T}_n)$ *where* $n \geq 0$*,* P *is an n-ary predicate, and each* T_i *(*$1 \leq i \leq n$*) is a term. When* P *is a 0-ary predicate, the fluent* $\mathsf{P}()$ *can be abbreviated to* P*. A fluent is either* **derived** *or* **primitive**.

Note that the truth value of primitive fluents is directly updated by the effects of actions, which will be defined soon, or by observation. On the other hand, the truth value of derived fluents is not directly updated, and is subject to the truth value of other fluents, as defined by clauses.

Definition 2.3. *A* **clause** *is of the form:* $\mathsf{F} \Leftarrow \mathsf{F}_1, \cdots, \mathsf{F}_n$ *where* $n \geq 0$*,* F *is a derived fluent called the* **head***, each* F_i *(*$1 \leq i \leq n$*) is a fluent, and the set of fluents* $\mathsf{F}_1, \cdots, \mathsf{F}_n$ *is called the* **body***. When* $n = 0$*, the clause* $\mathsf{F} \Leftarrow$ *is called a* **fact** *and can be expressed as the derived fluent* F*. The clause* $\mathsf{F} \Leftarrow \mathsf{F}_1, \cdots, \mathsf{F}_n$ **defines** *the fluent* G *if* F *is unifiable with* G*.*

The belief (corresponding to the program in Prolog) is defined using primitive fluents and clauses. Primitive fluents might be updated in the middle of plan execution. On the other hand, clauses defining derived fluents are never updated.

Definition 2.4. *A* **belief** *is of the form:* $\langle \mathsf{D}, \mathsf{S} \rangle$ *where* D *is a set of primitive fluents, and* S *is a set of clauses.*

Intuitively speaking, primitive fluents in D express the current state. The truth values of derived fluents are derived from the primitive fluents using the clauses in S. The separation of D and S is similar to the distinction between intensional and extensional programs in traditional logic programming.

A plan is defined as a list of tasks as follows. Abstract tasks are not directly executable. In order to execute an abstract task, it has to be decomposed to actions (= primitive tasks).

Definition 2.5. *A* **task** *is of the form:* $\mathsf{T}(\mathsf{X}_1, \cdots, \mathsf{X}_n)$ *where* $n \geq 0$, T *is an n-ary task symbol, and each* X_i $(1 \leq i \leq n)$ *is a term. When* T *is a 0-ary task symbol, the task* $\mathsf{T}()$ *can be abbreviated to* T. *A task is either* **abstract** *or* **primitive**. *An* **action** *is a primitive task.* **Cost**, *which is a non-negative number, is recorded in association with a task.*

Definition 2.6. *A* **plan** *is a list of tasks of the form:* $[\mathsf{T}_1, \cdots, \mathsf{T}_n]$ *where* $n \geq 0$ *and each* T_i $(1 \leq i \leq n)$ *is a task, which is called the i-th element of the plan. The* **cost** *of the plan* $[\mathsf{T}_1, \cdots, \mathsf{T}_n]$ *is the sum of each cost of* T_i $(1 \leq i \leq n)$.

In the planning phase, abstract tasks (task pluses) are decomposed to more concrete plans using the following HTN rules. It is possible to specify the precondition where an HTN rule is applicable. This precondition must be satisfied just before the execution of the abstract task.

Definition 2.7. *An* **HTN rule** *is of the form: htn*$(\mathsf{H}, \mathsf{C}, \mathsf{B})$ *where* H *is an abstract task called the* **head**, C *is a set of fluents called the* **precondition**, *and* B *is a plan called the* **body**. *The HTN rule htn*$(\mathsf{H}, \mathsf{C}, \mathsf{B})$ **defines** *the task* T *if* T *is unifiable with* H.

In order to express the effect of an action, the following action rules are used. Like HTN rules, it is possible to specify the precondition where the action rule is applicable. This precondition must be satisfied just before the execution of the action.

Definition 2.8. *An* **action rule** *is of the form: action*$(\mathsf{A}, \mathsf{C}, \mathsf{IS}, \mathsf{TS})$, *where* A *is an action,* C *is a set of fluents called the* **precondition**, IS *is a set of primitive fluents called the* **initiation set**, *and* TS *is a set of primitive fluents called the* **termination set**, *such that no primitive fluent is unifiable with both a primitive fluent in the initiation set and a primitive fluent in the termination set. The* **effect** *of an action rule refers to its initiation set and termination set. The action rule action*$(\mathsf{A}, \mathsf{C}, \mathsf{IS}, \mathsf{TS})$ **defines** *the action* B *if* B *is unifiable with* A.

Definition 2.9. **Planning knowledge** *is of the form:* $\langle \mathsf{AS}, \mathsf{HS} \rangle$ *where* AS *is a set of action rules, and* HS *is a set of HTN rules.*

2.5 Representing the Museum Guide Domain

This section expresses the domain of the scenario introduced in Section 2.3. The domain is expressed using the Prolog expression of belief and planning knowledge so that it is close[2] to the actual implementation of the planner which will be used for the experimental evaluation.

[2] The Prolog expression in the actual implementation is slightly different.

2.5.1 Belief of the Parent Planning Agent

This subsection shows how to express the belief of the parent planning agent. The parent planning agent needs the broad map information. However, it does not need detailed information inside areas.

The fact F is expressed by the Prolog fact: belief(F). Also, the clause $H \Leftarrow B_1, ..., Bn$ is expressed by the Prolog fact: belief(H, [B_1, ..., Bn]).

In Figure 2.3, the map is divided into eight areas. Each node belongs to an area. For example, n1 belongs to area1. This fact is expressed as follows:

```
belief(inArea(n1,area1)).
```

It is necessary to express the connection between areas. For example, an arc connects area1 and area2. This fact is expressed as follows:

```
belief(arc(area1,area2)).
```

The fact that an arc connects area5 and area7 through door1 is expressed as follows:

```
belief(arc(door1,area5,area7)).
```

The following clauses express that the directions of arcs that connect areas do not matter. However the door between the two areas has to be open. In the following clauses, note that each argument of the fluents is a variable.

```
belief(connectsAreas(Area1,Area2),[
  arc(Area1,Area2)
]).

belief(connectsAreas(Area1,Area2),[
  arc(Area2,Area1)
]).

belief(connectsAreas(Area1,Area2),[
  arc(Door,Area1,Area2),
  open(Door)
]).

belief(connectsAreas(Area1,Area2),[
  arc(Door,Area2,Area1),
  open(Door)
]).
```

In order to calculate the distance between areas, the location information of areas is necessary. For this purpose, for each area, the coordinate of the representative point is recorded. For example, the following fact expresses that the coordinate of area1 is (7,2):

```
belief(areaLocation(area1,7,2)).
```

Suppose that the robot is initially in area1, and that door1 and door2 are open. These facts are expressed as follows:

```
belief(in(area1)).
belief(open(door1)).
belief(open(door2)).
```

The truth value of the above fluents might change in the future. Therefore, these fluents are declared as follows where "_" is a variable:

```
dy(in(_)).
dy(open(_)).
```

The parent planning agent changes the belief of the child planning agent when the robot moves into the next area. This belief of the child planning agent is mainly about the detailed map information in each area. For this purpose, the parent planning agent records the path of the belief file in association with each area. For example, the following fact expresses the path of the belief file that the child planning agent will use in area1:

```
belief(localMap(area1,
  '/com/toshiba/dynagent/museum/axiomArea1.pl'
)).
```

2.5.2 Planning Knowledge of the Parent Planning Agent

This subsection shows how to express the planning knowledge for the parent planning agent. The parent planning agent needs to make plans for area movements.

First, the precondition and effect of actions are defined. Suppose that Area1 and Area2 are directly connected. When the robot is in Area1, it can move to the next area (Area2) by executing the action gotoNextArea(Area1,Area2). This can be expressed by the following action rule:

```
action(gotoNextArea(Area1,Area2),[
  in(Area1),
  connectsAreas(Area1,Area2),
],[
  in(Area2)
],[
  in(Area1)
]).
```

The precondition and effect of action suspension can be expressed using action rules. (It is possible to regard an action suspension as an action.) It is assumed that the robot can stop going to the next area and there is no effect[3] when suspending the action:

```
action(suspend(gotoNextArea(_,_)),[],[],[]).
```

[3] Note that the current area does not change before arriving in the next area.

The robot can move to the specified node (Node) in an area (Area) if the robot is in the area:

```
action(gotoNodeInArea(Node),[
  in(Area),
  inArea(Node,Area)
],[],[]).
```

It is assumed that the robot can stop going to the specific node in the current area and there is no effect when suspending the action:

```
action(suspend(gotoNodeInArea(_),[],[],[]).
```

The parent planning agent changes the belief of the child planning agent based on the current area (Area) by the following action where the path of the belief file is Path:

```
action(setLocalMap(Path),[
  inArea(Area),
  localMap(Area,Path)
],[],[]).
```

Next, it is necessary to specify how to decompose abstract tasks using HTN rules. The top-level abstract task is to go to a specified node (GoalNode). This abstract task can be done by first moving from the start area (StartArea) to the goal area (GoalArea), and then moving to the goal node as expressed by the following HTN rule:

```
htn(gotoNode(GoalNode),[
  inArea(GoalNode,GoalArea),
  in(StartArea)
],[
  gotoArea(StartArea,GoalArea),
  gotoNodeInArea(GoalNode)
]).
```

In the above HTN rule, gotoArea(StartArea,GoalArea) is an abstract task, and StartArea and GoalArea are not always directly connected. Therefore, the following HTN rules recursively decompose the tasks as follows:

```
htn(gotoArea(Area,Area),[],[]).

htn(gotoArea(StartArea,GoalArea),[
  connectsAreas(StartArea,NextArea)
],[
  gotoNextArea(StartArea,NextArea),
  setLocalMap(_),
  gotoArea(NextArea,GoalArea)
]).
```

Intuitively, the above HTN rules say that the robot can go from StartArea to GoalArea by moving to the directly connected next area (NextArea), updating the belief of the child planning agent, and then moving to the goal area. The robot does not have to move if the robot is already in the goal area.

Finally, the cost of an area movement task is estimated based on the straight-line distance as follows:

```
cost(gotoNextArea(Area1,Area2),Cost,[
  areaLocation(Area1,X1,Y1),
  areaLocation(Area2,X2,Y2),
  distance(Cost,X1,Y1,X2,Y2)
]).

cost(gotoArea(Area1,Area2),Cost,[
  areaLocation(Area1,X1,Y1),
  areaLocation(Area2,X2,Y2),
  distance(Cost,X1,Y1,X2,Y2)
]).
```

Here, `cost`(Task, Cost, [$B_1, ..., B_n$]) means that the cost of Task is Cost if $B_1, ..., B_n$ hold. Note that the undefined costs of the other tasks will be estimated to 0.

2.5.3 Belief of the Child Planning Agent

This subsection shows how to express the belief of the child planning agent. The child planning agent needs detailed information in the current area. First, it is necessary to express the nodes in the current area. Suppose that the robot is now in `area1`. Then, the child planning agent needs to know the nodes in `area1`. As in Subsection 2.5.1, the fact that `n1` belongs to `area1` is expressed as follows:

```
belief(inArea(n1,area1)).
```

The child planning agent needs to know how the nodes are connected. In addition, the child planning agent needs to know the arcs which lead to the next area. For example, `arc12` connects `n6` and `n10`. In addition, `arc12` connects `area1` and `area3`. These facts are expressed as follows:

```
belief(arcN(arc12,n6,n10)).
belief(arcA(arc12,area1,area3)).
```

As in Subsection 2.5.1, the directions of arcs do not matter:

```
belief(connectsNodes(Arc,Node1,Node2),[
  arcN(Arc,Node1,Node2)
]).

belief(connectsNodes(Arc,Node1,Node2),[
  arcN(Arc,Node2,Node1)
]).
```

```
belief(connectsAreas(Arc,Area1,Area2),[
  arcA(Area1,Area2)
]).

belief(connectsAreas(Area1,Area2),[
  arcA(Area2,Area1)
]).
```

In order to calculate the distance between nodes, the location information of nodes is necessary. For this purpose, the coordinate of each node is recorded. For example, the following fact expresses that the coordinate of `n1` is `(0,0)`:

```
belief(location(n1,0,0)).
```

Suppose that the robot is initially at `n1`. This fact is expressed as follows:

```
belief(at(n1)).
```

The current location is subject to change, which is declared as follows where "_" is a variable:

```
dy(at(_)).
dy(on(_)).
```

2.5.4 Planning Knowledge of the Child Planning Agent

This subsection shows how to express the planning knowledge of the child planning agent. The child planning agent needs to make detailed plans for going to nodes inside an area.

First, the precondition and effect of actions are defined. Suppose that the robot is at `Node1` or on `Arc`. The robot can move from the current location to `Node2` that is directly connected by `Arc`. This can be expressed by the following action rules:

```
action(gotoNextNode(Arc,Node2),[
  at(Node1),
  connectsNodes(Arc,Node1,Node2),
],[
  at(Node2)
],[
  at(Node1)
]).

action(gotoNextNode(Arc,Node2),[
  on(Arc),
  connectsNodes(Arc,Node1,Node2),
],[
  at(Node2)
],[
  on(Arc)
]).
```

It is assumed that the robot can stop going to the next node (`Node2`). If the robot suspends this action execution, the current location will be on the arc (`Arc`). This can be expressed by the following action rule:

```
action(suspend(gotoNextNode(Arc,Node2)),[
  connectsNodes(Arc,Node1,Node2),
],[
  on(Arc)
],[
  at(Node1)
]).
```

Next, it is necessary to express how to decompose the abstract tasks. The parent planning agent instructs the child planning agent to execute either the task to go to a node in the current area or the task to go to the next area.

The robot can move to the goal node (`GoalNode`) in `CurrentArea` by going to the next node (`NextNode`) and then moving to `GoalNode`. If the robot is already at the goal node, the robot does not have to move. The child planning agent recursively applies the following HTN rules:

```
htn(gotoNodeInArea(GoalNode)),[
  at(StartNode),
],[
  gotoNodeInArea2(StartNode,GoalNode)
]).

htn(gotoNodeInArea(GoalNode)),[
  on(StartArc),
  connectsNodes(StartArc,_,NextNode)
],[
  gotoNextNode(StartArc,NextNode),
  gotoNodeInArea2(NextNode,GoalNode)
]).

htn(gotoNodeInArea2(Node,Node)),[]).

htn(gotoNodeInArea2(CurrentNode,GoalNode)),[
  inArea(CurrentNode,CurrentArea),
  inArea(GoalNode,CurrentArea),
  connectsNodes(Arc,CurrentNode,NextNode),
  inArea(NextNode,CurrentArea)
],[
  gotoNextNode(Arc,NextNode),
  gotoNodeInArea2(NextNode,GoalNode)
]).
```

In order to move from `CurrentArea` to the next area (`NextArea`), it is necessary to go to a node (`GateNode`) which is directly connected to a node (`GoalNode`) in the next area. Afterwards, the robot can move to GoalNode in the next area. This strategy is expressed by the following HTN rules:

```
htn(gotoNextArea(CurrentArea,NextArea)),[
  at(StartNode),
  inArea(StartNode,CurrentArea),
  connectsAreas(Arc,CurrentArea,NextArea),
  connectsNodes(Arc,GateNode,GoalNode),
  inArea(GateNode,CurrentArea),
  inArea(GoalNode,NextArea)
],[
  gotoNodeInArea2(StartNode,GateNode),
  gotoNextNode(Arc,GoalNode)
]).

htn(gotoNextArea(CurrentArea,NextArea)),[
  on(StartArc),
  connectsAreas(StartArc,CurrentArea,NextArea),
  connectsNodes(StartArc,_,GoalNode)
  inArea(GoalNode,NextArea)
],[
  gotoNextNode(StartArc,GoalNode)
]).

htn(gotoNextArea(CurrentArea,NextArea)),[
  on(StartArc),
  connectsNodes(StartArc,_,NextNode),
  inArea(NextNode,CurrentArea),
  connectsAreas(Arc,CurrentArea,NextArea),
  connectsNodes(Arc,GateNode,GoalNode),
  inArea(GateNode,CurrentArea),
  inArea(GoalNode,NextArea)
],[
  gotoNextNode(StartArc,NextNode),
  gotoNodeInArea2(NextNode,GateNode),
  gotoNextNode(Arc,GoalNode)
]).
```

Finally, the cost of a movement task is estimated based on the straight-line distance as follows:

```
cost(gotoNextNode(Arc,Node2),Cost,[
  connectsNodes(Arc,Node1,Node2),
  location(Node1,X1,Y1),
  location(Node2,X2,Y2),
  distance(Cost,X1,Y1,X2,Y2)
]).
```

```
cost(gotoNodeInArea2(Node1,Node2),Cost,[
  location(Node1,X1,Y1),
  location(Node2,X2,Y2),
  distance(Cost,X1,Y1,X2,Y2)
]).
```

2.6 Terminologies for Online Planning Algorithms

For the purpose of planning and replanning, the notion of tasks, actions, and plans are extended to task pluses, action pluses, and plan pluses. Each task/action plus records two kinds of precondition: protected condition and remaining condition. These represent the fluents that have to be satisfied just before the execution. The satisfiability of the protected condition has been confirmed in the process of planning, and the protected condition will be used for plan checking when the belief is updated. On the other hand, the satisfiability of the remaining condition has not been confirmed yet. In addition, each action plus records the initiation set and the termination set. The initiation (termination) set records the primitive fluents which start (respectively, cease) to hold after the action execution.

Definition 2.10. *A* **task plus** *representing the task* T *is of the form:* $(\mathsf{T}, \mathsf{PC}, \mathsf{RC})$ *where* T *is a task,* PC *is a set of primitive fluents called the* **protected condition**, *and* RC *is a set of fluents called the* **remaining condition**. *The* **precondition** *of a task plus refers to its protected condition and remaining condition. The* **cost** *of the task plus representing the task* T *is the cost of* T.

Definition 2.11. *An* **action plus** *representing the action* A *is of the form:* $(\mathsf{A}, \mathsf{PC}, \mathsf{RC}, \mathsf{IS}, \mathsf{TS})$ *where* A *is an action,* PC *is a set of primitive fluents called the* **protected condition**, RC *is a set of fluents called the* **remaining condition**, IS *is a set of primitive fluents called the* **initiation set**, *and* TS *is a set of primitive fluents called the* **termination set**. *The* **precondition** *of an action plus refers to its protected condition and remaining condition. The* **effect** *of an action plus refers to its initiation set and termination set.*

A **solved action plus** *is an action plus whose remaining condition is empty. The* **cost** *of the action plus representing the action* A *is the cost of* A.

Definition 2.12. *A* **plan plus** *representing the plan:* $[\mathsf{A}_1, \cdots, \mathsf{A}_{n-1}, \mathsf{A}_n, \mathsf{T}_{n+1}, \cdots, \mathsf{T}_m]$ $(n \geq 0, m \geq n)$ *is of the form:* $[\mathsf{A}_1^+, \cdots, \mathsf{A}_{n-1}^+, \mathsf{A}_n^+, \mathsf{T}_{n+1}^+, \cdots, \mathsf{T}_m^+]$ *where each* A_i^+ $(1 \leq i \leq n-1)$, *which is called the i-th element of the plan plus, is a solved action plus representing the action* A_i, A_n^+, *which is called the n-th element of the plan plus, is an action plus representing the action* A_n, *and each* T_j^+ $(n+1 \leq j \leq m)$, *which is called the j-th element of the plan plus, is a task plus representing the task* T_j.

A **solved plan plus** *is a plan plus such that each element is a solved action plus. A* **supplementary plan plus** *is a plan plus of the form:* $[\mathsf{A}_1^+, \cdots, \mathsf{A}_n^+, \mathsf{TA}_{n+1}^+, \cdots, \mathsf{T}_m^+]$ *where* $n \geq 0$, $m \geq n+1$, *each* A_i^+ $(1 \leq i \leq n)$ *is a solved action plus, and* TA_{n+1}^+ *is a task plus or an action plus such that there exists only* *one* **marked**[4] *fluent, which is a primitive fluent, in the remaining condition of* TA^+_{n+1}. *The* **cost** *of the plan plus representing the plan* P *is the cost of* P.

2.7 Invalid Plans and New Valid Plans

In dynamic domains, while executing the plan, the environment might change, or the agent might find new information. When updating the belief, some plans might become invalid, and some new plans might become valid. Especially, when a plan depends on the deleted fluent, the plan is no longer valid. On the other hand, if a new fluent is added, it might be possible to make new valid plans. This section defines invalid plan pluses and new valid plan pluses.

Definition 2.13. *(Invalid Plan Pluses) Let* PLAN^+ *be a plan plus of the form:* $[\mathsf{T}^+_1, \cdots, \mathsf{T}^+_n]$ $(n \geq 0)$. PLAN^+ *becomes* **invalid** *after deleting the primitive fluent* F *from the belief iff:*

- *there exists* T^+_i $(1 \leq i \leq n)$ *such that* F *is unifiable with a primitive fluent* G *which belongs to the protected condition of* T^+_i,
- *and there does not exist* T^+_k $(1 \leq k \leq i-1)$ *such that* G *belongs to the initiation set of* T^+_k.

Note that fluents in the "protected conditions" are used to check the validity of plans at the time of belief updates.

Figure 2.4 shows that open(door1) is the precondition of task3. Suppose that the satisfiability of open(door1) has been confirmed in the process of planning. While executing the plan, if open(door1) is deleted from the belief just before task1, the plan becomes invalid unless task1 or task2 initiates open(door1).

On the other hand, even if the precondition (open(door1)) of task3 is invalid when planning, the plan is recorded as a supplementary plan, and open(door1) is marked. When later, open(door1) is added to the belief, the precondition becomes valid. In this way, it is possible to find new valid plans after belief updates.

Definition 2.14. *(New Valid Plan Pluses) Let* PLAN^+ *be a supplementary plan plus such that the marked primitive fluent* F *is unifiable with the primitive fluent* F_2 *using the mgu* θ. *The* **new valid plan plus** *made from* PLAN^+ *after adding* F_2 *to the belief is* $(\mathsf{PLAN}^+)\theta$ *such that the satisfiability of* F *in* PLAN^+ *is checked. (See Definition 2.17.)*

Note that supplementary plan pluses are used to create new valid plan pluses. Intuitively, a supplementary plan plus corresponds to an intermediate node of the search tree, and a new valid plan plus is made by adding a new branch to it.

[4] When trying to check the satisfiability of a primitive fluent in a plan plus, the primitive fluent is marked, and the plan plus is recorded as a supplementary plan plus. Supplementary plan pluses will be used to make new valid plans when the marked fluent becomes valid.

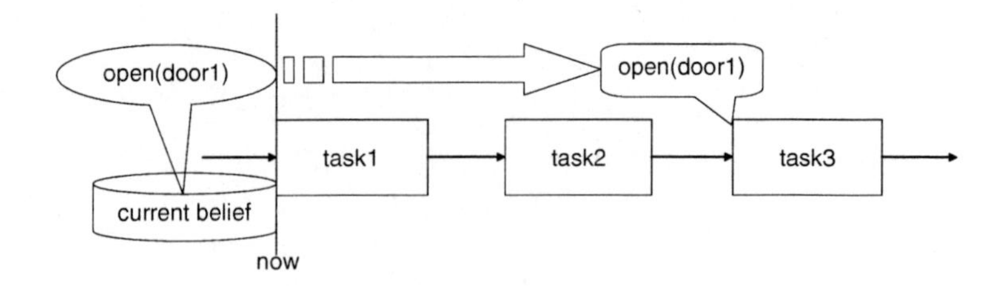

Fig. 2.4. Precondition Protection

2.8 Planning Algorithm

Using the belief and planning knowledge, this section now introduces the planning algorithm. The planner makes plans by decomposing abstract tasks and checking the satisfiability of the preconditions of tasks and actions. Using HTN rules, abstract tasks are decomposed as follows. In the following definition, the abstract task is decomposed to $(\mathsf{B}_1)\theta, \cdots (\mathsf{B}_\mathsf{k})\theta$, and extra precondition $(\mathsf{C})\theta$ is added.

Definition 2.15. *(Task Decomposition) Let* PLAN^+ *be the plan plus:* $[\mathsf{A}_1^+, \cdots, \mathsf{A}_\mathsf{n}^+, \mathsf{T}_{\mathsf{n}+1}^+, \mathsf{T}_{\mathsf{n}+2}^+, \cdots, \mathsf{T}_\mathsf{m}^+]$ *(*$n \geq 0, m \geq n+1$*) such that each* A_i^+ *(*$1 \leq i \leq n$*) is a solved action plus, and* $\mathsf{T}_{\mathsf{n}+1}^+$ *is the task plus:* $(\mathsf{T}_{\mathsf{n}+1}, \mathsf{PC}_{\mathsf{n}+1}, \emptyset)$.

- *When* $\mathsf{T}_{\mathsf{n}+1}$ *is an action, let* AR *be the action rule: action*$(\mathsf{A}, \mathsf{C}, \mathsf{IS}, \mathsf{TS})$ *such that* $\mathsf{T}_{\mathsf{n}+1}$ *is unifiable with* A *using the most general unifier (mgu)* θ. *The* **resolvent** *of* PLAN^+ *on* $\mathsf{T}_{\mathsf{n}+1}$ *by* AR *is the plan plus:* $([\mathsf{A}_1^+, \cdots, \mathsf{A}_\mathsf{n}^+, \mathsf{A}^+, \mathsf{T}_{\mathsf{n}+2}^+, \cdots, \mathsf{T}_\mathsf{m}^+])\theta$ *where* A^+ *is the action plus:* $(\mathsf{A}, \mathsf{PC}_{\mathsf{n}+1}, \mathsf{C}, \mathsf{IS}, \mathsf{TS})$.
- *When* $\mathsf{T}_{\mathsf{n}+1}$ *is an abstract task, let* HR *be the HTN rule: htn*$(\mathsf{H}, \mathsf{C}, [\mathsf{B}_1, \cdots \mathsf{B}_\mathsf{k}])$ *(*$k \geq 1$ *or* $\mathsf{PC}_{\mathsf{n}+1} = \mathsf{C} = \emptyset$*)*[5] *such that* $\mathsf{T}_{\mathsf{n}+1}$ *is unifiable with* H *using the mgu* θ. *The* **resolvent** *of* PLAN^+ *on* $\mathsf{T}_{\mathsf{n}+1}$ *by* HR *using* θ *is the following plan plus:* $([\mathsf{A}_1^+, \cdots, \mathsf{A}_\mathsf{n}^+, \mathsf{B}_1^+, \cdots, \mathsf{B}_\mathsf{k}^+, \mathsf{T}_{\mathsf{n}+2}^+, \cdots, \mathsf{T}_\mathsf{m}^+])\theta$ *where* B_1^+ *is the task plus:* $(\mathsf{B}_1, \mathsf{PC}_{\mathsf{n}+1}, \mathsf{C})$, *and each* B_i^+ *(*$2 \leq i \leq k$*) is the task plus:* $(\mathsf{B}_\mathsf{i}, \emptyset, \emptyset)$.

Figure 2.5 shows how taskA is decomposed into taskA1 and taskA2. The precondition of the task decomposition is added to the precondition of taskA1. It is easy to check the precondition because the planning agent knows what actions to execute before the task. When decomposing an action, in addition to the precondition, the effect of the action is recorded.

The satisfiability of a derived fluent depends on the satisfiability of the primitive fluents that imply the derived fluent. For this purpose, derived fluents are decomposed. In the following definition, the derived fluent F is decomposed to the fluents $(\mathsf{B}_1)\theta, \cdots, (\mathsf{B}_\mathsf{k})\theta$ which imply $(\mathsf{F})\theta$.

[5] When k is 0, it is still possible to relax this condition by merging $\mathsf{PC}_{\mathsf{n}+1}$ (and C), which will disappear otherwise, into the protected condition (the remaining condition) of T_{n+2}.

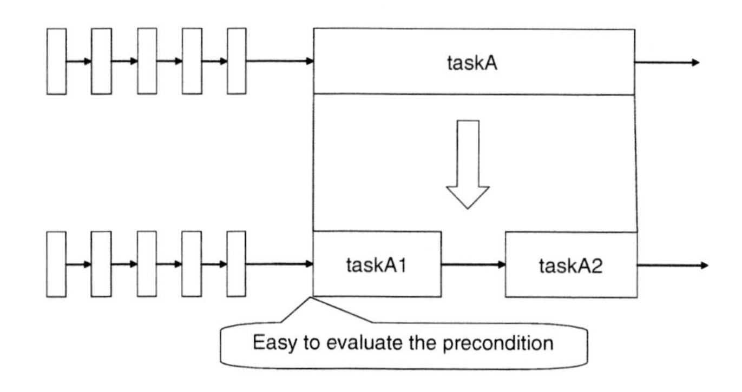

Fig. 2.5. Task Decomposition

Definition 2.16. *(Derived Fluent Decomposition) Let* PLAN^+ *be the plan plus:* $[\mathsf{A}_1^+, \cdots, \mathsf{A}_n^+, \mathsf{TA}_{n+1}^+, \mathsf{T}_{n+2}^+, \cdots, \mathsf{T}_m^+]$ *(*$n \geq 0$*,* $m \geq n+1$*) such that each* A_i^+ *(*$1 \leq i \leq n$*) is a solved action plus,* TA_{n+1}^+ *is either the task plus* $(\mathsf{T}_{n+1}, \mathsf{PC}_{n+1}, \mathsf{RC}_{n+1})$ *or the action plus* $(\mathsf{A}_{n+1}, \mathsf{PC}_{n+1}, \mathsf{RC}_{n+1}, \mathsf{IS}_{n+1}, \mathsf{TS}_{n+1})$*, the derived fluent* F *belongs to* RC_{n+1}*, and each* T_i^+ *(*$n+2 \leq i \leq m$*) is a task plus. Let* CL *be the clause of the form:* $\mathsf{H} \Leftarrow \mathsf{B}_1, \cdots, \mathsf{B}_k$ *(*$k \geq 0$*) such that the derived fluent* F *is unifiable with* H *using the mgu* θ*. The* **resolvent** *of* PLAN^+ *on* F *by* CL *using* θ *is the following plan plus:* $([\mathsf{A}_1^+, \cdots, \mathsf{A}_n^+, \mathsf{TA}^+, \mathsf{T}_{n+2}^+, \cdots, \mathsf{T}_m^+])\theta$ *where if* TA_{n+1}^+ *is a task plus, then* TA^+ *is the task plus:* $(\mathsf{T}_{n+1}, \mathsf{PC}_{n+1}, \mathsf{RC}_{n+1} \setminus \{\mathsf{F}\} \cup \{\mathsf{B}_1, \cdots, \mathsf{B}_k\})$*, otherwise* TA^+ *is the action plus:* $(\mathsf{A}_{n+1}, \mathsf{PC}_{n+1}, \mathsf{RC}_{n+1} \setminus \{\mathsf{F}\} \cup \{\mathsf{B}_1, \cdots, \mathsf{B}_k\}, \mathsf{IS}_{n+1}, \mathsf{TS}_{n+1})$*.*

The satisfiability of primitive fluents in the remaining condition of a task plus in a plan plus is checked based on the belief, the initiation sets and termination sets mentioned in the action pluses. When decomposing a task in a plan, the planner knows what actions to execute before the task. Therefore, it is possible to check the condition of the task plus when decomposing it. This is one of the advantages of forward-chaining HTN planning. In the following definition, it is checked that the primitive fluent F (precondition of a task) is true now or initiated before the task execution. We also confirm that the primitive fluent F is not terminated before the task execution.

Definition 2.17. *(Primitive Fluent Checking) Let* PLAN^+ *be the plan plus:* $[\mathsf{A}_1^+, \cdots, \mathsf{A}_n^+, \mathsf{TA}_{n+1}^+, \mathsf{T}_{n+2}^+, \cdots, \mathsf{T}_m^+]$ *(*$n \geq 0$*,* $m \geq n+1$*) such that each* A_i^+ *(*$1 \leq i \leq n$*) is a solved action plus,* TA_{n+1}^+ *is either the task plus:* $(\mathsf{T}_{n+1}, \mathsf{PC}_{n+1}, \mathsf{RC}_{n+1})$ *or the action plus:* $(\mathsf{A}_{n+1}, \mathsf{PC}_{n+1}, \mathsf{RC}_{n+1}, \mathsf{IS}_{n+1}, \mathsf{TS}_{n+1})$*, and the primitive fluent* F *belongs to* RC_{n+1}*. Given the belief* $\langle \mathsf{D}, \mathsf{S} \rangle$*, if*

- F *is unifiable with a primitive fluent mentioned in* D *using the mgu* θ*,*
- *and there does not exist an action plus* A_i^+ *(*$1 \leq i \leq n$*) such that* $(\mathsf{F})\theta$ *is unifiable with a primitive fluent in its termination set,*

or

- *there exists an action plus* A_i^+ *(*$1 \leq i \leq n$*) such that* F *is unifiable with a primitive fluent in its initiation set using the mgu* θ*,*

- *and there does not exist an action plus* A_j^+ $(i+1 \leq j \leq n)$ *such that* $(\mathsf{F})\theta$ *is unifiable with a primitive fluent in its termination set,*

then F *is satisfiable, and after checking the satisfiability of* F, PLAN^+ *is updated to:* $([\mathsf{A}_1^+, \cdots, \mathsf{A}_n^+, \mathsf{TA}^+, \mathsf{T}_{n+2}^+, \cdots, \mathsf{T}_m^+])\theta$ *where if* TA_{n+1}^+ *is a task plus, then* TA^+ *is the task plus:* $(\mathsf{T}_{n+1}, \mathsf{PC}_{n+1} \cup \{\mathsf{F}\}, \mathsf{RC}_{n+1} \setminus \{\mathsf{F}\})$, *otherwise* TA^+ *is the action plus:* $(\mathsf{A}_{n+1}, \mathsf{PC}_{n+1} \cup \{\mathsf{F}\}, \mathsf{RC}_{n+1} \setminus \{\mathsf{F}\}, \mathsf{IS}_{n+1}, \mathsf{TS}_{n+1})$.

The planning algorithm in this chapter is similar to the planning algorithm of SHOP [26]. The agent makes several alternative plans (= current set of plan pluses) by decomposing a task in a plan into subtasks. The main difference between SHOP and the planning algorithm in this section is that extra information is recorded for the purpose of replanning. The protected condition recorded in each action plus and each task plus is used to detect invalid plans when deleting a fluent from the belief. The supplementary plan pluses are used to make new valid plans when adding a new fluent to the belief. The following algorithm will be used not only for initial planning but also for replanning. In the case of initial planning, given the task G as the goal, the current set of plan pluses is set to $\{[(\mathsf{G}, \emptyset, \emptyset)]\}$, and the current set of supplementary plan pluses is set to $\emptyset$. The purpose of planning is to decompose the goal G to a sequence of actions. In the agent algorithm which will be introduced later, even after the initial planning, the agent keeps and continuously modifies the current belief, the current set of plan pluses, and the current set of supplementary plan pluses.

Algorithm 1. *(Planning)*

1. *Given a set of plan pluses (***current set of plan pluses***), a set of supplementary plan pluses (***current set of supplementary plan pluses***), a belief (***current belief***), and the planning knowledge, repeat the following procedure until there exists at least one solved plan plus* PLAN^+ *in the current set of plan pluses and the first element of each non-empty plan plus is a solved action plus:*
   a) *(Plan Selection for Task Decomposition) From the current set of plan pluses, select a plan plus* PLAN^+ *of the form:* $[\mathsf{A}_1^+, \cdots, \mathsf{A}_n^+, \mathsf{T}_{n+1}^+, \mathsf{T}_{n+2}^+, \cdots, \mathsf{T}_m^+]$ $(n \geq 0,\ m \geq n+1)$ *such that each* A_i^+ $(1 \leq i \leq n)$ *is a solved action plus representing* T_i, *and* T_{n+1}^+ *is a task plus representing the task* T_{n+1}.
   b) *(Task Decomposition) Replace the occurrence of* PLAN^+ *in the current set of plan pluses with its resolvents* $\mathsf{R}_1, \cdots, \mathsf{R}_k$ $(k \geq 0)$ *on* T_{n+1} *by the HTN rules that define* T_{n+1} *in the planning knowledge. (See Definition 2.15.)*
   c) *Repeat the following procedure until each remaining condition of action pluses and task pluses mentioned in the current set of plan pluses*[6] *becomes empty.*

[6] It is necessary to check the remaining condition of the resolvents produced at Step 1b and Step 1(c)ii.

i. *(Plan Selection for Fluent Decomposition) From the current set of plan pluses, select a plan plus* PLAN^+ *of the form:* $[\mathsf{A}_1^+, \cdots, \mathsf{A}_n^+, \mathsf{TA}_{n+1}^+, \mathsf{T}_{n+2}^+, \cdots, \mathsf{T}_m^+]$ $(n \geq 0,\ m \geq n+1)$ *such that each* A_i^+ $(1 \leq i \leq n)$ *is a solved action plus,* TA_{n+1}^+ *is an action plus or a task plus, and the fluent* F *is mentioned in the remaining condition of* TA_{n+1}^+.

ii. *(Derived Fluent Decomposition) If the fluent* F *is a derived fluent, replace the occurrence of* PLAN^+ *in the current set of plan pluses with its resolvents* $\mathsf{R}_1, \cdots, \mathsf{R}_k$ $(k \geq 0)$ *on* F *by the clauses that define* F *in the current belief. (See Definition 2.16.)*

iii. *If the fluent* F *is a primitive fluent, do the following procedure:*
   A. *(Supplementary Plan Recording) Add* PLAN^+ *to the current set of supplementary plan pluses, marking the occurrence of* F *in* PLAN^+.
   B. *(Primitive Fluent Checking) Check the satisfiability of the primitive fluent* F *in* PLAN^+ *based on the current belief. (See Definition 2.17.) If* F *is not satisfiable, delete* PLAN^+ *from the current set of plan pluses.*

2. *Return the solved plan plus* PLAN^+*, the current set of plan pluses, and the current set of supplementary plan pluses.*

A* [20, 21] is a well-known heuristic graph search algorithm for finding the shortest path. Suppose that A* has found a route from the starting point to another point p. The distance $g(p)$ of the route to p can be calculated easily. A* estimates the distance $h(p)$ between the point p and the destination. In this way, A* estimates the distance $(f(p) = g(p) + h(p))$ from the starting point to the destination via the point p. Afterwards, A* picks up the best already computed path to p in terms of $f(p)$, expands the path to the next points, and continues the search in the same way.

In Algorithm 1, when decomposing a task in a plan, all the tasks before the task have been decomposed into actions. The cost of an action is the exact cost, and the cost of the abstract task is its estimation. (There are many ways to execute the abstract task, and the actual cost of the task is unknown until the task is decomposed to actions, which are primitive tasks.) Like A*, if the estimated cost is always lower than the actual cost, the plan with the lowest cost is found in the following way:

Algorithm 2. *(Heuristic Planning Using Cost Information) The algorithm is a special case of Algorithm 1 where the cost of the plan plus selected at Step 1a is the lowest among the non-solved plan pluses in the current set of plan pluses.*

2.9 Stratified Multi-agent Planning and Replanning

In stratified multi-agent planning, the parent planning agent executes a rough plan by giving subgoals to its child planning agents. For the parent planning

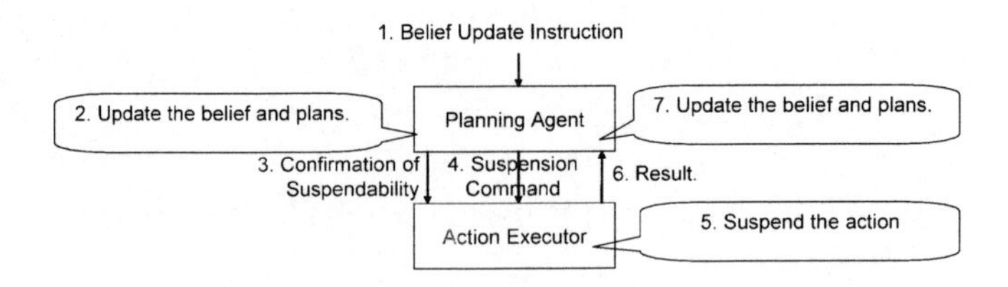

Fig. 2.6. Action Suspension

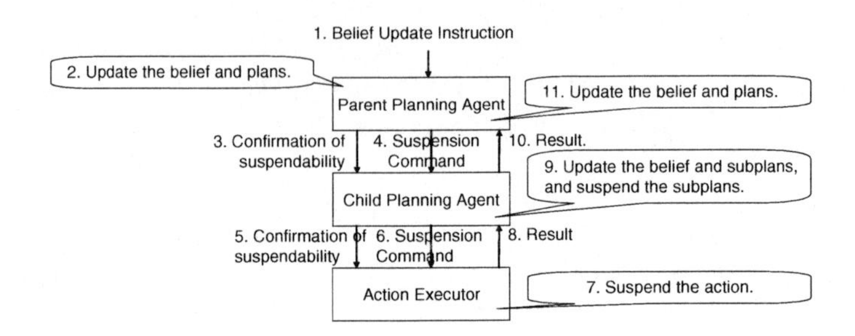

Fig. 2.7. Action Suspension in Stratified Multi-Agent Replanning

agent, its child planning agents are just action executors, and the parent planning agent does not know how its action executors or child planning agents are implemented. Replanning during a plan execution would be easy if the planning agent could wait for an action executor (or a child planning agent) to finish the current action execution. As in the museum guide scenario, however, it is meaningless and time-consuming to continue the current action execution before replanning when using child planning agents. This section intuitively shows how to start replanning quickly by stopping the current action execution.

As shown in Figure 2.6, replanning is triggered by a belief update. After replanning, if the plan is modified and the current action execution becomes no longer relevant with regard to the new plan, then the agent should stop the action execution. However, it is not always possible to stop the current action execution. Therefore, the planning agent asks the action executor if it is possible to suspend the current action execution. If yes, the planning agent tells the action executor to suspend the current action execution. After stopping the current action execution, the planning agent updates its belief and plans considering the effect of the action suspension. For example, if the action to go to $n6$ along $arc5$ is suspended, then the location of the robot becomes on $arc5$. It assumed that the planning agent knows the effect of action suspension.

In stratified multi-agent systems, action suspension can be implemented in a similar way. As shown in Figure 2.7, when the parent planning agent suspends the current action execution after replanning, it tells the child planning agent to stop the current action execution. (Note that a child planning agent is just an action executor from the viewpoint of its parent planning agent.) When the child planning agent receives the suspension command from the parent planning agent, it tells the action executor to stop the current action execution. After suspending the current action execution, all the planning agents update the belief based on the effect of action suspension, and the parent planning agent reupdates its plans, if necessary. In Figure 2.7, two planning agents are stratified. More planning agents can be stratified in a similar way.

2.10 Agentization

Planning agents receive three kinds of inputs: 1. goals; 2. belief update instructions; 3. suspension commands. Given a goal, planning agents make and execute a plan. During the plan execution, if a planning agent receives a belief update instruction, it updates the belief, replans, and suspends the action execution if necessary. If a planning agent has a parent planning agent, it might receive a suspension command from its parent planning agent during the plan execution, in which case it has to stop the plan execution. Note that more than two planning agents can be stratified. However, it is assumed that planning agents do not execute actions concurrently. The following subsections introduce the algorithms that handles these three kinds of inputs.

2.10.1 Combining Planning and Execution

The planning agent keeps several alternative plans. Each task in an alternative plan is not necessarily an action (= a primitive task). The agent decomposes a task into subtasks when necessary. However, the first task in each plan must be an action before selecting a plan and executing an action. After successfully executing an action, the agent updates the belief based on the effect of the action, and removes the executed action from each plan if the first action in the plan is the executed action. When the agent fails to execute an action, it removes each plan such that the first action in the plan is the action that is not executable. After the belief update, it is easy to find and remove invalid plans. Invalid plans can be found by rechecking the (protected) precondition of each action in the plans. Some preconditions must be protected if they might become unsatisfiable when the belief is updated. Even if the precondition of an action is unsatisfiable when planning initially, it might become satisfiable later because of a belief update. Therefore, the agent keeps such invalid plans (= supplementary plans) and makes new valid plans from them when unsatisfiable preconditions become satisfiable.

The following algorithm takes a goal as an input and executes a plan for the goal. Note that when the planning agent receives a suspension command from its parent planning agent, this algorithm is finished by another thread (Algorithm 5.)

Algorithm 3. *(Planning and Plan Execution) Given the task* G (**goal**)*, the belief* (**current belief**)*, the effects*[7] *of action suspension, and the planning knowledge, the* **agent** *starts the following procedure:*

[7] It is assumed that the planning agent knows the effects of action suspensions.

1. *Set the status of the goal to "active."*
2. *Set the* **current set of plan pluses** *to* $\{[(\mathsf{G}, \emptyset, \emptyset)]\}$.
3. *Set the* **current set of supplementary plan pluses** *to* $\emptyset$.
4. *Repeat the following procedure while the status of the goal is "active:"*
   a) *(Planning) Following Algorithm 1 (or its special case: Algorithm 2), try to make a solved plan plus, and update the current set of plan pluses and the current set of supplementary plan pluses.*
   b) *(Successful Plan Execution) If there exists an empty plan plus [] in the current set of plan pluses, then change the status of the goal to "success" and break.*
   c) *(Plan Execution Failure) If the current set of plans is empty, then change the status of the goal to "failure" and break.*
   d) *(Plan Selection) Select a solved plan plus from the current set of plans pluses.*
   e) *(Action Execution) Following the selected plan, tell the action executor to execute the next action and wait for the result ("success", "failure", or "suspended") that is reported from it.*
   f) *If the result of the action execution*[8] *of* $(\mathsf{A}_1)\theta$ *is "success", do the following procedure:*
      i. *From the current set of supplementary plan pluses, delete*[9] *each plan plus such that the first element is not a solved action plus.*
      ii. *(Removal of Executed Action) For each plan plus in the current set of plan pluses and in the current set of supplementary plan pluses, if the plan plus is of the form:* $[\mathsf{B}_1^+, \mathsf{B}_2^+, \cdots, \mathsf{B}_\mathsf{m}^+]$ $(m \geq 1)$*, and* B_1^+ *represents an action that is unifiable with* $(\mathsf{A}_1)\theta$ *using the mgu* ρ*, then replace the plan plus with:* $(([\mathsf{B}_2^+, \cdots, \mathsf{B}_\mathsf{m}^+])\theta)\rho$.
      iii. *(Addition of Fluents) For each primitive fluent* F *that belongs to the initiation set of* $(\mathsf{A}_1^+)\theta$*, add* F *to the set of primitive fluents in the current belief.*
      iv. *(Removal of Fluents and Invalid Plans) For each primitive fluent* F *that belongs to the termination set of* $(\mathsf{A}_1^+)\theta$*, delete* F *from the set of primitive fluents in the current belief, and delete all the invalid plan pluses*[10] *from the current set of plan pluses and the current set of supplementary plan pluses. (See Definition 2.13.)*
   g) *If the result of the action execution of* A_1 *is "failure"*[11]*, do the following procedure:*
      i. *(Removal of Nonexecutable Plan) For each plan plus in the current set of plan pluses (and for each plan plus in the*

[8] After the action execution, some variables might be bound by the substitution θ.
[9] Note that this operation is similar to the cut operation of Prolog.
[10] The effect of an action in a plan might invalidate the other plans.
[11] It is assumed that action failure does not have any effect.

current set of supplementary plan pluses), if the plan plus is of the form: $[\mathsf{C}_1^+, \cdots, \mathsf{C}_m^+]$ *($m \geq 1$) and* C_1^+ *is an action plus which represents an action that is unifiable with* A_1*, then delete the plan plus from the current set of plan pluses (respectively, from the current set of supplementary plan pluses).*

h) (Plan Suspension) If the result of the action execution is "suspended," then update the belief considering the effect of the action suspension. If the belief is updated, then remove all the invalid plan pluses from the current set of plan pluses and add all the new valid plan pluses to the current set of plan pluses. (See Definition 2.13 and 2.14.)

5. Output the status of the goal ("success", "failure", or "suspended.")

Figure 2.8 shows how plans are modified after successful execution of a, and already executed a is deleted from the first element of each plan. In Algorithm 3, this is done at step 4(f)ii.

Figure 2.9 shows how plans are modified after unsuccessful execution of a, and each plan whose first task is a is deleted because these plans cannot be executed. In Algorithm 3, this is done at step 4(g)i.

As in [22, 23], it is possible to use "undoing actions." For example, in Figure 2.8, if the undoing action (undo(a)) of a is used, the plans [b,c], [b,b,c], [d,c], and [d,a,c] would be [b,c], [undo(a),b,b,c], [d,c], and [undo(a),d,a,c]. In this case, unlike [22, 23],

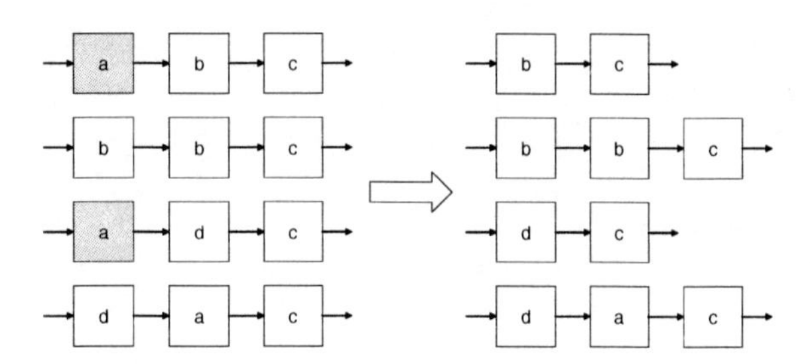

Fig. 2.8. Plans after Successful Action Execution

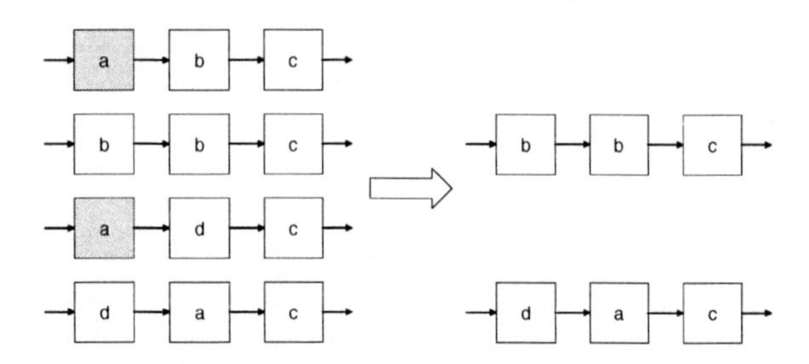

Fig. 2.9. Plans after Action Execution Failure

it is necessary to check that the effect of undo(a) does not invalidate the plans. For simplicity, the concept of undoing actions is not included in this chapter. However, this is an important technique for avoiding side-effects of already executed actions. For example, before changing a travel plan, all the hotel reservations should be canceled. Undoing actions can be used in this case. The importance of undoing is recognized not only in the area of planning [14] but also in the area of web service composition [5].

2.10.2 Belief Updates, Replanning, and Action Suspension

The following algorithm takes a belief update instruction as an input and updates the belief and plans. Following the new plan, it tries to suspend the meaningless current action execution.

Algorithm 4. *(Belief Update and Replanning)*

1. *(Input) A belief update instruction is given as an input. (Belief updates are addition and deletion of fluents.)*
2. *(Belief Update) Update the current belief following the instruction.*
3. *If a plan is being executed by Algorithm 3, execute the following procedure:*
   a) *(Addition of New Valid Plans) For each fluent* F *that has been added to the current belief, make the new valid plan pluses from the supplementary plan pluses in the current set of supplementary plan pluses, and add them to the current set of plan pluses. (See Definition 2.14.)*
   b) *(Deletion of Invalid Plans) For each fluent* F *that has been removed from the current belief, delete each invalid plan plus from the current set of plan pluses and the current set of supplementary plan pluses. (See Definition 2.13.)*
   c) *(Replanning) Following Algorithm 1 (or its special case: Algorithm 2), make a solved plan plus and update the current set of plan pluses and the current set of supplementary plan pluses.*
   d) *(Action Suspension) If the current action execution becomes no longer relevant to the current plan execution and it is possible to stop the action execution, then give the suspension command to the action executor[12] (or its child planning agent that is executing the action) and wait for the result.*

2.10.3 Plan Suspension

In stratified multi-agent planning, the parent planning agent might suspend the plan execution of the child planning agent. Therefore, the child planning agent need to be able to stop the current plan execution. The following algorithm takes a suspension command from its parent planning agent or the user as an input and stops the execution of the action and the plan. Note that when the following

[12] Note that a child planning agent is an action executor of the parent planning agent.

algorithm changes the status of the goal to "suspended," then the plan execution process (Algorithm 3) is finished.

Algorithm 5. *(Goal Suspension)*

1. *(Input) A suspension command is given as an input.*
2. *(Action Suspension) If it is possible to stop the current action execution, then give the suspension command to the action executor (or its child planning agent that is executing the action) and wait for the result.*
3. *(Plan Suspension) Change the status of the goal to "suspended."*

2.11 Experiments

This section evaluates the efficiency of replanning by means of experiments based on the museum guide scenario explained in Section 2.3.

Initially, the robot is at $n1$ in $area1$, and $door1$ is open. (See Figure 2.3.) The destination is $n40$ in $area8$. If the goal is given when $door2$ is open, then the parent planning agent tries to execute the plan to go from $n1$ to $n40$ via $area1$, $area3$, $area5$, $area7$, and $area8$. If the goal is given when $door2$ is closed, then the parent planning agent tries to execute the plan to go to $n40$ via $area1$, $area3$, $area5$, $area6$, and $area8$. In any case, in order to move from $area1$ to $area3$, the child planning agent starts the plan execution to go from $n1$ to $n10$ via $n6$. When the robot arrives at $n6$, in order to move from $area3$ to $area5$, the child planning agent starts planning and plan execution to go from $n10$ to $n17$. While the robot is moving in $area3$ ($arc17$) or $area5$ ($arc23$ or $arc28$), the robot finds that $door2$ is closed or opened and the planning agents replan.

The time for replanning is evaluated. The online replanning algorithm in this chapter is compared with the naive method to replan from scratch. It is not the aim of these experiments to evaluate the efficiency of pathfinding because HTN planners can be used for other purposes. The agent system is implemented in Java and the planners that the planning agents use are implemented in Prolog. Prolog used for this experiment is implemented in Java. The PC (Windows XP) which was used for the experiments is equipped with a Pentium4 2.8GHz and 512MB of RAM.

Table 2.1 shows the replanning time of the parent planning agent. The online replanning is $3 \sim 4$ times as fast as the naive replanning. This is because the online planning agents reuse old plans when replanning.

Table 2.2 shows the replanning time of the child planning agent. When replanning in $area3$, the replanning time is 0.0 sec. This is because the child planning agent does not change the plan. On the other hand, when replanning in $area5$, the online replanning is as fast as the naive replanning. This is because the child planning agent changed the next destination area (from $area7$ to $area6$ or from $area6$ to $area7$) and has to make plans from scratch.

Table 2.3 shows the total replanning time. The online replanning is $2 \sim 5$ times as fast as the naive replanning.

Table 2.1. Replanning Time (Parent Planning Agent)

Place of Replanning	door2	Online Replanning	Naive Replanning
arc17(area3)	*open* → *closed*	0.1 sec	0.4 sec
arc28(area5)	*open* → *closed*	0.1 sec	0.4 sec
arc17(area3)	*closed* → *open*	0.1 sec	0.4 sec
arc23(area5)	*closed* → *open*	0.1 sec	0.4 sec

Table 2.2. Replanning Time (Child Planning Agent)

Place of Replanning	door2	Online Replanning	Naive Replanning
arc17(area3)	*open* → *closed*	0.0 sec	0.1 sec
arc28(area5)	*open* → *closed*	0.1 sec	0.1 sec
arc17(area3)	*closed* → *open*	0.0 sec	0.1 sec
arc23(area5)	*closed* → *open*	0.1 sec	0.1 sec

Table 2.3. Total Replanning Time

Place of Replanning	door2	Online Replanning	Naive Replanning
arc17(area3)	*open* → *closed*	0.1 sec	0.5 sec
arc28(area5)	*open* → *closed*	0.2 sec	0.5 sec
arc17(area3)	*closed* → *open*	0.1 sec	0.5 sec
arc23(area5)	*closed* → *open*	0.2 sec	0.4 sec

2.12 Efficiency (General Case)

In the previous section, the efficiency of stratified multi-agent replanning was evaluated by means of experiments. In general, when a planning agent A replans, the total replanning cost largely depends on whether its child planning agent needs to change the goal or not. If there is no need for the child planning agent to change the goal when A replans, there is no need for the descendants of A to change their plans. In this case, it is clearly possible to save the replanning cost which is roughly proportional to the number of the descendants of A that are executing a plan. On the other hand, if the child planning agent needs to change the goal, the descendants of A cannot reuse their plans. Even in such case, the planning agent A normally replans efficiently because Dynagent is an online planning agent which reuses its plans when replanning.

2.13 Related Work

In order to detect an error in a plan, it is necessary to confirm that each action precondition in the plan is not affected. For this purpose, PLANEX [16] uses triangle tables. (PLANEX is the execution monitoring system of the well-known classical planner STRIPS [17].) Also, causal links of partial-order plans which were introduced first in NONLIN [32] can be used to detect an error in a plan. Causal links are used in many backward-chaining partial-order planners such as TWEAK [8] and SNLP [25]. As explained in the standard textbook [28] of AI, IPEM [1] is the first online planner which smoothly integrates partial-order backward-chaining planning and execution. SIPE [33] is known as an online HTN planner. The plan repairing strategy based on O-Plan [10] and NONLIN [32], both of which are partial-order HTN planners, is explained in [13]. Dynagent [24] is an online forward-chaining HTN planning agent that extends the algorithm of SHOP [26]. There is another attempt [2] to make an online planner based on SHOP. As explained before, we modified the algorithm of Dynagent so that unnecessary action execution can be suspended after replanning, which is especially important to stratify planning agents. Dynagent and the algorithm in this chapter dynamically selects the optimal plan from alternative plans. Similarly, the plan monitoring algorithm in [18] checks not only the validity of plans but also optimality of plans. This algorithm can be applied to backward-chaining planners.

As surveyed in [11, 12, 15], there are mainly two kinds of multi-agent planning:

- Planning for task distribution to agents;
- Coordination of plans of different agents.

Typically, when planning for task distribution, the parent agent makes a rough plan and distributes tasks to its child agents following the plan. Stratified multi-agent planning is related to planning for task distribution. Coordination of plans is necessary because each agent has a different goal and plan execution of one agent affects plan execution of another agent. Dynamic recoordination of plans is researched in [3]. A good survey on distributed and continual (= online) planning can be found in [12].

Stratified multi-agent systems, which are not necessarily planning systems, are often used for the adaptation to the dynamic world. For example, Mobilespaces [30] is a mobile agent system such that the parent mobile agent changes its child mobile agents according to the computer environment. Mobile agents are software agents that move from one computer to another through the network. Therefore, when moving to the next computer, the mobile agent needs to adapt to the new environment. The layered architecture [6] of robotics, where the agents in higher levels control the agents in lower levels, makes mobile robots robust to the environment.

2.14 Conclusion

This chapter has presented a new online forward-chaining HTN planning agent. This chapter has also shown how to stratify these online planning agents. They can replan even before finishing the current action execution. After replanning, if the current action execution becomes meaningless, the planning agent tries to stop it. In particular, after replanning, the parent planning agent tries to stop the unnecessary plan execution of its child planning agent, and then rechecks the new plan considering the side-effects of the plan suspension of the child planning agent. Compared with the naive replanning method that makes plans from scratch, the stratified online planning agents replan efficiently when modifying the belief as confirmed by the experiments.

References

1. Ambros-Ingerson, J., Steel, S.: Integrating planning, execution and monitoring. In: AAAI 1988, pp. 735–740 (1988)
2. Fazil Ayan, N., Kuter, U., Yaman, F., Goldman, R.P.: HOTRiDE: Hierarchical ordered task replanning in dynamic environments. In: ICAPS Workshop on Planning and Plan Execution for Real-World Systems: Principles and Practices for Planning in Execution (2007)
3. Bartold, T., Durfee, E.: Limiting disruption in multiagent replanning. In: AAMAS 2003, pp. 49–56 (2003)
4. Belker, T., Hammel, M., Hertzberg, J.: Learning to optimize mobile robot navigation based on HTN plans. In: ICRA 2003, pp. 4136–4141 (2003)
5. BPEL4WS v1.1 specification (2003)
6. Brooks, R.: A robust layered control system for a mobile robot. IEEE Journal of Robotics and Automation 2(1), 14–23 (1986)
7. Cavazza, M., Charles, F., Mead, S.: Planning characters' behavior in interactive storytelling. The Journal of Visualization and Computer Animation 13(2), 121–131 (2002)
8. Chapman, D.: Planning for conjunctive goals. Artificial Intelligence 32(3), 333–377 (1987)
9. Clocksin, W., Mellish, C.: Programming in Prolog: Using the ISO Standard, 5th edn. Springer, Heidelberg (2003)
10. Currie, K., Tate, A.: O-plan: The open planning architecture. Artificial Intelligence 52(1), 49–86 (1991)
11. de Weerdt, M., ter Mors, A., Witteveen, C.: Multi-agent planning: An introduction to planning and coordination. In: Handouts of the European Agent Summer School, pp. 1–32 (2005)
12. desJardins, M., Durfee, E., Ortiz, C., Wolverton, M.: A survey of research in distributed, continual planning. AI Magazine 20(4), 13–22 (1999)
13. Drabble, B., Tate, A., Dalton, J.: Repairing plans on-the-fly. In: NASA Workshop on Planning and Scheduling for Space (1997)
14. Eiter, T., Erdem, E., Faber, W.: Plan reversals for recovery in execution monitoring. In: NMR 2004, pp. 147–154 (2004)
15. Ferber, J.: Multi-Agent Systems: An Introduction to Distributed Artificial Intelligence. Addison-Wesley, Reading (1999)
16. Fikes, R., Hart, P., Nilsson, N.: Learning and executing generalized robot plans. Artificial Intelligence 3(4), 251–288 (1972)
17. Fikes, R., Nilsson, N.: STRIPS: A new approach to the application of theorem proving to problem solving. Artificial Intelligence 2(3-4), 189–208 (1971)
18. Fritz, C., McIlraith, S.A.: Monitoring plan optimality during execution. In: ICAPS 2007, pp. 144–151 (2007)
19. Ghallab, M., Nau, D., Traverso, P.: Automated Planning: Theory and Practice. Morgan Kaufmann, San Francisco (2004)

20. Hart, P., Nilsson, N., Raphael, B.: A formal basis for the heuristic determination of minimum cost paths. IEEE Transactions on Systems, Science, and Cybernetics 4(2), 100–107 (1968)
21. Hart, P., Nilsson, N., Raphael, B.: Correction to a formal basis for the heuristic determination of the minimum path costs. SIGART Newsletter 37, 28–29 (1972)
22. Hayashi, H., Cho, K., Ohsuga, A.: Mobile agents and logic programming. In: MA 2002, pp. 32–46 (2002)
23. Hayashi, H., Cho, K., Ohsuga, A.: Speculative computation and action execution in multi-agent systems. Electronic Notes in Theoretical Computer Science 70(5) (2002)
24. Hayashi, H., Tokura, S., Hasegawa, T., Ozaki, F.: Dynagent: An incremental forward-chaining HTN planning agent in dynamic domains. In: Baldoni, M., Endriss, U., Omicini, A., Torroni, P. (eds.) DALT 2005. LNCS, vol. 3904, pp. 171–187. Springer, Heidelberg (2006)
25. McAllester, D., Rosenblitt, D.: Systematic nonlinear planning. In: AAAI 1991, pp. 634–639 (1991)
26. Nau, D., Cao, Y., Lotem, A., Mũnoz-Avila, H.: SHOP: simple hierarchical ordered planner. In: IJCAI 1999, pp. 968–975 (1999)
27. Obst, O., Maas, A., Boedecker, J.: HTN Planning for Flexible Coordination of Multiagent Team Behavior. Technical report, Universität Koblenz-Landau (2005)
28. Russell, S., Norvig, P.: Artificial Intelligence: A Modern Approach. Prentice Hall, Englewood Cliffs (1995)
29. Sacerdoti, E.: A Structure for Plans and Behavior. American Elsevier, Amsterdam (1977)
30. Satoh, I.: Mobilespaces: A framework for building adaptive distributed applications using hierarchical mobile agent system. In: ICDCS 2000, pp. 161–168 (2000)
31. Sirin, E., Parsia, B., Wu, D., Hendler, J., Nau, D.: HTN planning for web service composition using SHOP2. Journal of Web Semantics 1(4), 377–396 (2004)
32. Tate, A.: Generating project networks. In: IJCAI 1977, pp. 888–893 (1977)
33. Wilkins, D.: Practical Planning. Morgan Kaufmann, San Francisco (1988)
34. Wilkins, D., desJardins, M.: A call for knowledge-based planning. AI Magazine 22(1), 99–115 (2001)

3

Mobile Agent-Based System for Distributed Software Maintenance

Gordan Jezic, Mario Kusek, Igor Ljubi, and Kresimir Jurasovic

University of Zagreb, Faculty of Electrical Engineering and Computing,
Department of Telecommunications
{gordan.jezic,mario.kusek,igor.ljubi,kresimir.jurasovic}@fer.hr
http://agents.tel.fer.hr

Abstract. The evolution of today's computing environments towards heterogeneous distributed systems introduces specific challenges to the area of software maintenance and testing. Software maintenance on emerging telecommunication systems that are distributed over a wide area is a hard task because it is not easy or even possible to perform final testing on a remote target system, as well as on system in operation.

Experiences show that it is possible for new software running on a target system to give a result different from the one obtained on test system. The reasons are mostly the structural and/or functional differences between both systems. Therefore, only implementation and testing on the actual target system can give the answer whether the new software solves the problem (i.e., error, new operational circumstances, enhancement, and maintainability improvement) or not. Service management and software configuration operations in distributed systems become very demanding tasks as the number of computers and/or geographical distances between them grow. The situation gets worse with an increase in the complexity of the network and the number of nodes.

This chapter describes Multi-Agent Remote Maintenance Shell (MA-RMS) as a mobile agent–based system for distributed software maintenance. It represents a protected environment for the software management without suspending or influencing regular operation. It enables the distributed service management operations (deployment, configuration, control, monitoring, upgrading, and versioning), as well as the advanced features related to the verification of the actual target system. MA-RMS is based on the operations performed by mobile agents that act within an agent team. Presented case studies elaborate three scenarios in which MA-RMS prototype is used for managing software employed in the distributed environment. The first scenario represents the maintenance process for two applications specifically written to be used with MA-RMS. The second one illustrates the procedures needed to be done when the software can be modified to use the advanced features of the RMS prototype, while the third scenario shows how the software which cannot be modified, can yet be maintained and thus adapted to MA-RMS.

3.1 Introduction

The maintenance of the communication software that is distributed over a network is a complex task due to the fact that it is difficult and sometimes even impossible to perform modification and verification on a target system which is remote, and especially

L.C. Jain, N.T. Nguyen (Eds.): Knowl. Proc. & Dec. Mak. in Agent-Based Sys., SCI 170, pp. 43–69.
springerlink.com

on a target system which is currently in operation. Software installation, modification and verification without suspending regular operation, isolation of software under maintenance in order to prevent side-effects influencing regular operation, as well as the support for performing all these operations remotely, are serious problems.

The mobile software agents model, due to its characteristics (reactivity, autonomy, proactivity, persistence, and mostly important – mobility), has been proven good for solving problems in remote management of software systems located on various distributed nodes of the telecommunications network.

This chapter describes Multi-Agent Remote Maintenance Shell (MA-RMS), which represents a protected environment for software verification without suspending or influencing regular operation. The described MA-RMS supports the maintenance by managing of working and/or newly installed software on a remote system.

It includes the following remote operations that support software maintenance: delivering software to a remote system, remote installation/uninstallation, program starting/stopping, tracing and trace data collection, maintaining several versions of software, selective or parallel execution of two versions, and version replacement.

The chapter is organized as follows: software maintenance and the role of Remote Maintenance Shell in software maintenance, as well as related work are described in Section 3.2. The architecture of the MA-RMS and details about the prototype implementation are given in Section 3.3. Section 3.4 presents the case study of remotely management of a web application, while Section 3.5 concludes the chapter, giving directions for future research.

3.2 Software Maintenance

The software maintenance is a modification of a software product after delivery, according to IEEE definition. The purpose of the software maintenance is to correct the faults, improve performance or other attributes, adapt the product to environmental changes, or improving the product maintainability [1]. The maintenance of the communication software that is distributed over the network is a complex task due to the fact that it is difficult and sometimes even impossible to perform modification and verification on a target system which is remote, and especially on a target system which is currently in operation. Software installation, modification and verification without suspending regular operation, isolation of software under maintenance in order to prevent side-effects influencing regular operation, as well as the support for performing all these operations remotely, are serious problems.

3.2.1 Definition

During the late 1980's there was a need to standardize the processes of software maintenance. The Institute of Electrical Engineers, Inc. (IEEE), specifically Software Engineering Standards Committee of the IEEE Computer Society has issued the "IEEE standard for software maintenance" in 1993 [2]. It defines the repetitive process for management of software maintenance activities. The IEEE standard organizes the maintenance process in seven phases, as demonstrated in Fig. 3.1. In addition to identifying the phases and their order of execution, the standard indicates input and output deliverables, the activities grouped, related and supporting processes, the

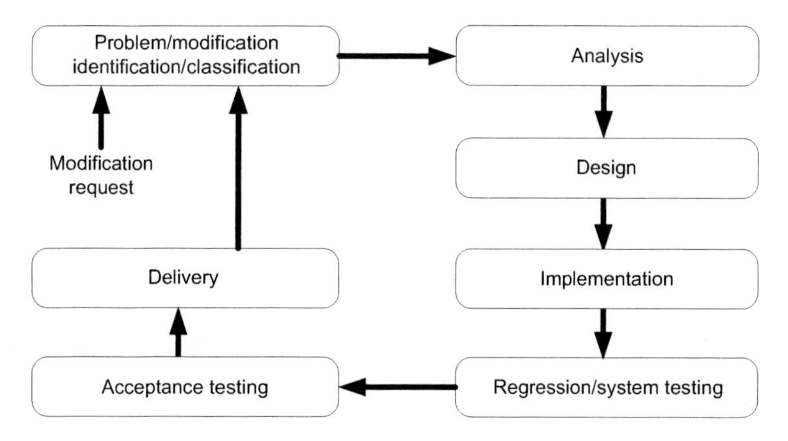

Fig. 3.1. The IEEE Maintenance Process

control, and a set of metrics for each phase [3]. One of the setbacks of this model is that it doesn't consider the activities done prior to application's delivery, which may be extremely important for an efficient and cost-effective maintenance.

Problem/modification identification, classification, and prioritization. This is the phase in which the request for a change (MR – modification request) is usually issued by a user. Each MR is given a unique identifier and is stored in a database for tracking. This phase also includes activities to determine whether it is possible to fulfil the request, or to include the solution for an upcoming patch.

Analysis. This phase devises a preliminary plan for design, implementation, test, and delivery. This step should come up with possible answers to what the alternative solutions of the problem are and how much the implementation of such a solution would cost. It also defines the requirements for the modification, devises a test strategy, and develops an implementation plan. It is important to determine what parts of the system could be modified and what parts must not be modified due to system stability issues.

Design. During this phase the actual modification of the system are designed. Designers use all current system and project documentation, existing software and databases, and the output of the analysis phase. Activities include the identification of affected software modules, the modification of software module documentation, the creation of test cases for the new design, and the identification of regression tests.

Implementation. This phase includes the activities of coding and unit testing, integration of the modified code, integration and regression testing, risk analysis, and review. All of the documentation concerning the system, including the application information and new design issues, are aligned with the new implementation.

Regression/system testing. This is the phase in which the entire system is tested to ensure compliance to the original requirements plus the modifications (consolidated in new documentation). This phase includes regression testing to validate that no new faults have been added.

Acceptance testing. This level of testing concerns the fully integrated system and involves users, customers, or a third party designated by the customer. Acceptance testing comprises functional tests, interoperability tests, and regression tests.

Delivery. This is the phase in which the modified systems are released for installation and operation. It includes the activity of notifying the user community, performing installation and training, and preparing an archival version for the backup.

The Remote Maintenance Shell is positioned in the last three parts of the described process (i.e. regression/system testing, acceptance testing and delivery). While it fully covers the delivery segment of the software maintenance process, only parts of the regression, system and acceptance tests have been implemented in the MA-RMS, mainly those required for remote testing on actual systems.

3.2.2 Positioning of Remote Maintenance Shell in Maintenance Scheme

The motivation for this work can be found in software maintenance experience [4, 5], where it can be found that the same software run on a target system gives results different from the ones obtained on a test system. The usual reasons are hidden in structural and/or functional differences between both systems. Therefore, verification on a target system is necessary to answer the question of whether a software modification solves a problem (fault, new operational circumstances, performance issue, maintainability) [6, 7].

The purpose is to develop a method of remote software maintenance suitable for telecommunication systems. It is aimed towards creating architecture capable of remote software installing, starting, stopping, tracing and testing, and capable to implement a prototype which would prove proposed method. The method does not include performance analysis because it is clear that real-time maintenance slows the performance. The main concern is to make the remote software operation capable of maintenance and testing in a real operating environment.

RMS is the system that employs mobile and stationary agents to perform distributed service maintenance. The position of the MA-RMS in a network node of the distributed system is shown in Fig. 3.2. The main role of the MA-RMS system is to provide the requirements for service management in the distributed system and handle several versions of services (verX) on multiple network nodes simultaneously, without suspending or influencing normal services operation MA-RMS system uses an agent platform as a base for agent creation and management, and adopts all security mechanisms, as well as other basic features from the agent platform.

In the environment with a large number of network nodes and different kinds and versions of the heterogeneous services placed on the network nodes, MA-RMS system is capable to remotely control and manage the service. Mobile agents are the carriers of all operations. They offer several important advantages that make them especially suitable for implementation in the distributed systems, as agents can be employed to perform tasks as a team.

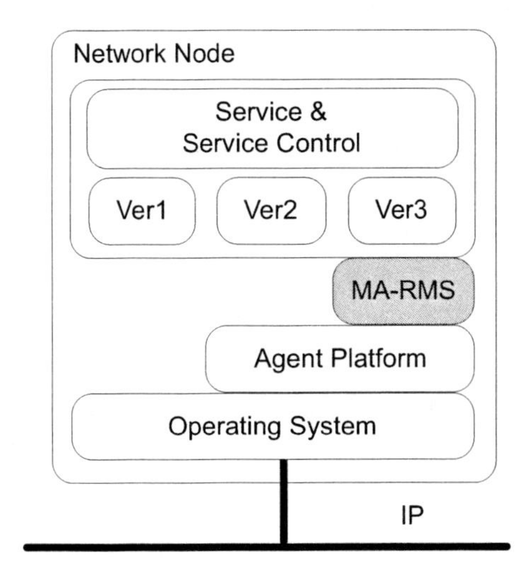

Fig. 3.2. MA-RMS Position

The limitations of this approach can be found in the fact that agent platform must be installed and always started on all network nodes in the system. This will additionally increase the load on the node, which can lead to some stability issues (particularly on the nodes which are already running at or very near their projected capacity). Moreover, Java Virtual Machine must be run on all of the nodes supporting MA-RMS.

MA-RMS is based on three principles [8], as follows:

- Design for remote maintenance: An application should be designed according to specific rules in order to fit the requirements for remote maintenance;
- Low resource implementation: Only MA-RMS parts needed for a specific maintenance job are activated, all other remain inactive in order to save the system resources for regular operation;
- Agent-supported maintenance session: Software agents support all remote operations requested by a maintenance session and guided by user.

3.2.3 Related Work

No one likes the idea of manual installations of new applications or updates of the existing ones. This is the reason why several frameworks for software maintenance have been developed; to make the process easier.

One of them is a framework from Microsoft based on Active Directory infrastructure [9]. The framework allows two types of deployments: user based and computer based [10]. User based deployment allows the administrator to define which application has to be installed for every user individually. The application is not installed until the user tries to use the application for the first time. The second type of deployment is computer based. The administrator defines a group policy for a computer within a certain domain which defines applications that have to be installed on it. Applications are installed next time the system is rebooted.

There are several disadvantages of this kind of approach. The policies for software deployment have to be defined on a very low level (department or less) to avoid conflicts between different policies. The practise has also shown that mixing two deployment types can also lead to conflicts. All install packages for applications must be made in the MSI (Microsoft Installer Package) format. Since the framework is made mostly for organisations it would be very hard to use this system for remote software deployment [11]. Furthermore, since the installation process is rather strait forward there are very few possibilities for intelligent deployment or optimisation.

IBM Tivoli framework offers several components that are used for software maintenance. The component that deals with deployment is called Tivoli Provisioning Manager [12]. The framework is made up of three components: a provisioning server, web-based console and Automation Package Development Environment. The provisioning server contains information about all the computers, switches, applications and security policies present in the system it controls. It consists of the provisioning database used to store all the information, automation package which contains scripts, workflows and tools used that apply to the operation of a specific type of software component or hardware, compliance management used to examine security setup on a target group or computer as well as reporting and discovery management. A web-based console provides an interface towards the provisioning server while Automation Package Development Environment provides an Eclipse based plug-in for automation package development. Due to the fact that Tivoli Provisioning Manager is a propriety framework, we were unable to test its capabilities.

3.3 Architecture of Multi-Agent Remote Maintenance Shell

MA-RMS is designed as an active shield used for protecting the system from possibly faulty functionality of the newly created and installed software. It is also an interface between the environment (system) and the new software. It is made of three main parts: Application Testbed, RMS Console and Maintenance Environment.

MA-RMS manager is a user that determines agents' tasks, injects agents into a network allowing them to roam toward the target node, and return to their home node to report results.

The basic MA-RMS concept is shown in Fig. 3.3. MA-RMS consists of a management station and of the remote systems distributed over the network. The management station is responsible for software delivery to remote systems and for remote operations on them. MA-RMS user connects to management station through RMS Console, the client part of MA-RMS. The software under maintenance (i.e. application) must be adapted for MA-RMS. Service management operations are performed by multi-operation mobile agents, which migrate from the console to the remote system. The Application Testbed, an application-dependent part, which has to be built along with the application, provides the design for remote maintenance. When the application is ready for the delivery, it is moved together with the Application Testbed to the target system.

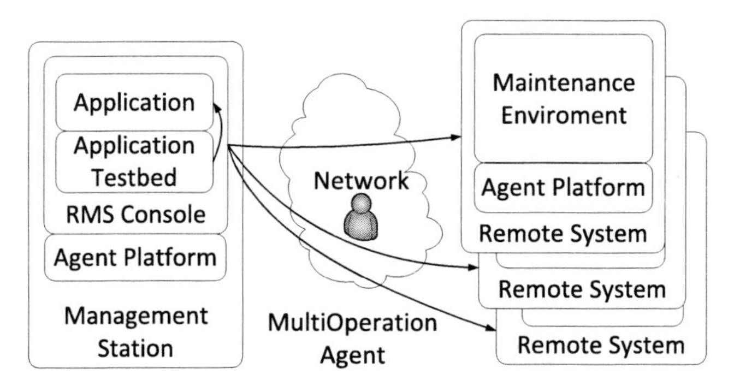

Fig. 3.3. MA-RMS concept

The Maintenance Environment is a common MA-RMS part, pre-installed on the target remote system(s), which enables maintenance actions. It is application independent. The Maintenance Environment is responsible for communications with the Management station. Its main tasks are enabling remote operations and storing the installed software's data.

3.3.1 Maintenance Environment

The Maintenance Environment (Fig. 3.4), which is implemented as MA-RMS server located on the remote system, is responsible for communications with the management station. Its main task is to enable remote operations. It also handles local database which stores data on the installed software and their properties e.g. activity status [13].

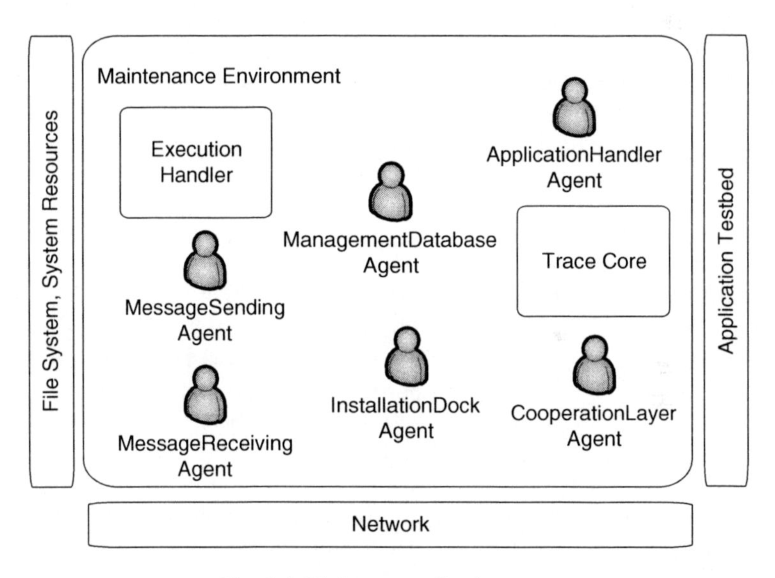

Fig. 3.4. Maintenance Environment

Stationary agents in Maintenance Environment (ME agents) are responsible for providing services to the *Multi-Operation Agent*, whose functionality will be explained later. Due to security restrictions, only local agents are able to perform certain deployment and maintenance operations.

Stationary agents in MA-RMS are: RemoteRequestListenerAgent, CooperationUnitAgent, ApplicationHandlerAgent, InstalationDockAgent and ManagementDatabaseAgent. Trace Core and Execution Handler module are non-agent functional units.

MessageReceivingAgent waits for the requests from the management station. That message is transferred to the MessageSendingAgent.

MessageSendingAgent transfers the message to the appropriate API in the Application Testbed (Resource Manager and VersionHandler).

CooperationUnitAgent is responsible for data interchange between the remote system and management stations. It should be used for receiving new applications.

ApplicationHandlerAgent should be responsible for activating (starting), pausing, stopping, trace activation and other actions over tested application.

InstallationDockAgent is responsible for support of the new application installation. The InstallationDockAgent must be able to receive and place installation files (in cooperation with the CooperationUnitAgent), create directory structure (previously examine existing structure), maintain installation with installation scripts and finally test installation (according received instructions).

ManagementDatabaseAgent should keep data about environment on the remote site. It includes information about installed applications, versions. Also, if something changes on the node, all subscribed consoles are notified about the event.

3.3.2 Multi-Operation Agents

Multi-Operation Agents are capable of executing multiple operations. The MA-RMS user inputs only the desired end state of the software through the GUI. This end state is passed over to the Management Console agent, which generates the necessary operations and assigns them to the cooperative Multi-Operation Agent(s). Each multi-operation agent is assigned one or several operations.

The MA-RMS user only gives a desired end state of service, without specifying particular operations or the order in which they have to be performed. The MA-RMS system automatically determines the operations needed for the specified transition within the service state, as well as the order in which the operations will be executed by the agents. Each operation has some input and output variables. The input variables specify necessary preconditions for the operation execution to begin. The output variables specify the operations that depend on this one, i.e. the operations that have to be notified when this one completes its execution. In this way a graph of interdependencies is created between the operations. The operations do not communicate directly, but through their enclosing agents [14].

Besides the input and output variables each operation has a variable defining the network node on which it has to be executed. When the input data for that operation are available (i.e. when all preconditions are satisfied), the agent migrates to the node where it must execute the operation and initiates its execution there. When the operation is completed, the system notifies of all the operations defined by the output variables and announces the completion of the execution to the agent. Then the agent checks if there is any other operation ready for execution. If affirmative, the whole process is repeated. If some operation is next in line for execution but its preconditions are not satisfied yet, the agent simply waits. When a message arrives, the agent dispatches it to the appropriate operation and tries to execute it again. In case of multiple unsatisfied preconditions this process may have to be repeated. When all operations are completed, the agent destroys itself.

3.3.3 Application Testbed

The Application Testbed (Fig. 3.5) serves as an interface between the software (application) and the Maintenance Environment. It conveys all data from the outer world to the application and vice versa. The application can have multiple versions, and each version is somewhat different from the other, and it is a result of each cycle of the software maintenance process.

Design for maintainability requirements means practically that software input and output have not to be handled directly, but through the Application Testbed and its components. The Application Testbed includes Resource Manager, Version Handler, Trace support API.

Resource Manager represents the input-output layer and provides connection to system resources. It also handles the communication between the software and the environment.

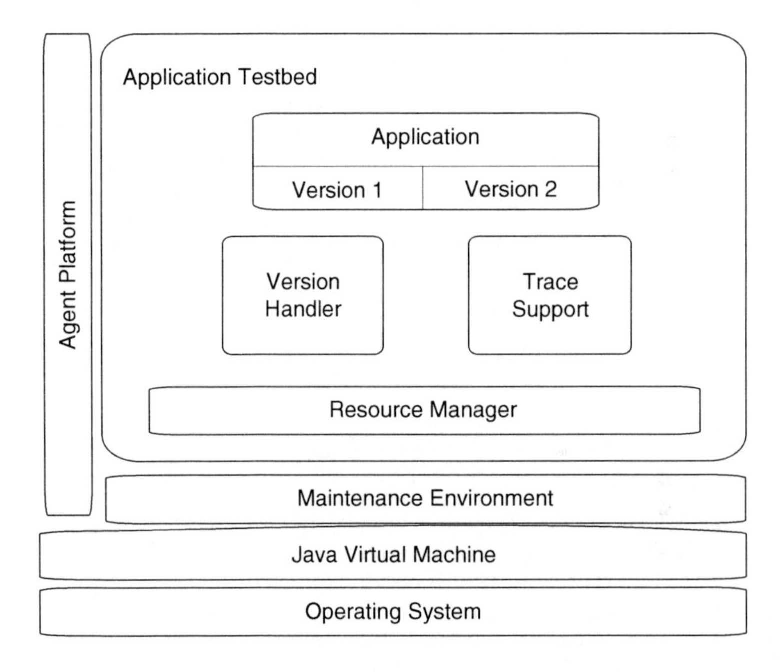

Fig. 3.5. Application Testbed

Version Handler is responsible for software versions starting, selecting (for selective mode) and timing (for parallel mode).

Trace Support API is designed as a set of utilities that should be used for collecting, modeling and delivering trace data to the Maintenance Environment. Software designer should be aware of these utilities used for trace activities.

3.3.4 Remote Software Operations

All maintenance actions are started and managed from the *RMS* Console on Management station, which creates corresponding agents and sends them to remote systems to perform defined actions. *RMS* Console represents a graphical interface between the human manager and the MA-RMS, and it observes the software and agents execution.

In the case that remote software maintenance is needed on the target system, it is necessary to make possible remote software operations. Usage of MA-RMS as a tool for remote software maintenance provides following remote software operations: software migration, installation, starting, stopping, uninstallation, and finally running and tracing.

Remote software operations are started on the client side (Management station) and most of the work is done on the server side (Remote system). Since software under maintenance can be found in different states, maintenance actions must be defined (Fig. 3.6).

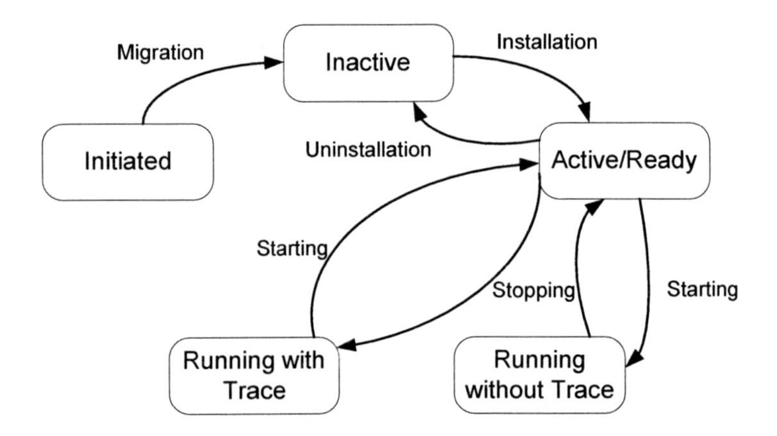

Fig. 3.6. Remote maintenance operations with mobile agents

Software Migration operation includes the actions for software transfer from the management station to a remote system. Before the migration the software is in the *initiated* state. After the transfer has been completed, a delivery report must be sent.

Together with the testbed, installation archives for every migrated version have to be created, too. The basic idea is the following: each service version must be packaged into an installation archive, which contains a special installation script. That installation script describes the configuration details that have to be changed when installing the service. When an agent executes the installation it extracts the script from the archive, loads it with local specific parameters and initiates its execution.

After that the installation script takes over and does all the work. The parameters passed to the script include the parameters such as local host name, the exact path that service will be installed to etc.

It should be noted that the installation scripts for the two versions used in this case study are somewhat different, due to the fact the configuration requirements have changed between the versions. All the knowledge required to write these scripts should be obtained from the monitoring service documentation.

In addition to specifying configuration details, Ant scripts [15] can be used for other purposes as well. In the case study, a solution that enables the migration of large service is tested. Commonly, the agent loads the whole installation archive into the memory and takes it to the HE. However, in the case of a really large service, this can cause heavy memory load and lead to system instability. For that reason an alternative approach has been developed, in which the agents transfer only the Ant script. During the installation, the script downloads the actual service from a HTTP server to the HE. Both approaches are depicted on Fig. 3.7.

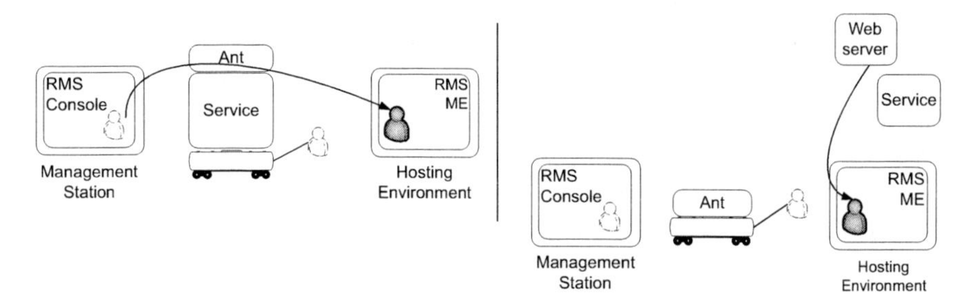

Fig. 3.7. Alternative service migration scenarios

After the migration, the software has switched from the *initiated* to the *inactive* state, and it is ready for installation.

Software Installation operation includes actions for software installation on the remote system. Upon completion of this operation, the software state is changed to *active/ready* state. In that state the software is active and ready for the execution.

An important action that precedes software installation is the checking of the target system for installation capability. The usual problem is the lack of resources on the remote systems and corrupted installation as a consequence. The remote system checking should include available disk space, installed services, free ports, percentage of processor and memory utilization, etc.

The installation of simple software is easy and it includes the following actions:

– Creation of directory structure, and
– Software placement to the newly created directory structure.

In case of large and complex software, with many interfaces towards the environment, additional actions should be included:

– Software registration, and
– Connections to the environment (e.g. to other applications).

Software Starting operation includes actions that change the software state from *ready* to *running*. In the *running* state, software is being executed. In this state software can be stopped and it has to be restarted to reach the *running* state again.

Possible errors and faults generated during the starting operation can be dangerous. In case of malfunctions, errors should be detected, the software execution stopped, and finally, a report sent to the user that initiated the starting operation.

Software Stopping operation can be activated only from the *running* state, and in this case, the software returns to the *ready* state. The software can be stopped in case of problems with the operating system or hardware (memory, processor). Also, the user can initiate software stopping.

Software Uninstallation operation can be started only from the *ready* state, and after the uninstallation, software returns to the *inactive state*. Before the uninstallation, it is important to determine relationships between the software that will be uninstalled and other applications installed in the same system. Some software units could be used by other applications, and their removal would affect regular operations. If such relationships exist, the common parts must be retained.

Software Tracing operation (passive logging) includes actions for trace data collecting during execution of the software. Tracing can be initiated from the *ready* state and performed during the *running* state, or dynamically turned on during software execution. The main idea of tracing is to collect input and output data from maintained software in order to analyze correctness.

Comparing log files from remote and testing systems can be useful for debugging. It is important to underline that the first execution of the new software should always be started with the tracing turned on.

For the purpose of future trace data analysis, the organization of collected data is very important. Each trace data entry contains the following encapsulated information:

- Server name – the name that represents remote system;
- Module source – the name of the unit that generates trace information;
- Module destination – in case of communication between modules, it represents the unit where the message is directed to;
- Time – when the trace information is generated;
- Note – special note about trace information;
- Type – type of trace information (e.g. receiving message, send message, internal information, fatal error, module out of synchronization or module is stopped).

3.3.5 Execution Modes

The software on a remote system can be executed in three modes: normal, selective and parallel. Selective and parallel modes require two versions of the software.

Normal Mode. The normal mode includes the execution of a well-tested version of the software. This mode is used in regular condition when (if) the software works correctly.

Remote software maintenance requires the coexistence of two versions of the software, the old one (input to software maintenance) and the new one (produced during

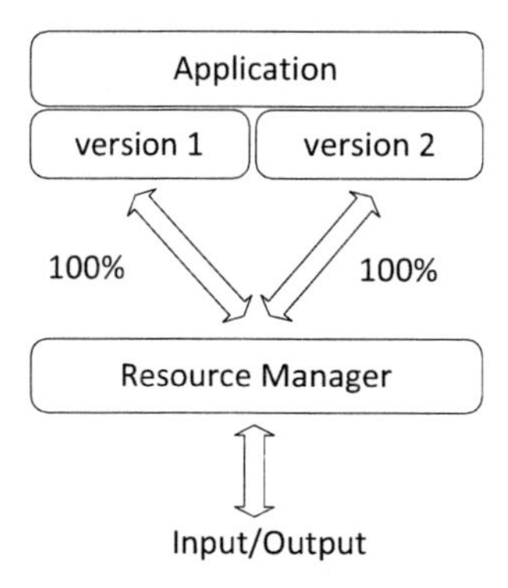

Fig. 3.8. Parallel execution

software maintenance). The replacement of the old version by the new one and the execution of both versions at the same time should be provided.

Parallel Mode. The parallel mode of execution allows the execution of two versions in a pseudo-parallel way (Fig. 3.8). In parallel mode, both versions are executed with the same inputs. The old version is designated as the main version and output messages from that version are sent to the environment. It should be always performed with tracing in order to allow comparison, i.e. checking whether the outputs are equivalent or not.

Collected trace results can then be compared. After the comparison, there are three events that require automatic actions as follows:

- Stopping the new version after receiving different messages,
- Stopping the version that sends the message after defined timeout period, and
- Stopping the version that never responds (message is lost).

For a real-time comparison it is very important to synchronize the versions. When the software consists of a few units and all units execute in parallel mode, it is strictly required to determine which messages of a particular unit must be synchronized.

When the main version finishes its job, it should forward the output to the environment. Before the new execution cycle, the new version must reach the same state, i.e. the main version must wait until the new version finishes its job (Fig. 3.9). If the new version has not finished before the timeout expires, the version is marked as rejected.

There are different equivalence checking models that can be performed, including strong, weak and observational ones. Strong equivalence includes the checking of all messages during execution; the final result, as well as all generated messages. Weak equivalence checks the set of messages and all other messages produced during execution are neglected. In observational equivalence, one specific message can be defined for checking.

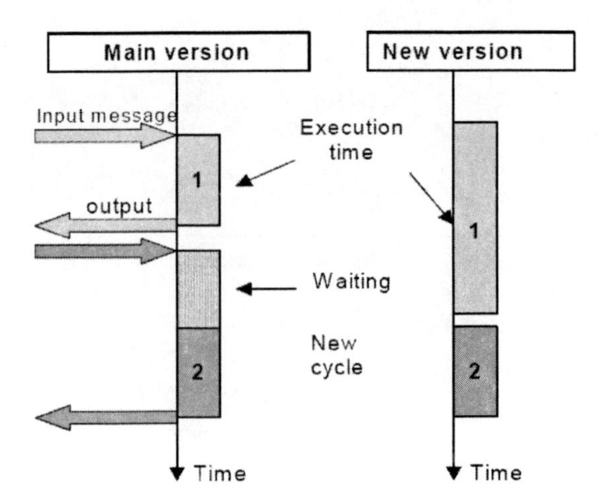

Fig. 3.9. The synchronization of versions

Selective Mode. The selective mode allows the execution of two versions with predefined distribution (Fig. 3.10). Each input message is sent to only one version and this version sends the output message as a result back to the environment. The amount of the requests given to each version is determined by an execution probability, which defines the probability that the request will be given to the new version of the software.

The main role of this mode is the experimental execution of the new software. During the experimental phase, the execution probability for the new software should be kept low in order to observe its behaviour in real environment, without stronger influence on regular operations. When obtaining expected results, the execution probability could be gradually increased as long as complete load is finally turned to the new software.

Within the environment of several application versions running, messages could be easily routed to a version that is not expecting any message. For example, in case there are two versions of some application (i.e. web server) (Fig. 3.11). When a client sends a request to the application, the MA-RMS environment (Resource Manager) will route the request according to the rules of the selective execution mode. For example, the first client's request will be sent to the version 1 and that version will create a session and will store user data in the session. If the second request is routed to the version 2 then the session data will not be available in the version 2, for only the version 1 knows that the session is created.

An even bigger problem occurs in the case when one application (A) uses another and both are running in selective execution mode (Fig. 3.12). In such a case, when the first request to the application A is received, it will send a new request message to application B. The application B responds to application A. If the application B does not respond to the same version in application A then the version in application A does not expect and does not know what to do with such message. The MA-RMS system must ensure that return message will use the same path as an original message which means that the responding and the following new requests must use the same set of application versions.

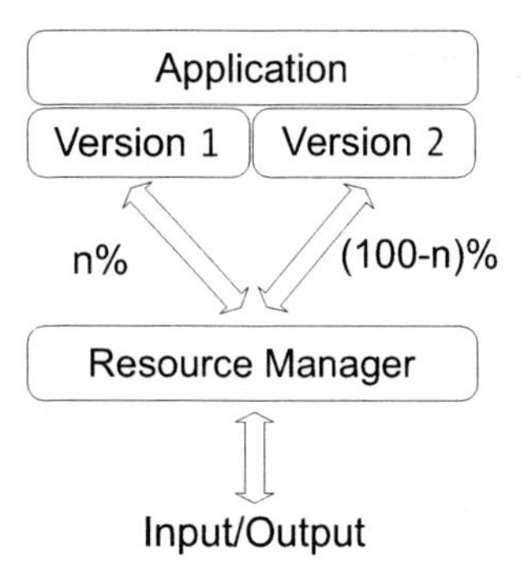

Fig. 3.10. Selective execution

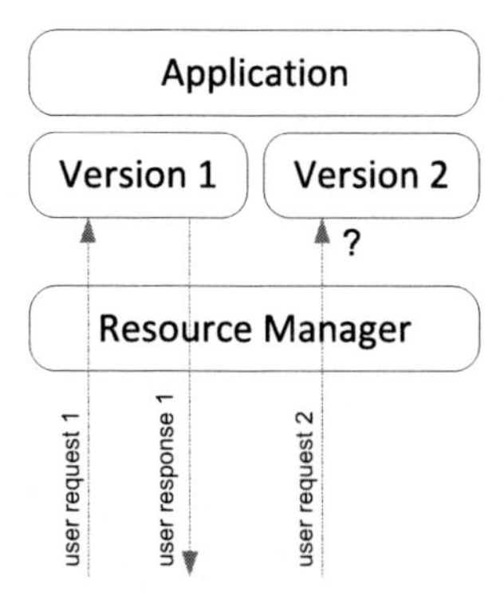

Fig. 3.11. Wrong Routing with One Application

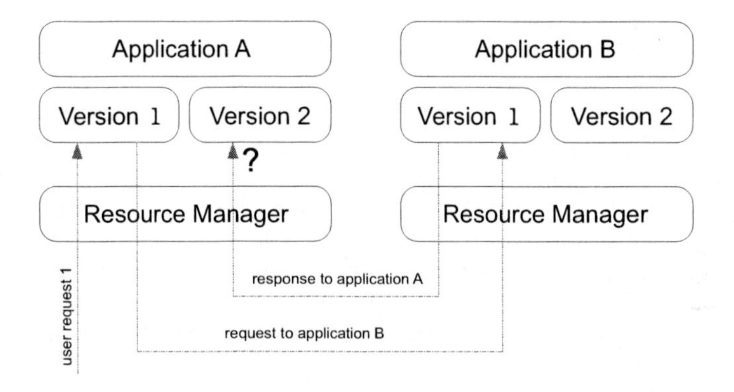

Fig. 3.12. Wrong Routing with Two Applications

This problem is solved by using version signatures within a message. For example, if the first request that needs to establish the context is received by Application A, version 1, the new request to application B will have the signature of the source of this message (application A, version 1). When the version 1 in the application 2 responds to the request, it will send a response to the version 1 in application 1. In this case there are only two elements in the path, but if there are more applications in the path all versions and the applications must be put in the message (the whole path). A similar approach is now used in the SIP (Session Initiation Protocol)[16].

3.3.6 Prototype Implementation

All maintenance operations are performed by mobile agents [17] and each agent executes actions defined for particular functionality [18]. When an agent arrives to the remote system, it sends a request to the agent responsible for processing that request depending on the operation the agent wants to perform. At that point the agent starts further actions.

The mobile agent platform used for MA-RMS prototype is JADE [19], while agents are designed in the Java programming language which includes secure mechanisms. Security issues include secure communication between management station and remote systems, secure agent transfer and execution, the protection of the agents and controlled access to MA-RMS resources. The MA-RMS prototype enables the following functionalities:

- Software delivery to remote system,
- Remote software installation and uninstallation,
- Remote software starting and stopping,
- Trace collecting on remote system,
- Remote execution of two software versions in:
  - selective mode, or
  - parallel mode, and
- Remote versions replacement.

The prototype is implemented as a framework and any object-oriented software (application) written in Java can be adapted for maintenance by using MA-RMS.

Software Delivery to Remote System. Management station creates a MultiOperationAgent (MOA) with the assigned migrate operation. The software that needs to be delivered to the remote system is stored within the operation. The agent then migrates to the remote location and sends a request to the CooperationLayerAgent (CLA). CLA stores the software in its software repository and sends a message to the ManagementDatabaseAgent (MDA) with the information about the migrated software. When all operations are completed, MOA sends a report to the RMS Console.

Remote Software Installation. Management station creates a MOA with the assigned installation operation containing the ID of the software on the remote location. The MOA then migrates to the remote location and sends a request to the InstallationDockAgent (IDA). This agent then sends a message to the MDA requesting information on the software based on the provided ID. In case of version installation it also requests information about the version's testbed. After retrieving the information the IDA starts the installation process. When all operations are completed IDA sends the updated application status to the MDA. MOA sends a report to the RMS Console.

Remote Software Uninstallation. Management station creates a MOA with the assigned uninstallation operation. The uninstallation operation contains the ID of the software on the remote location. MOA agent then migrates to the remote location and sends a request to the IDA. Upon receiving the request the IDA first retrieves the information about the software from the MDA, then checks whether a version can be uninstalled and whether it can start the uninstallation process. When the process is finished the information about the software is updated at the MDA and a report is sent by the MOA to the RMS Console.

Remote Software Starting. Management station creates a MOA with the assigned start operation. The start operation contains the ID of the software on the remote location. After migrating to the remote location MOA sends a request to the ApplicationHandlerAgent (AHA). This agent first retrieves information about the software from the MDA, then checks if the software can be started. If all the prerequisites are fulfilled the application is started and the MDA is updated with the new information. When all operations are completed the MOA sends a report to the RMS Console.

Remote Software Stopping. Is performed on the same way as in the remote software starting functionality. The only difference is that the requested operation is stop operation.

Collecting Trace Data on Remote System. There are two modes of tracing software execution: collected trace and interactive trace. In the collected trace mode a Trace Agent (TA) is created on the remote location. After creation, this agents registers on the Trace Core and starts receiving inputs and outputs data of the software being traced. When the software execution is stopped the TA sends the trace data to the RMS Console. In the interactive mode two agents are created: Interactive Agent (IA) and Trace Agent (TA). The first agent is created on the RMS Console side while the second agent is created on the Maintenance Environment side. After both agents are created the TA agent registers on the Trace Core and starts receiving inputs and output data of the software. Trace data is sent to the IA when those are received so the administrator can observe the execution of the software. Collecting trace data is stopped when the software is stopped or when the trace is turned off.

Remote Versions Replacement. Version replacement allows the old version of the software under maintenance to be replaced with the new one. The replacement is possible with or without the service stoppage. In case of service stoppage, the situation is easier, because only one version is in active/ready state. In this case, MOA performs the following operations: first the MOA migrates to the remote location and delivers the software. Then another MOA stops the execution of the old version and uninstalls it at the remote location. After the uninstallation process ends the new version is installed by the second MOA agent. The final step is to start the new version. At this point only the new version is registered by the Version Handler and all requests go to it. The replacement without service stoppage is a more complex task because in some situations two versions can be in active/ready state. The procedure is the following: first the new version is installed and started. The application is reconfigured in a way that the current version becomes old, and the new version becomes current. All requests are now redirected by the Version Handler to the current version. The old version is working till it has active jobs, after that it is stopped and uninstalled.

In order to perform any action on the remote system, the management station first checks the current software state (active version, execution mode). The software state is obtained by registering on to the remote location. The agent responsible for management station registration on remote system is RegisterDeregisterMCAgent (RA). This agent registers the RMS Console on the remote location. Every time the software state is updated the MDA agent sends a notification message, to all the RMS consoles currently registered, containing the new status of software. This information is used on the management station to define which operations have to be performed at the remote location for it to reach the desired state.

3.4 Case Studies

Three different scenarios for software maintenance using the developed MA-RMS prototype have been conducted. The first scenario is the maintenance process for two applications specifically written to be used with MA-RMS. The second one illustrates the procedures needed to be done when the software can be modified to use the advanced features of the RMS prototype; while the third scenario shows how the software which cannot be modified, can yet be maintained and thus adapted to MA-RMS (i.e. source of the software is not available).

Each case study is used for deploying a new application to the remote systems and upgrading it as part of software maintenance. The first step is to deploy a new application, configure it and start it. The administrator has to specify the desired end state of the application and Management Console agent generates necessary operations and establishes interdependencies between them. The operations include: application and testbed migration, installation, setting execution parameters and starting. Fig. 3.13 shows a dependency graph for one remote system.

The first operation is testbed migration (migrate TB), followed by testbed installation (install TB) and version (application) migration (migrate version) that depends on

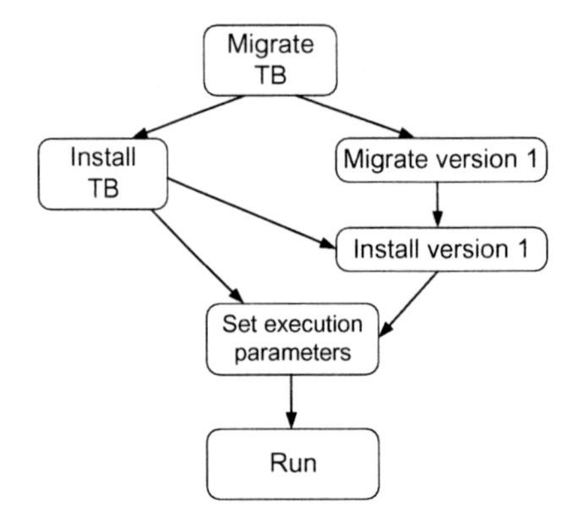

Fig. 3.13. Execution Graph for Application Deployment

testbed migration. The version installation (install version) depends on the version migration and the operation for setting execution parameters depends on both install testbed and install version. The last operation is starting the application (run), which can be executed after setting execution parameters. When the MA-RMS administrator initiates task execution, Multi-Operation Agents are created as a team.

The second part deals with application upgrading, which means deploying, installing and starting newer version of application (Fig. 3.14). When the upgrade is initiated the agents are created. Once the migration of a new application is completed at all servers the new version installation, stopping of the old application, setting execution parameters and starting is executed.

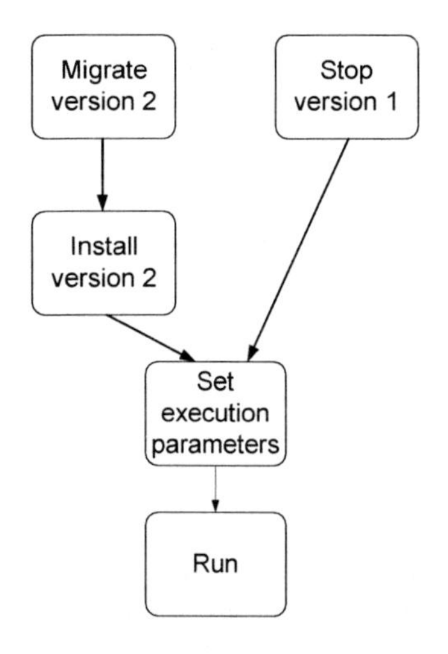

Fig. 3.14. Execution Graph for Application Upgrading

The advanced version handling mechanisms provides the support for gradual introduction of a newer version so that the remote system can remain operational during the testing phase. It is based on the execution modes. When the new application is started for the first time it is started in the parallel execution mode.

In the parallel mode, both versions are executed with the same inputs. The old version is designated as the main version and the output messages from that version are sent to the environment, but the outputs from both versions are collected and compared. Thus, correct functioning of the new version is verified.

After a certain amount of time running without flaws the parallel execution mode is replaced with the selective one. The selective execution mode allows simultaneous execution of two versions with predefined distribution of the incoming user's requests. Each input message is sent to one version only, and as a result this version sends the output message back to the environment. At the beginning of running in this mode the execution probability for a new version must be kept low in order to monitor its behaviour in a real environment. If the predicted results are obtained, the execution probability can be gradually increased, until the complete load is finally turned to a new service and then the selective mode can be replaced with the normal execution mode but with the new version. After that the old version can be uninstalled.

3.4.1 Application Designed for the MA-RMS

Two demo applications are developed specifically for the demonstration of usage of the MA-RMS advanced features [20]. The first one is USI (User Side Interface) and the second is NSI (Network Side Interface). They are installed and maintained as two separate applications, but only when put together they form a simple network chat program. As their names suggest, USI provides a GUI for user interaction, while NSI takes care of the network communication. Fig. 3.15 depicts USI/NSI and their relationship.

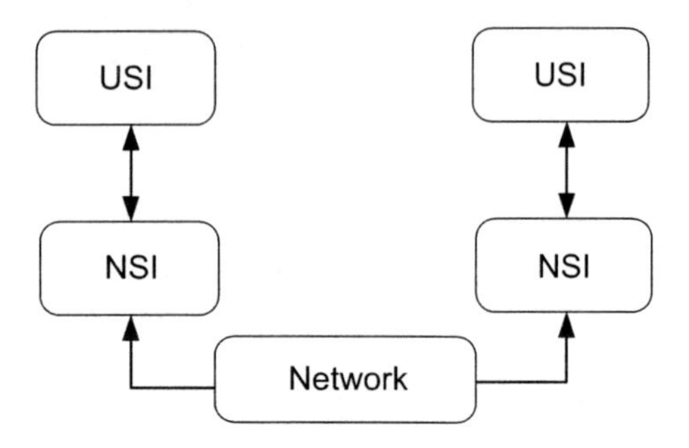

Fig. 3.15. RMS demo applications – USI and NSI

Both USI↔NSI and NSI↔NSI communications utilize a strictly defined protocol, exchanging only messages from a predefined set, which are used to establish, maintain and close the connection between two hosts.

The result of designing a new application for MA-RMS environment is an application that supports all operations provided by MA-RMS environment: migration, installation, starting (normal, parallel, selective and testing mode), stopping and uninstallation. Two parts of the application should be created. The first is the application testbed and the second is one version of the application. The application testbed consists of Resource Manager, Trace API, Version Handler and interface specific to the application. This interface is used for starting/stopping the actual version of application and for receiving the messages from other applications. The application version must have at least one class that implements interface from the testbed.

After creating the testbed and the version they must be packaged into jars (Java archive), the testbed into one jar and the version into another. The application's testbed contains VersionHandler, ResourceManager, Trace API and interface (i.e. for the NSI application it is called Insi) and one additional file named SWInfo.properties. The SWInfo.properties. contains the application's name, the developer's name, a short description of the application, the name of testbed, and the full qualified name of ResourceManager class in application testbed. The version jar contains the actual application and the SWInfo.properties for the version. When the application is going to be installed it is extracted from the file system. If just extracting files is not enough for application to be installed then the additional Build.xml file must be packaged into the jar file. This Build.xml contains Apache Ant script that will be called during the process of installation. The install target in this script will be called and some system variables will be set (related to the RMS environment).

The first version of application is deployed and run. While in operation, a malfunction in USI application (version 1) which prevents the user to successfully disconnect chat session has been found. The developer of the application has found a bug in the application and has issued a new version of the USI application (version 2).

After the new version has been built, it is migrated and installed on the remote system using the management station. During software starting, the administrator will use a parallel mode of execution, for testing the functionalities of the new version as explained earlier within the maintenance procedure.

3.4.2 MA-RMS Compliant Web Application

To remotely manage the Web application, some of its components should be redesigned in order to be compliant with the MA-RMS specification [21]. The problem was how to access the same resource (i.e. network connection) from different versions of the application. The solution is to divide the application to three separate entities: user front-end, routing component and serving component. The user front-end is realized as a servlet and is deployed on the Web server. The routing component and the serving component are realized as the MA-RMS applications and can be deployed as separate entities (Fig. 3.16). The difference from the user front-end and the routing component is the same as between Apache Web server (in our case user front-end) redirecting requests to the Tomcat application server (in our case routing component). The web applications inside the Tomcat application server are similar to the serving component in our case.

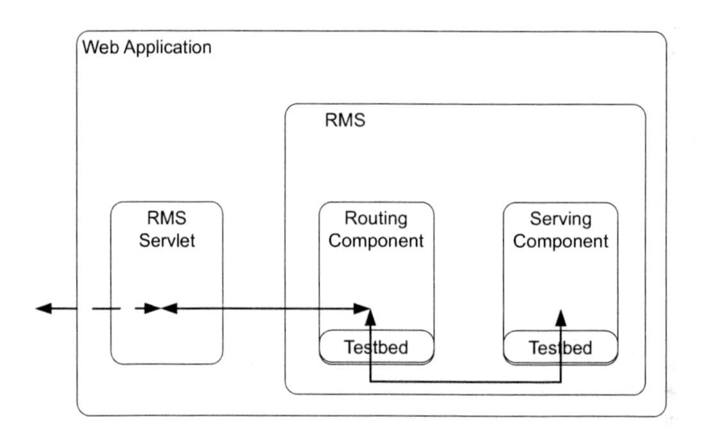

Fig. 3.16. Web application structure

User Front-End. The user front-end is implemented as a servlet on the web server. The MA-RMS servlet receives the requests from the clients who are communicating with the MA-RMS-based application. The RMS servlet analyzes the received request, extracts the URL query parameters and parameter values, and sends them to the routing component. The MA-RMS servlet communicates with a routing component using the appropriate protocol. The MA-RMS servlet can communicate with the systems that are implementing the servlet communication protocol.

For each client request the MA-RMS servlet establishes the communication with the routing component as shown in Fig. 3.17. The first message is a sending integer value that represents the number of strings (one per line), which the routing component must read. The first string is the URL and the other strings are parameters and parameter values extracted from the client request. Having sent the data to the routing component, the servlet waits to receive the data. The routing component sends the integer value, which represents the number of bytes that the servlet must read. The servlet reads the bytes and closes the connection.

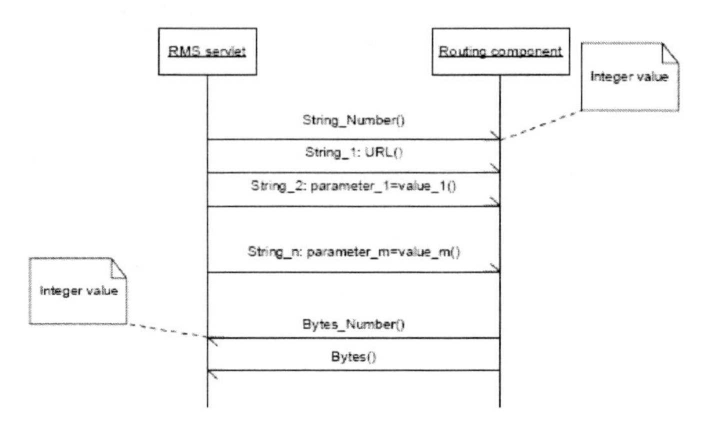

Fig. 3.17. Communication between MA-RMS Servlet and Routing Component

Routing Component. The role of the routing component is to accept the requests sent by the MA-RMS servlet and to forward them to the serving component. Also, it must properly send the response from the serving component back to the MA-RMS servlet. The communication between MA-RMS servlet and the routing component is based on the sockets. To enable the connections with the servlet, a server socket has to be created and kept permanently active for a request to come, for which a new socket must be opened. Each request has a separate socket and is processed in a separate thread.

The received request is parsed and packed into a message appropriate for the serving component. After sending the message to the serving component, the thread ends, but the socket stays opened until the serving component sends a response back. The socket must not be closed, because it is needed to send a response back to the servlet. It is, therefore, stored until this message arrives. A message received from the serving component is a response to some recent requests. The application has to find an appropriate socket to send the response back, after what the socket is closed. The serving component always sends the MA-RMS compliant messages. Therefore, the routing component has first to make the message appropriate for being sent to the servlet.

The routing application consists of three classes: Transporter, Server Listener and Sender. The Server Listener is responsible only for waiting for the requests and opening of a new socket every time a request comes. That thread is ended when the application stops. A message received from the servlet must always include the name of the serving component so that the routing component sends the message further to that application.

Serving Component. The serving component is the actual application. The role of the serving component is to receive the request messages from the routing component through the MA-RMS, to serve the request and to send a response message back to the routing component using an appropriate protocol. This component should be a strict MA-RMS application in order to support all MA-RMS modes of execution. There can be multiple versions of a serving component.

3.4.3 Using MA-RMS with Third-Party Application

A case study presented here attempts to show how MA-RMS can be used to perform Grid service management operations [22] in multiple hosting environments (HEs). The goal is to install and run third-party application (i.e. monitoring service) on three HEs, and later replace it with a newer version. Since the monitoring software in our case study was already in a production phase, and since it was not possible to obtain the source code of the application, it was necessary to wrap the monitoring service into the application that conforms to MA-RMS environment. In order to do that a testbed and a version was crated.

The testbed employed in this case study supports only basic MA-RMS features. Only one testbed needs to be made for a particular service, irrespective of how many versions of that service will be used.

In addition to the testbed, installation archives for both versions had to be created, too. The basic idea is the following one: each service version must be packaged in an installation archive, which contains a special installation script. That installation script describes the configuration details that have to be changed when installing the service. When an agent executes the installation it extracts the script from the archive, loads it with local specific parameters and initiates its execution. After that the installation script takes over and does all the work. The parameters passed to the script include the parameters such as the local host name, the exact path that the service will be installed to etc.

It should be noted that the installation scripts for the two versions used in this case study are somewhat different. This is so because the configuration requirements have changed between the versions. All the knowledge required to write these scripts should be obtained from the monitoring service documentation.

Imagine a Grid environment with three simple hosting environments, referred to as HE1, HE2 and HE3. All of them should contain the same user application. In the first part of the scenario, Grid administrator wants to run the service in the HEs still lacking the service. It is supposed that RMS Maintenance Environment is started on all three HEs. At the beginning of the scenario, the MA-RMS user starts the console and registers at all three HEs. MA-RMS Maintenance Environment was started on all three HEs, none containing the monitoring service. Grid administrator started the RMS Console and registered at all three HEs. To achieve a desired end state, a Management Console agent generates necessary operations and establishes interdependencies between them. The operations are the following ones: service and testbed migration, installation, setting of the execution parameters and starting.

The service containing testbed has to be picked up and set for start on one of the HEs. After that, the configuration can be easily copied on all other systems. Each operation has two variables, one for the input conditions and the other for the output notification. Input variables define the conditions to be fulfilled so that the operation begins the execution. After completion, the operation has to notify the operations that are waiting it to complete the job whether it was successful or not. The variables have two values: report for successful completion of the operation, and abort for unsuccessful operation. Once the operation sends the abort variable as its output, that variable propagates through the entire operations communication tree. Thus, all remaining operations fail and the users must be notified accordingly.

In the first part of the scenario, the service was transferred over the network in the usual way – the agent loaded the whole installation archive into the memory. Since the monitoring service used in the scenario is about 5 MB in size, there were no significant problems during the service migration.

The second part of the scenario deals with service upgrading with a newer version. A Grid administrator wants to terminate the existing user application and to install and run a new version of the service in this hosting environment. The following operations must be executed: migrate and install the new version, stop the old one, set the new version to be the active one and start the new version.

This concept, in which the service is downloaded directly from the web, is particularly powerful when it comes to updating multiple HEs running monitoring service as soon as the new version gets released. The important fact to notice is that the service can be downloaded directly from Web server (i.e. directly from third-party developer). That implies that the downloaded archive file is not modified for MA-RMS. It is the same publicly available archive file that can be downloaded and manually installed; only RMS does it automatically. So when a new version of some service gets published on the web, the only thing that has to be done is to write a relatively simple Ant script and pack it into an MA-RMS installation archive. The service can then be transferred to the HEs in the same way it would have been done manually, without providing any special distribution channels.

3.5 Conclusions

When developing a new software product, every development team or a company is confronted with the fact that the product life cycle does not end when the software is delivered to the users. It is often necessary to modify the software at a later time in order to fix bugs that have been discovered by the users or simply because new functionality has to be added into the software due to the change in requirements. Because of this the software maintenance process is a very important step in software life cycle.

By IEEE definition, the software maintenance is a modification of a software product after the delivery. The purpose of software maintenance is to correct the faults, to improve the performance or other attributes, to adapt the product to environmental changes, or to improve the product maintainability. This process is divided into several phases: problem/modification identification, classification and prioritization, analysis, design, implementation, regression/system testing, acceptance testing and delivery. Each of these steps is needed to effectively find the problem, solve it and finally deliver the new version of software to the user. The last step is very complicated when it is required to deploy new version of the software to a large number of geographically distributed network nodes. Another problem with the software development is that often the software on the targeted node does not work equally as during the testing phase on the testing environment.

For that purpose the Mobile-Agent Remote Maintenance Shell (MA-RMS) as a mobile agent–based system for distributed software maintenance is described. It is designed as an active shield used for protecting the system from faulty functionality of a newly created and installed application as well as an interface between the application and new operating system it is run on. The operations provided by the MA-RMS are: software migration, installation, starting, stopping, uninstallation, and finally running and tracing. MA-RMS has several execution modes which enable the gradual introduction of a new application into the remote system. This allows the application to be fully tested on the remote network node(s) and after it is verified that the application works properly it is fully deployed. All operations are performed by mobile software agents that allow the framework to intelligently perform the maintenance process.

This framework is currently used as part of an effort to create an agent based framework for service provisioning in telecommunication systems that would allow the telecommunication providers to efficiently deploy new services, adopt them and configure according to the providers business strategies in a cost effective way. The other very important factor is to enable service personalization and to provide deploying an application based on user profiles. Every user has its profile which defines user preferences. Those preferences define which types of applications the user prefers, the service quality he expects as well as the content types he likes. Based on his profile the framework triggers the automated deployment of application that offer new services to the users and installs the service on network nodes and user terminals according to provider's business rules. Future work will include modifying the MA-RMS with goal oriented concepts introduced in the Belief, Desire and Intention (BDI) paradigm. The goal is to use these concepts to increase the adaptability of MA-RMS to any changes in the software maintenance process. It will also decrease the complexity of adopting third-party application to MA-RMS thus allowing the software maintenance process to be performed on a wider range of application [23, 24].

Acknowledgments

This work was carried out within the research project 036-0362027-1639 "Content Delivery and Mobility of Users and Services in New Generation Networks", supported by the Ministry of Science, Education and Sports of the Republic of Croatia, and "Agent-based Service & Telecom Operations Management", supported by Ericsson Nikola Tesla, Croatia. Results from this work were patented with the European Patent Office under EP1518175 and Patent Nr. EP1306754.

References

1. Pigosky, T.M.: Practical Software Maintenance. Wiley, New York (1996)
2. IEEE Std. 1219: Standard for Software Maintenance. IEEE Computer Society Press, Los Alamitos (1993)

3. Canfora, G., Cimitile, A.: Software Maintenace. In: Handbook of Software Engineering and Knowledge Engineering, Ch:2, 1st edn., vol. 1. World Scientific Publishing Company, Singapore (2002)
4. Jezic, G., Kusek, M., Ljubi, I.: Mobile Agent Based Distributed Web Management. In: Proc. 4th Int. Conference on Knowledge-Based Intelligent Engineering Systems & Allied Technologies, Brighton, vol. 2, pp. 679–682 (2000)
5. Lovrek, I., Kos, M., Mikac, B.: Collaboration between academia and industry: Telecommunications and informatics at the University of Zagreb. Computer communications 26(5), 451–459 (2003)
6. Mikac, B., Lovrek, I., Sinković, V., Car, Ž., Podnar, I., Pehar, H., Carić, A., Burilović, A., Naglić, H., Sinovčić, I., Visković-Huljenić, T.: Assessing the Process of Telecommunications Software Maintenance. In: Proceedings of the combined 10th European Software Control and Metrics conference and 2nd SCOPE conference on Software Product Evaluation ESCOM-SCOPE 1999, Herstmonceux, pp. 267–275 (1999)
7. Podnar, I., Mikac, B., Carić, A.: SDL Based Approach to Software Process Modeling. In: Conradi, R. (ed.) EWSPT 2000. LNCS, vol. 1780, pp. 190–202. Springer, Heidelberg (2000)
8. Lovrek, I., Caric, A., Huljenic, D.: Remote Maintenance Shell: Software Operations using Mobile Agents. In: ICT 2002, International Conference on Telecommunications, Beijing (2002)
9. Windows Server Active Directory (2003), http://www.microsoft.com/windowsserver2003/technologies/directory/activedirectory/default.mspx
10. Using Group Policy to Deploy Office, http://office.microsoft.com/en-us/ork2003/HA011402011033.aspx
11. Using Group Policy to Deploy Applications, http://www.windowsnetworking.com/articles_tutorials/Group-Policy-DeployApplications.html
12. IBM Redbooks, Deployment guide series: IBM Tivoli Provisioning Manager Version 5.1, http://www.redbooks.ibm.com/abstracts/sg247261.html?Open
13. Kusek, M., Jezic, G., Ljubi, I., Mlinaric, K., Lovrek, I., Desic, S., Labor, O., Caric, A., Huljenic, D.: Mobile Agent Based Software Operation and Maintenance. In: Proceedings of the 7th International Conference on Telecommunications ConTEL 2003, Zagreb, Croatia, pp. 601–608 (2003)
14. Jezic, G., Kusek, M., Desic, S., Caric, A., Huljenic, D.: Multi-Agent System for Remote Software Operation. In: Palade, V., Howlett, R.J., Jain, L. (eds.) KES 2003. LNCS, vol. 2774, pp. 675–682. Springer, Heidelberg (2003)
15. The Apache Ant Project, http://ant.apache.org
16. RFC 3261, SIP: Session Initiation Protocol, http://www.ietf.org/rfc/rfc3261.txt
17. Cockayne, W.R., Zyda, M.: Mobile Agents. Prentice Hall, Englewood Cliffs (1997)
18. Lovrek, I., Jezic, G., Kusek, M., Ljubi, I., Caric, A., Huljenic, D., Desic, S., Labor, O.: Improving Software Maintenance by using Agent-based Remote Maintenance Shell. In: Proceedings of International Conference on Software Maintenance, pp. 440–449. IEEE Computer Society, Amsterdam (2003)
19. Java Agent DEvelopment Framework – JADE, http://jade.tilab.com/

20. Marenic, T., Jezic, G., Kusek, M., Desic, S.: Using Remote Maintenance Shell for Software Testing in the Target Environment. In: Proceedings of 26th International Conference on Software Engineering (2nd International Workshop on Remote Analysis and Measurement of Software Systems), pp. 19–23, Edinburgh (2004)
21. Zivic, M., Rac, L., Medar, A., Kusek, M., Jezic, G.: Designing of a Distributed Web Application in the Remote Maintenance Shell Environment. In: Proceedings of the 12th IEEE Mediterranen Electrotechnical Conference, MELECON 2004, Zagreb, The Institute of Electrical and Electronics Engineers, pp. 709–712 (2004)
22. Jezic, G., Kusek, M., Marenic, T., Lovrek, I., Desic, S., Trzec, K., Dellas, B.: Grid service management by using remote maintenance shell. In: Jeckle, M., Kowalczyk, R., Braun, P. (eds.) GSEM 2004. LNCS, vol. 3270, pp. 136–149. Springer, Heidelberg (2004)
23. Jurasovic, K., Jezic, G., Kušek, M.: Using BDI Agents for Automated Software Deployment in Next Generation Networks. In: Proceedings of the 11th International Conference on Software Engineering and Applications, pp. 423–428 (2007)
24. Podobnik, V., Petric, A., Jezic, G.: The crocodileAgent: Research for efficient agent-based cross-enterprise processes. In: Meersman, R., Tari, Z., Herrero, P. (eds.) OTM 2006 Workshops. LNCS, vol. 4277, pp. 752–762. Springer, Heidelberg (2006)

4

Software Agents in New Generation Networks: Towards the Automation of Telecom Processes

Vedran Podobnik[1], Ana Petric[1], Krunoslav Trzec[2], and Gordan Jezic[1]

[1] University of Zagreb, Faculty of Electrical Engineering and Computing, Department of Telecommunications, Unska 3, HR-10000 Zagreb, Croatia
{vedran.podobnik,ana.petric,gordan.jezic}@fer.hr
http://www.tel.fer.hr

[2] Ericsson Nikola Tesla, R&D Center, Krapinska 45, HR-10000 Zagreb, Croatia
krunoslav.trzec@ericsson.com
http://www.ericsson.hr

Abstract. Today, there is a wide variety of telecom services available provided by a number of competing stakeholders in the telecom environment. Consequently, consumers in such an environment require efficient mechanisms to match their demands (requested services) to supplies (available services). Businesses, on the other hand, aim to achieve efficient allocation of their resources. This chapter introduces a combination of Artificial Intelligence mechanisms and Computational Economics concepts as a means of solving the above mentioned problems. These mechanisms and concepts are implemented in agent-based electronic markets. Electronic markets function as digital intermediaries that create value by bringing consumers and providers together to create transactional immediacy and supply liquidity, by supporting the exchange of demand and supply information, and reducing transaction costs. The efficiency of the created multi-agent system is realized by applying novel discovery and/or negotiation models.

4.1 Introduction

We are entering a period of time when everything is becoming digitized and almost all software and devices are innately network-aware [1]. The amazing advances in the Information and Communication Technology (ICT) industry over the past 60 years have enabled the advent of the ubiquitous *Network,* aimed at employing goal-directed applications which can intelligibly and adaptively coordinate information exchange and actions [2, 3]. However, this new *Network* is not a result of the latest paradigm shift enforced by ICT industry leaders, but an outcome of the natural fusion of three different technologies. These technologies are computers, the Internet and mobile networks.

Computers have come a long way, from the gigantic machines of the early 1950s to the today's micro-scale devices [4]. Throughout the history of computing, three main eras can be identified [5]. The first era was the era of mainframe computing, when large and powerful computers were shared by many people. The second era was the era of personal computing, when there was one computer per person. In the upcoming third era, we as humans will interact no longer with one computer at a time,

L.C. Jain, N.T. Nguyen (Eds.): Knowl. Proc. & Dec. Mak. in Agent-Based Sys., SCI 170, pp. 71–99.
springerlink.com

but rather with a dynamic set of small networked computers, often invisible and embodied in everyday objects in the environment (e.g., temperature sensor weaved into the clothes) [6]. This third era is the era of *ubiquitous computing* (now also called *pervasive computing*), or the age of *calm technology*, when technology recedes into the background of our lives [5].

The Internet emerged in the early 1970s, as a small network interconnecting just a few computers. As the Internet grew through the 1970s and 1980s, many people started to realize its potential. Nevertheless, the Internet did not experience real proliferation until the invention of the World Wide Web (WWW or simply Web 1.0), a service provisioned through the Internet infrastructure [7]. Web 1.0, as a global information medium enabling users to read and write via computers connected to the Internet, became the bearer of the digital revolution in the 1990s which was a major catalyst of globalization and an important driver of economic prosperity. Consequently, all further Internet evolution after the invention of Web 1.0, is characterized as Web X.0, in spite of the fact that the WWW is just one of many Internet services. Web 2.0, also called "the Social Web", is no longer simply about connecting information, but also about connecting people through various forms of social networks (e.g., Facebook[1], MySpace,[2] or LinkedIn[3]). The phrase "Web 2.0" was coined a couple years ago when the social networking phenomenon was recognized, now having tens of millions of users world-wide employing it on a daily basis for both personal and businesses uses. Web 3.0, also called "the Semantic Web", is the next stage in the evolution of the Internet in which it will become a platform for connecting knowledge. Web 3.0 is an evolutionary path for the Internet which will enable people and machines to connect, evolve, share, and use knowledge on an unprecedented scale and in many new ways make our experience of the Internet better[4]. One of the most promising Web 3.0 technologies, besides the Semantic Web [8], are intelligent software agents which can utilize semantically annotated information and reason in a quasi-human fashion.

Mobile network evolution started in the 1980s when the network was designed merely to provide voice communication. The first and second generations (1G and 2G) of mobile networks have enabled circuit-switched voice services to go wireless. As the Internet grew, it became necessary to ensure mobile Internet access. The 2G GSM (*Global System for Mobile communications*) system was enhanced to 2.5G by introducing data communication and packet-switched services into the GSM network. The technologies of 2.5G, GPRS (*General Packet Radio Service*) and EDGE (*Enhanced Data Rate for GSM Evolution*) were the first step towards creating a mobile Internet. The third generation (3G) system, known as UMTS (*Universal Mobile Telecommunications System*), has introduced higher data rates which enable multimedia communications. This has made mobile Internet access available to users, including a wide spectrum of Internet-based data services, better coverage and multiple services in a terminal. The development of mobile networks has continued in both access and core networks. The UMTS access network has been improved by HSPA (*High Speed*

---

[1] http://www.facebook.com
[2] http://www.myspace.com
[3] http://www.linkedin.com
[4] Project10X's Semantic Wave 2008 Report (http://www.project10x.com)

Packet Access) technology which enables very high bit rates and throughput focusing on streaming and interactive services. The core network incorporates IMS (*IP Multimedia Subsystem*) [9] which integrates mobile communications and the Internet, enabling the convergence of existing networks with the Internet in mobile broadband networks. Long Term Evolution (LTE) is working on the evolution of mobile communication systems beyond GSM-UMTS-HSPA systems (B3G) which introduce higher levels of capacity, bit rates and performance, and support new services and features.

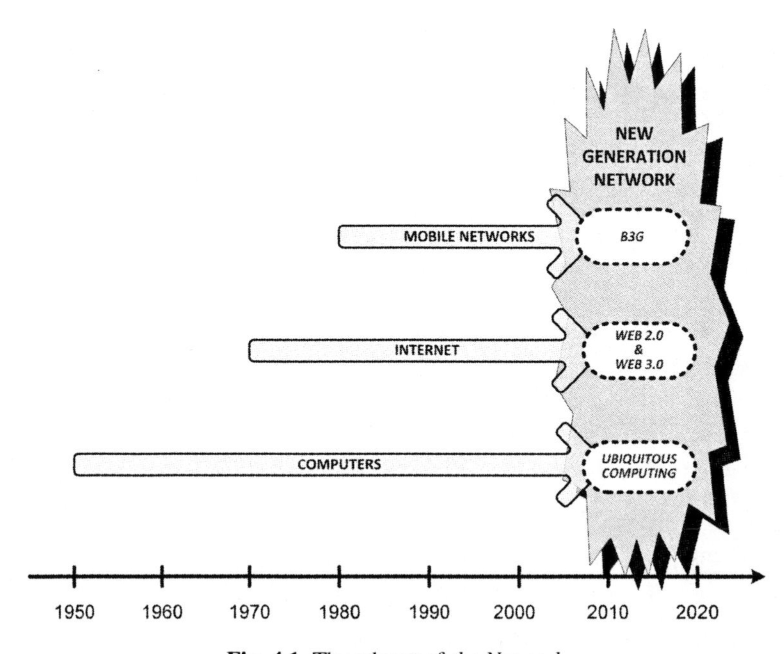

Fig. 4.1. The advent of *the Network*

As shown in Figure 4.1, computers, the Internet and mobile networks are technologies invented at different times with different purposes, each of them having their own individual evolution. Furthermore, they have all individually made an indelible footprint in recent human history. Nevertheless, today we are witnessing their fusion into a single, but extremely prominent and globally ubiquitous, technology: *the Network*. *The Network* will enable the transformation of physical spaces into computationally active and intelligent environments [10], characterized with ambient intelligence where devices embedded in the environment provide seamless connectivity and services all the time. This is aimed at improving the human experience and quality of life without explicit awareness of the underlying communication and computing technologies. Tremendous developments in wireless technologies and mobile telecommunication systems, as well as rapid proliferation of various types of portable devices, have significantly amended computing lifestyle, thus advancing the vision of ubiquitous computing toward technical and economic viability [11]. The vision of *the*

Network is becoming a reality with the new generation of communication systems: the New Generation Network (NGN) [12].

This chapter is organized as follows. First, we present the NGN, the telecom environment of the future that we are trying to automate with the help of the electronic market (e-market) concept and the agent-based computing paradigm. Section 4.3 briefly describes e-markets as an organizational framework for the future telecom domain, agents as representatives of all stakeholders participating in such an e-market, and auctions as an efficient mechanism for allocating resources/services in such an e-market. In Section 4.4, processes in telecom e-markets are described and the solutions for their automation are proposed. Next, agent-based B2B e-markets for information resources and transport capacities, which respectively use the multi-attribute auction and continuous double auction as negotiation protocols, are described. Finally, a multi-agent system (MAS) [13] implementing the agent-mediated B2C e-market is presented: the architecture of the designed e-market is described, as well as an overview of market mechanisms. Section 4.5 concludes this chapter.

4.2 New Telecom Environment

In the rapidly evolving telecom market, the upcoming NGN concept introduces multiple convergences, as presented in Figure 4.2:

- *network convergence*, describing the integration of wireline and wireless access technologies;
- *terminal convergence*, describing the introduction of adaptive services which can be delivered anytime, anywhere, and to any device the consumer prefers;
- *content convergence*, describing the ever-growing tendency of digitizing various forms of information;
- *business convergence*, describing the fusion of the ICT and media industries.

The number of telecom value-added service consumers is rising continuously. Moreover, the competition among stakeholders in the telecom market, as well as the NGN concept which introduces a whole spectrum of new services, enables consumers to be very picky. Consequently, realization of the full potential of the NGN will make it necessary for service providers to offer dynamic, ubiquitous and context-aware personalized services. Moreover, a large part of the services offered by the NGN will contain multimedia sessions which will be composed of a different number of audio and/or video communications with a certain quality of service (QoS) [14]. Providing such services to consumers transparently is challenging from the technical, business and social points of view. Consequently, when modeling novel relationships between stakeholders in the telecom market, it is important to analyze the new telecom environment with regard to the technical standpoint, but also to investigate the social and the economic perspectives.

There are a number of different stakeholders present in the telecom market. They need to establish strategic partnerships in order to provide end-users (i.e., consumers) with converged services, integrating information resources and transport capacities. In

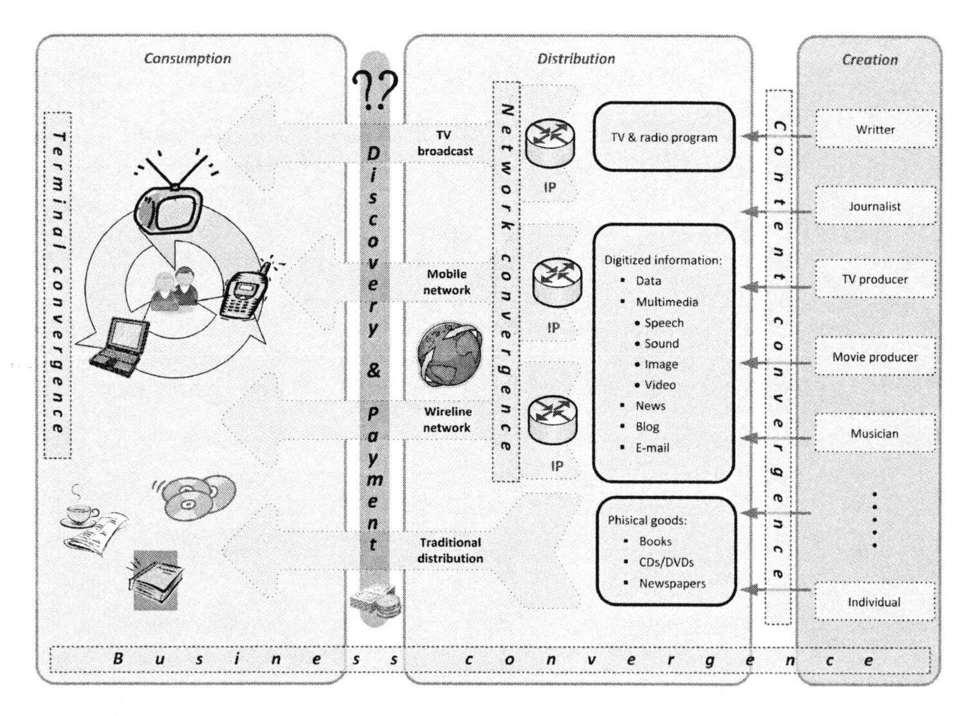

Fig. 4.2. The multiple convergences characteristic of the NGN environment

general, a business model is a model comprising stakeholders present in a market and their roles and relationships. A stakeholder may take on a number of roles in a particular scenario, and furthermore, a number of stakeholders can play the same role. Examples of roles include the following [15]:

- *Consumer:* A telecom service user, having at his disposal one or more devices (e.g., mobile phone, laptop, digital TV receiver) attached to the network (e.g., the Internet).
- *Access Provider:* Provides access to service consumers.
  - *Line Provider:* Provides physical access to service consumers (e.g., an operator providing wireline access through a local loop, a mobile/wireless access operator).
  - *Connection Provider:* Provides network layer access to services (e.g., an operator with entry points to the Internet).
- *Service Broker:* Provides simplified filtering and access to a vast amount of services available in the network (e.g., an Internet search engine).
- *Service Provider:* Facilitates integrated services for consumers (e.g., a company offering IPTV (*Internet Protocol Television*) or a telecom operator offering MMS (*Multimedia Messaging Service*)).
- *Information Enabler:* The enabler of information resources.
  - *Content Owner:* The owner of information in its original form (e.g., a movie producer or the stock exchange).

  - *Content Enabler:* Converts the information into a format suitable for network-based transmission (e.g., digitized information suitable for transmission over the Internet).
  - *Server Infrastructure Owner:* Provides storage capacity and server functionality (e.g., a company owning a "server farm").
  - *Wholesaler of Content:* Provides lower-cost content.
- *Transport Enabler:* The enabler of information resource transport through the network.

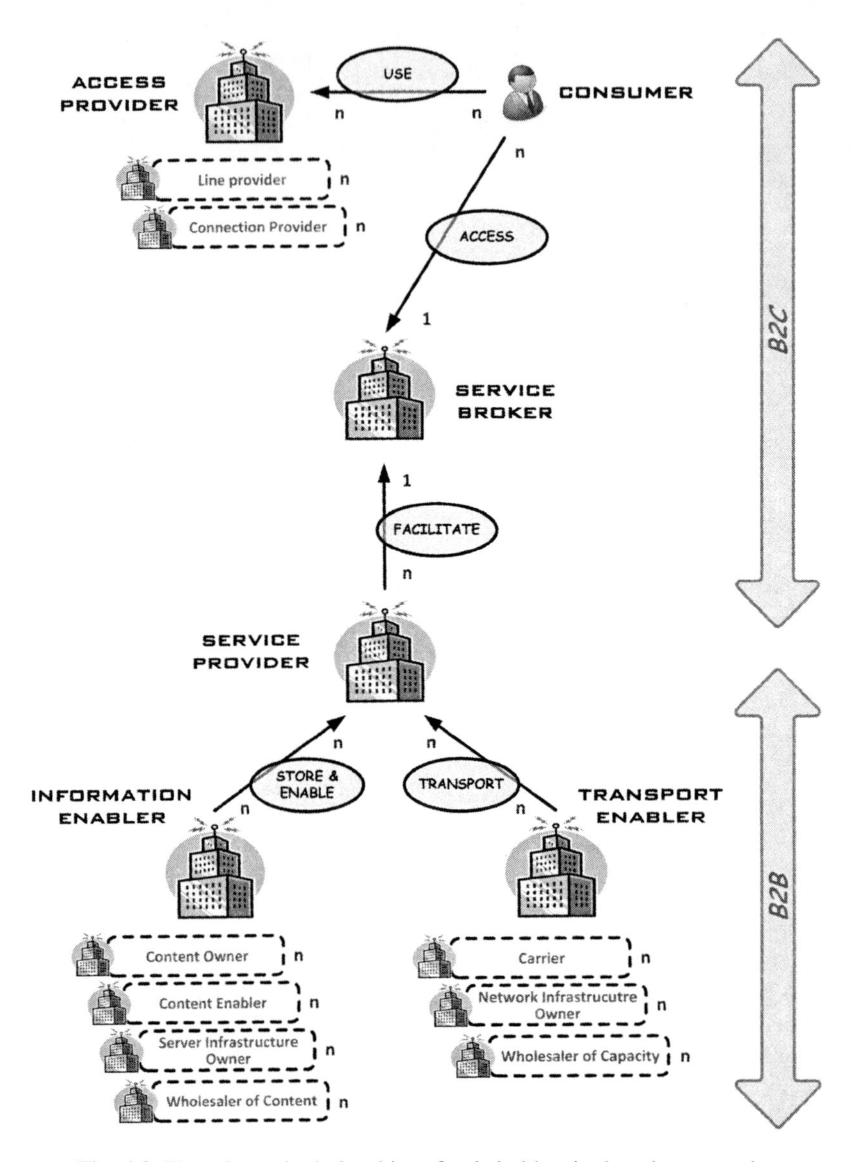

Fig. 4.3. The roles and relationships of stakeholders in the telecom market

- *Network Infrastructure Owner:* Provides transmission lines (e.g., telecoms or cable TV operators).
- *Carrier*: Provides a transport service for data traffic (e.g., companies which buy bandwidth from the Network Infrastructure Owner).
- *Wholesaler of Capacity:* Provides lower-cost transport capacity (e.g., large ISPs (*Internet Service Providers*) selling capacity to smaller ISPs).

Figure 4.3 illustrates the roles and relationships of stakeholders in the telecom market. The relationships may be divided into two broad categories: Business-to-Consumer (B2C) and Business-to-Business (B2B).

The NGN will create heterogeneous environments populated with diverse types of interconnected user devices that will cooperatively and autonomously collect, share, and process information aimed at adapting the associated context, as well as providing the user with unobtrusive connectivity and services at all times. Concurrently, a variety of enablers/providers will offer a remarkable selection of information resources through a set of services, while continuously competing with each other to improve market share and increase profit. In such an environment, consumers should be able to access all of the available services anytime, anywhere, with any device, regardless of the type of access network. Taking into consideration the growing portfolio of attractive new services, the existence of user related constraints, and the importance of business decisions and mechanisms to charge users for provided services, developing an automated service provisioning process is critical [16].

In this chapter, we propose an agent-based approach to enabling autonomous coordination between all stakeholders across the telecom value chain, thus enabling context-aware B2C service provisioning for consumers. Moreover, the proposed model also implements an automated mechanism for B2B trading between service providers and information/transport enablers.

4.3 Electronic Markets

One of the most important consequences of the NGN advent is the rapid proliferation of e-markets. E-markets function as digital intermediaries that create value by bringing consumers and providers together in order to create transactional immediacy and supply liquidity, as well as reducing transaction costs.

A variety of telecom services, information resources and transport capacities are currently being provided by a number of competing enablers/providers. Consequently, the telecom market is facing market saturation and fierce competition among telecom operators and other such providers [17]. Thus, it has become necessary to allocate services/resources dynamically and efficiently in order to be competitive. To meet this requirement, novel business relationships among telecom market stakeholders have emerged through the deployment of B2C and B2B e-markets.

4.3.1 Agents

The dynamic and distributed nature of services/resources in the NGN requires telecom stakeholders, not only to respond to requests, but also to intelligently anticipate and adapt to their environment. Software agents are the computing paradigm, which is very convenient for the creation of programs able to conform to the requirements set forth. A software agent is a program which autonomously acts on behalf of its principal, while carrying out complex information and communication tasks that have been delegated to it. From the owner's point of view, agents improve his/her efficiency by reducing the time required to execute personal and/or business tasks. A system composed of several software agents, collectively capable of reaching goals that are difficult to achieve by an individual agent or a monolithic system, is called a multi-agent system (MAS) [18].

In an MAS implementing the proposed B2C and B2B e-market models, intelligent software agents are used to impersonate businesses, consumers and mediators in the environment of the NGN in order to enable automated interactions and business transactions.

4.3.1.1 The Intelligent Software Agent Model

An agent must possess some intelligence grounded on its *knowledge base*, *reasoning mechanisms* and *learning capabilities*. The intelligence of an agent is a prerequisite for all its other characteristics. Depending on the assignment of a particular agent, there are differences in types of information contained in its knowledge base. However, generally this information can be divided into two parts – the *owner's profile* and the agent's *knowledge about its environment*. It is very important to notice that the agent's knowledge base does not contain static information. Adversely, the agent continuously updates its owner's profile according to its owner's latest needs. This allows the agent to efficiently represent its owner, thus realizing the *calm technology* concept. Calm technology is that which serves us, but does not demand our focus or attention [5]. Furthermore, the agent also updates knowledge regarding its environment with the latest events from its ambience and the current state of observed parameters intrinsic to its surroundings, thus realizing *context-awareness*. Context-awareness describes the ability of an agent to provide results that depend on changing context information [19]. In our model, we differentiate between *situation context* (e.g., user location and environment temperature) and *capability context* (e.g., features of a user's terminal). An agent executes tasks *autonomously* without any interventions from its owner, making it an invisible servant. An agent must be *reactive*, so it can properly and timely respond to impacts from its environment. An agent not only reacts to excitations from its environment, but also takes initiatives coherent to its tasks. A well-defined objective is an inevitable prerequisite for *proactivity*. An efficient software agent collaborates with other agents from its surroundings: i.e.; it is *cooperative*. If an agent is capable of migrating between heterogeneous network nodes, it is classified as a *mobile* software agent [20]. An agent has a lifetime throughout which the persistency of its identity and its states should be retained. Thus, it is characterized by *temporal continuity*.

Figure 4.4 visualizes a generic model of an intelligent software agent [21, 22, 23, 24, 25], which we used while designing an MAS employing the B2C and B2B telecom e-markets.

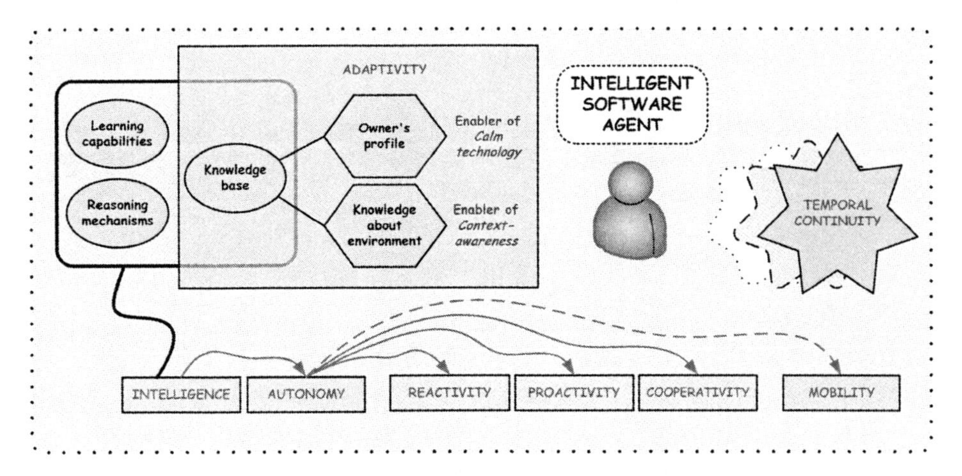

Fig. 4.4. A visualization of intelligent software agent characteristics

4.3.1.2 Ontology-Based Communication

Ontologies provide a shared vocabulary to represent the meaning of entities, while knowledge representation provides structured collections of information and inference rules for automated reasoning. As a result, intelligent software agents can interpret and exchange semantically enriched knowledge for users (Web 3.0) [26]. Ontologies represent a formal and explicit Artificial Intelligence (AI) tool for facilitating knowledge sharing and reuse [27]. An ontology refers to a description of concepts and relationships between these concepts in an area of interest. Therefore, an ontology is the terminology used for a given domain of interest. Ontologies can refer to other ontologies, and thus create domain-dependent terminologies which describe certain concepts and relationships in more detail.

"*People can't share knowledge if they don't speak a common language*" [28]. This statement can be transposed from "people" to "machines". The exchange of knowledge is possible only if the communication participants speak a common language, i.e., they are able to map a sign to an object in the same way. This means that both evoke the same concept when using the same sign (see Figure 4.5). Ontologies should enable machines to achieve such a common understanding.

In our MAS implementing the proposed B2C e-market model, the Semantic Web technology [1, 29] is used to represent telecom services on the market. By applying OWL-S (*Web Ontology Language for Services*), every telecom service can be described with an ontology. Each OWL-S ontology utilizes one or more domain ontologies which define the concepts important for a particular domain of interest. Concepts in domain ontologies, as well as the relations between the concepts themselves, are specified with OWL (*Web Ontology Language*), a semantic markup language based on description logics (*DL*) and used for publishing and sharing ontologies on the Internet.

4.3.2 Auctions

The range and value of objects sold in auctions today are taking astonishing proportions. Auctions are defined as a market institution that acts in pursuant of a set of

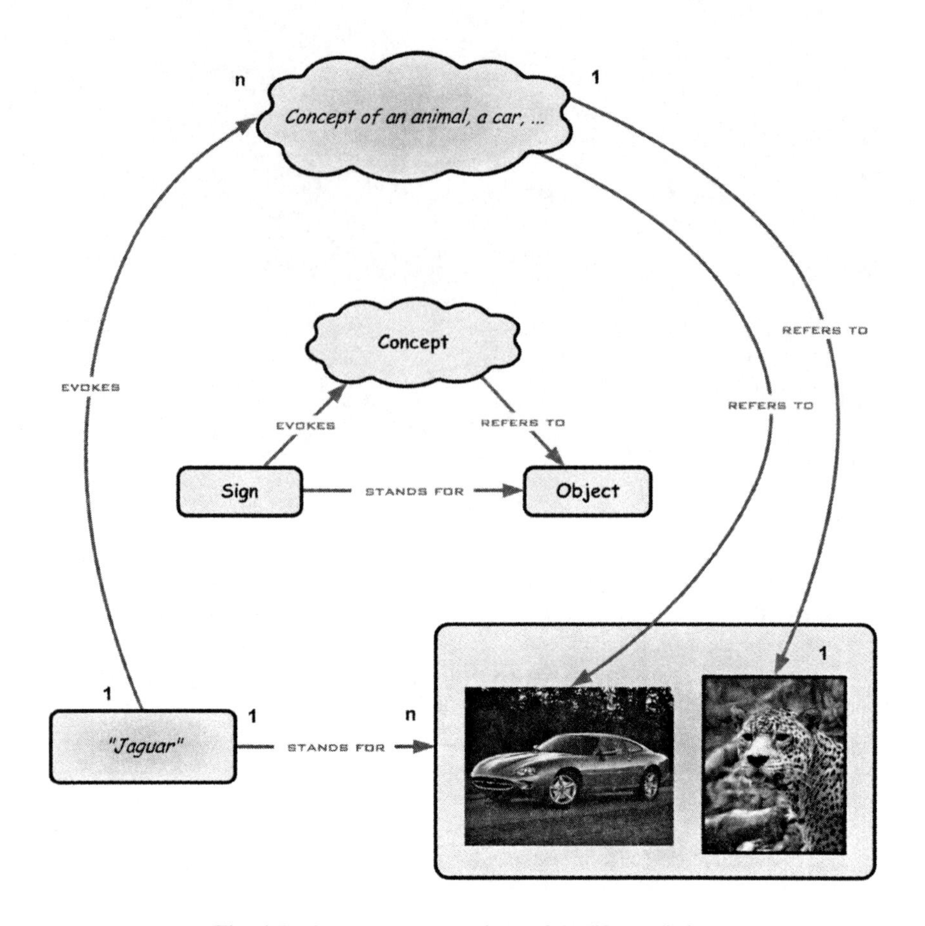

Fig. 4.5. Multiple interpretations of the "Jaguar" sign

predefined rules. Based on offers placed by market participants, resource allocation and prices are determined [30]. Due to their well defined negotiation protocols, auctions are also suitable enablers of negotiations in agent-based e-markets.

An auction defines an explicit set of rules for determining resource allocation and prices on the basis of traders' messages. These messages include an offered price, called a *bid*, for an offer to buy and a query, called an *ask*, for an offer to sell. This competitive process serves to aggregate scattered information regarding traders' declared valuations and to dynamically set a transaction price. An auction is one-sided if only bids or only asks are permitted, and two-sided (double) if both are permitted. Moreover, an auction can be either one-shot or repeated. A repeated auction consists of several rounds. In a multi-object auction, multiple units of the same or different items are on sale. Such auctions can be individual or combinatorial, depending on whether offers are restricted only to individual items or whether they can include a combination of several items. In a homogeneous multi-object auction, a number of identical units of a standardized item (i.e., commodity) can be auctioned, thus referred to as a multi-unit auction. Multi-unit auctions can be discriminatory or uniform, depending on whether units are sold at different or equal prices. They are of great practical importance, and have been applied to selling transport capacity in telecom networks, MWs of electric power, as well as capacity units of natural gas and oil pipelines [31]. In the simplest multi-unit auction scenario, each buyer is interested in only one unit. Multi-attribute auctions are an extension to standard auction theory [32], which enables negotiation regarding several attributes in addition to the price of the resource.

4.4 The Automation of Processes in the Telecom Environment

The immense growth of the Internet and the rapid proliferation of e-markets have provoked dynamic and extensive research aimed at developing efficient e-market models [33, 34, 35, 36]. The most prominent model for systematic analysis of processes in B2C e-markets is the CBB (*Consumer Buying Behavior*) model, while the BBT (*Business-to-Business Transaction*) model systematically analyses processes in B2B e-markets.

4.4.1 Processes in a B2B e-Market

The proliferation of auctions on the Internet, and the dynamic nature of auction interactions, argues for the development of intelligent trading agents which act on behalf of human traders (i.e., buyers and sellers). The nature of buying and selling tasks makes trading agents one of the most relevant applications of agent technologies. Namely, an agent, i.e. a software component, can monitor and participate in the market continuously. Furthermore, an agent has the autonomy to make decisions according to the preferences of the user it represents, thus enabling it to autonomously place offers on an e-market.

From the BBT model perspective [37], we can formally identify six fundamental steps which must be executed in order to successfully complete one transaction in a B2B environment. These steps are as follows (Figure 4.6): 1) *partnership formation*, 2) *brokering*, 3) *negotiation*, 4) *contract formation*, 5) *contract fulfillment*, and 6) *service and evaluation*. Even though B2B e-markets have significantly less offered resources than B2C markets, B2B negotiation is much more complex. Typically, for example, it involves larger volumes, repeated transactions and more complex contracts. This is the reason why most researchers have concentrated on the negotiating phase of B2B market transactions.

The *partnership formation* phase usually includes forming of a new virtual enterprise or finding partners to form a supply chain. A virtual enterprise represents a form of cooperation of independent stakeholders which combine their competencies in order to provide a service [38]. On the B2B e-market, *content owners*, *content enablers, server infrastructure owners* and *wholesalers of content* can form a *virtual enterprise* in order to successfully place and sell information resources to various service providers. Moreover, *carriers*, *network infrastructure owners* and *wholesalers of capacity* may also form a *virtual enterprise* to enhance trading with transport capacity.

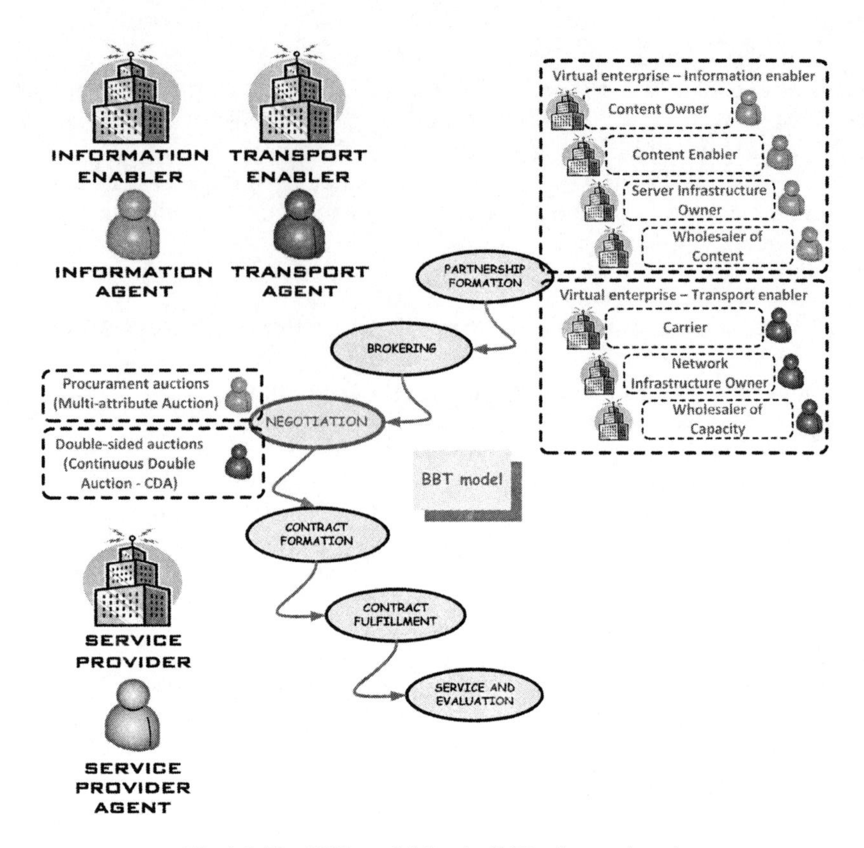

Fig 4.6. The BBT model for the B2B telecom domain

With the expansion of the e-market, the number of buyers and sellers grows accordingly making it more difficult to find all potential business partners trading a requested service/resource. The main role of the *brokering* phase is to match service providers with information/transport enablers that sell resources/capacities needed for the creation of a new service or improvements on an old one.

Negotiation is a process which tries to reach an agreement regarding one or more resource attributes (e.g., price, quality, etc.). The trading agent uses a negotiation strategy suitable for the type of auction applied (i.e., negotiation protocol) on the market. The negotiation protocol defines the rules of encounter between trading agents. It should ensure that the negotiation's likely outcome satisfies certain social objectives, such as maximizing allocation efficiency (i.e., ensuring that resources are awarded to the participants who value them the most) and achieving market equilibrium [39]. The negotiation strategy represents a set of rules that determines the behavior of a trading agent. It is defined by the resource provider and aims to help buy or sell a particular resource optimally (i.e., trying to maximize the expected payoff).

The negotiation process can be either distributive or integrative [40]. In distributive negotiations, one issue (i.e., resource attribute) is subject to negotiation while the

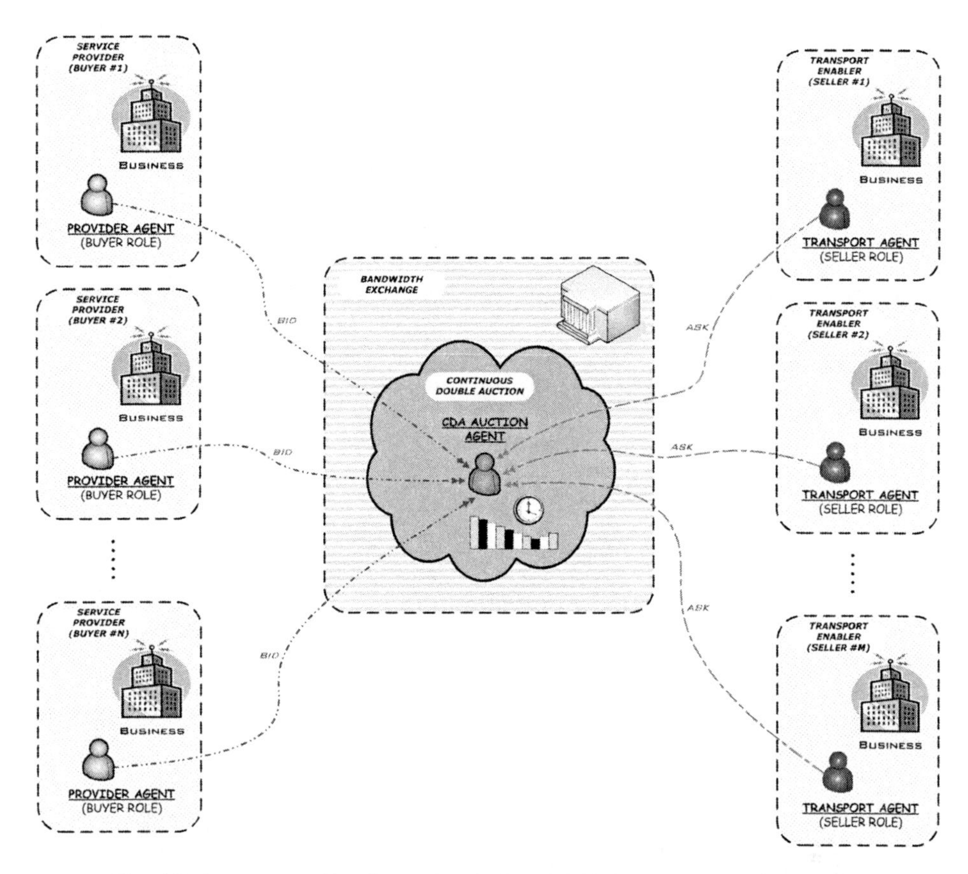

Fig. 4.7. An agent mediated e-market for bandwidth (transport capacity) trading

parties involved have opposing interests. One party tries to minimize loss (to give as little as possible) and the other party tries to maximize gain (to receive as much as possible). Distributive negotiations are also characterized as "win-lose" negotiations. The continuous double auction (CDA), which is suitable for optical transport capacity (i.e., bandwidth) trading in a B2B e-market for transport capacities (see Figure 4.7), represents a distributive type of negotiation in a multi-unit auction with multiple buyers and sellers [35, 36]. The CDA is one of the most common trading mechanisms and is used extensively in stock and commodity exchanges. It yields good profits for traders, as well as efficient market outcomes [39].

In integrative negotiations, multiple issues are negotiated while the parties involved have different preferences towards these issues. For example, two information enablers may want to sell multimedia information resources to a portal provider, but one is primarily interested in the sale of news (video with voice), whereas the other is interested in the sale of movie clips. These variant valuations can be exploited to find an agreement resulting in mutual gain. If their preferences are the same across multiple issues, the negotiation remains integrative until opposing interests are identified. In such a case, both parties can realize gains: consequently, another name for this class of negotiations is "win-win" negotiations. A multi-attribute auction represents an integrative negotiation process which can be used after information resource discovery in the e-market.

Item characteristics (attributes) represent an important factor in deciding which auction should be used in the negotiation phase. Negotiation on commodities, such as transport capacities, focuses mainly on the price of the item. These items are mostly sold in conventional single-attribute auctions. On the other hand, complex items such as information resources often require negotiation of several attributes, and not just the price [41]. They are sold in multi-attribute auctions [42, 43], which are a special

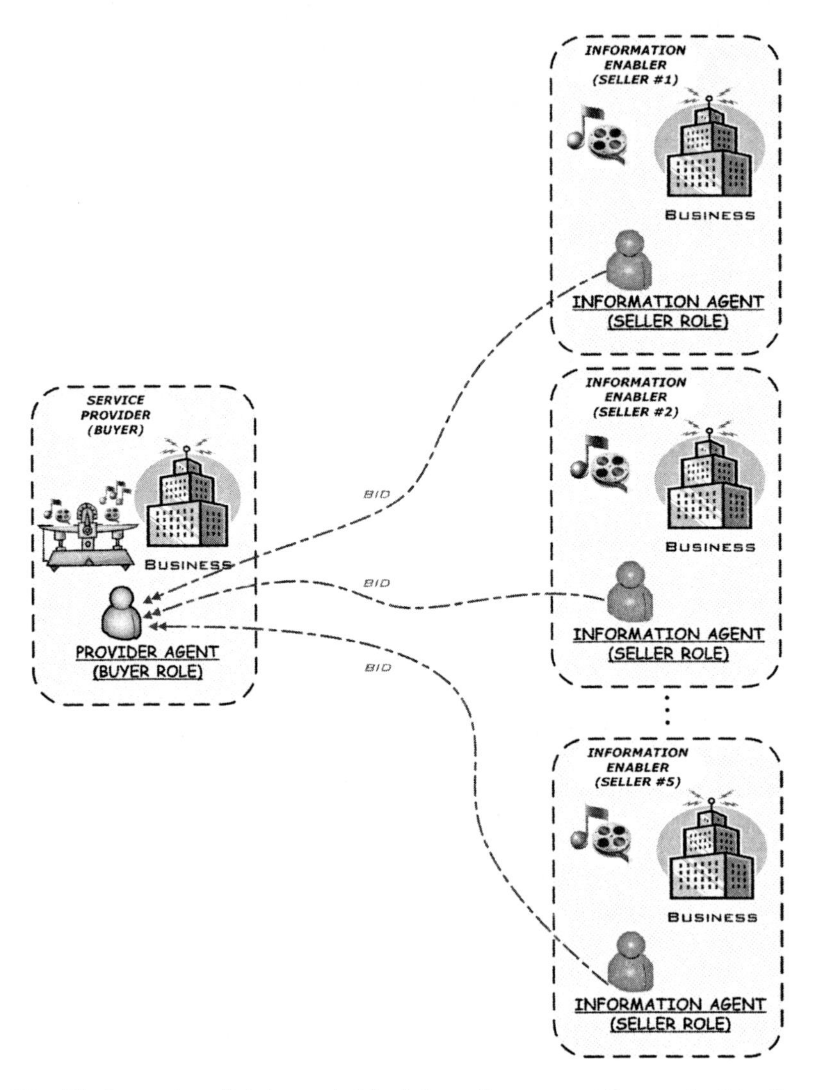

Fig. 4.8. An agent mediated e-market for information resource (i.e., content) trading

case of procurement auctions. Procurement auctions are also called reverse auctions since there are multiple sellers (e.g., information enablers) and only one buyer (e.g., service provider) that purchases items (e.g., information resources). Multi-attribute auctions have been attracting more and more attention in B2B markets since the price is not the only important attribute considered in the decision making process[5].

The first step in a multi-attribute auction is for the buyer to specify his preferences regarding the item he wishes to purchase. Preferences are usually defined in the form of a scoring function based on the buyer's utility function [44]. In order to familiarize sellers with buyer's valuations of relevant attributes, the buyer usually publicly announces his scoring function. Sellers are not obligated to disclose their private values of an item. The winner of the multi-attribute auction is the seller that provided the highest overall utility for the buyer. The buyer sends a request to all interested sellers which than reply by sending bids. The buyer selects the bid with the highest overall utility. If the auction is one-shot, this bid is declared the winning one, otherwise it is declared as the currently leading bid and the new round of the auction begins. The buyer can also define the bid increment or minimum requirements the bid has to fulfill in order to compete in the next round. Figure 4.8 shows a multi-attribute auction between a service provider and several information enablers. Information enablers offer multimedia content composed of video and audio streams with different performances. Based on its utility function, the service provider reaches an agreement with the information enabler whose information resource has the highest overall utility.

4.4.2 Processes in a B2C e-Market

When designing a solution for the automation of a B2C telecom e-market, one important phenomenon of telecom supply chains should be considered:

- in classical supply chains a factory cannot use the same purchased component in more than one product;
- in service supply chains, a service provider can reuse some purchased components in multiple services: for instance, when a service provider purchases a movie clip from an information enabler, the service provider can reuse the same movie clip in several different services, whereas every service can be provisioned to an unlimited number of consumers.

Additionally, B2C telecom markets have a huge number of players on the consumer side, but just a few service providers on the business side.

Figure 4.9 shows the introduction of software agents into a B2C telecom market. The entities marked as the *Consumer Agent (CA)*, *Broker Agent (BA)* and *Provider Agent (PA)* impersonate consumers, service brokers, and service providers, respectively.

The model we use for systematic analysis of processes in B2C e-markets is based on the Consumer Buying Behavior (CBB) model [45], adapted for the B2C telecom domain [46]. The model identifies six fundamental steps, shown in Figure 4.10, which need to be executed in order to successfully complete one transaction in the B2C

[5] http://www.cindywaxer.com/viewArticle.aspx?artID=149 (Business 2.0 magazine)

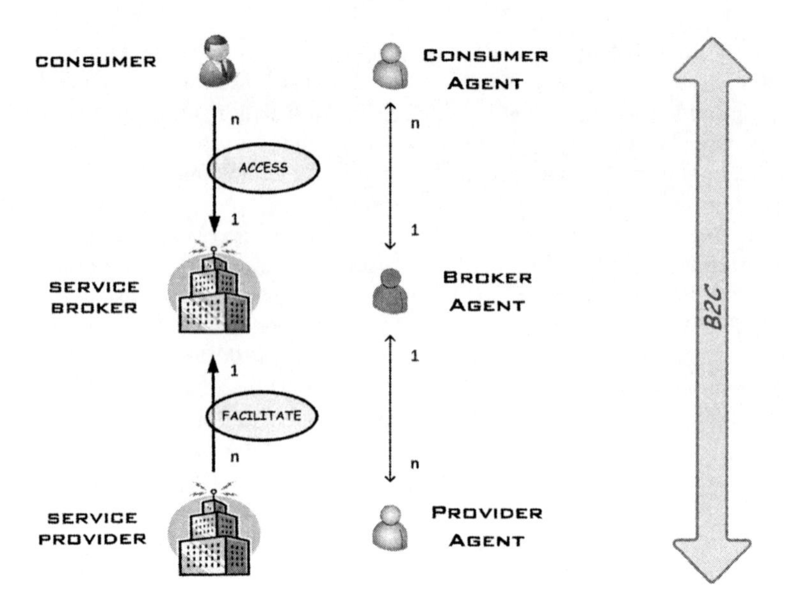

Fig. 4.9. A multi-agent system for automation of processes in a B2C telecom market

telecom environment: 1) *identifying needs*, 2) *service brokering*, 3) *provider brokering*, 4) *negotiation*, 5) *service provisioning and charging*, and 6) *provider evaluation*. A brief description of these steps follows.

Identifying needs: Prior to service utilization, the consumer must become aware of the existence of a service which conforms to his/her needs. One of the possible mechanisms for achieving this is using recommendations from consumers with similar profiles.

Service brokering: This step is responsible for determining *what* (which service) to buy, which encompasses an evaluation of service alternatives based on consumer-provided criteria. The output of this stage is called the *consideration set* of services.

Provider brokering: This step combines the derived *consideration set* with service provider-specific information to help determine which service provider to buy the service from. This includes an evaluation of service provider alternatives based on consumer-selected criteria (e.g., price, reputation, etc.).

Negotiation: After the consumer has selected the service provider which offers the desired service, (s)he must negotiate the conditions (e.g., price, QoS, etc.) for service utilization.

Service provisioning & charging: Once service utilization is arranged, the service provider provisions the service with the agreed level of QoS, while the consumer is charged accordingly.

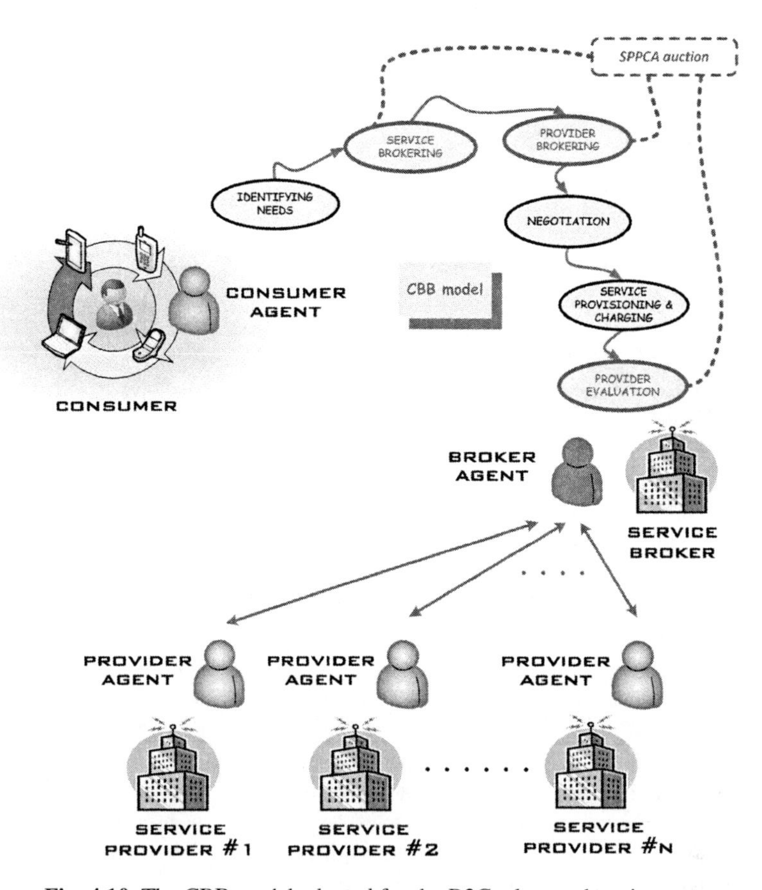

Fig. 4.10. The CBB model adapted for the B2C telecom domain

Provider evaluation: This step evaluates of the consumer's level of satisfaction with respect to the service provided by the selected service provider and its conformance to the agreed QoS level specified in a Service Level Agreement (SLA)[6].

In B2C e-markets, almost an unlimited quantity of available services is offered to each consumer. Consequently, identifying a small set of services which best satisfy his/her needs in order to negotiate the utilization of one of them is a crucial problem for consumers.

Here we propose an integrated solution for the *service* and *provider brokering* phases[7]. Our model is also partially based on information acquired in the provider evaluation phase, thus emphasizing the integrity of the CBB model. The final ranked set of eligible services, recommended to consumers in response to their requests, is highly-relevant input for the *negotiating* phase.

---

[6] ITU-T Recommendation E.860, Framework of a service level agreement, June 2002.

[7] The Nicosia model of B2C e-market transactions merges both brokering phases into one *Search Evaluation* stage [45]. Hereafter we use the term *discovery* when referencing both brokering phases in the CBB model.

Discovery is the process of searching for possible matches between requested and available services. The objective of this process is not simply to find all the available services which match a requester's demand. Namely, efficient discovery processes should identify all the supplies that can fulfill a given demand to some extent, and then propose just the most promising ones [47]. A number of discovery infrastructures have been proposed. However, most approaches either lack a mechanism for ranking matches or rank potentially suitable services according to only their semantic attributes (i.e., semantic matchmaking).

In this chapter, in order to rank potentially adequate services offered to consumers in the B2C e-market, we use a combination of Artificial Intelligence (AI) mechanisms and Computational Economics (CE) concepts. Firstly, by introducing an auction model based on Pay-Per-Click (PPC) advertising auctions[8], we enable service providers to dynamically and autonomously advertise semantic descriptions of the services they provide. We call this new auction model the Semantic Pay-Per-Click Agent (SPPCA) auction.

4.4.2.1 The Architecture of the Implemented B2C e-Market

A description of the implemented agent-mediated B2C e-market architecture follows, along with a demonstration of how it operates. Figure 4.11 illustrates the e-market supported by a discovery mediator (i.e., service broker). Here, we consider the discovery mediator to be a "black box". The specific mechanisms enabling the functionalities of the discovery mediator are described in the next subsection.

Our proof-of-concept prototype is implemented as a JADE (*Java Agent DEvelopment Framework*) MAS [51]. In the prototype, agents communicate by exchanging ACL (*Agent Communication Language*) messages. Efficient coordination between agents is achieved by applying FIPA (*Foundation of Intelligent Physical Agents*) interaction protocols. Two types of pre-defined FIPA conversation protocols – FIPA Request and FIPA Contract-Net are used.

As shown in Figures 4.9, 4.10 and 4.11, we introduce three types of agents in the B2C e-market: *Consumer Agents (CA)*, *Provider Agents (PA)* and *Broker Agents (BA)*. These agents enable automated service discovery. A more detailed description of the above mentioned agents and their interactions follows.

The Broker Agent: A BA represents a discovery mediator between the remaining two types of agents, i.e., PAs and CAs. A BA enables PAs to advertise their service descriptions and recommends ranked sets of eligible services to CAs in response to their requests. It is assumed that the BA is a trusted party which is fairly neutral between

---

[8] By virtue of Pay-Per-Click (PPC) auctions (also called paid or sponsored search mechanisms), content providers pay Web search engines to display sponsored links in response to user queries alongside the algorithmic links, also known as organic or non-sponsored links [48]. Pay-Per-Click auctions are an indispensable part of the business model of modern Web search engines and are responsible for a significant share of their revenue [49]. PPC auctions run continuously for every possible character sequence. In each auction, a competitor c enters a bid $b_k(c)$ which is the amount (s)he is willing to pay should a customer click on his/her advertisement in the search results for keyword k. The auctioneer (e.g., Google) sorts the bids for keyword/auction k and awards position 1 to the highest bidder, position 2 to the second highest bidder, and so on. Each participant pays an amount equal to the number of customers that visit their web-site multiplied by their bid price [50].

service requesters and service providers. When a BA is contacted by another agent, it must first determine whether the other agent is a PA or a CA. If the other agent is a PA, the BA includes it into the SPPCA auction. As a result, the PA is allowed to advertise its services at the discovery mediator the BA represents. If the other agent is a CA, then the BA provides it with a ranked list of eligible services in response to its discovery request.

The Provider Agent: A PA represents a business which offers a certain service. Initially, each PA wishes to advertise its service at the discovery mediator entity. A PA accomplishes this by participating in the SPPCA, which enables businesses to dynamically and autonomously advertise semantic descriptions of their services. After successfully advertising its service, the PA waits to be contacted by a CA which is interested in the service it is providing.

The Consumer Agent: A CA acts on behalf of its human owner (consumer) in the discovery process of suitable services and subsequently negotiates the utilization of these services. A CA wishes to get an ordered list of ranked advertised services which are most appropriate with respect to its needs. After a CA receives recommendations from the BA, it tries to contact a desired number of proposed PAs and find the one

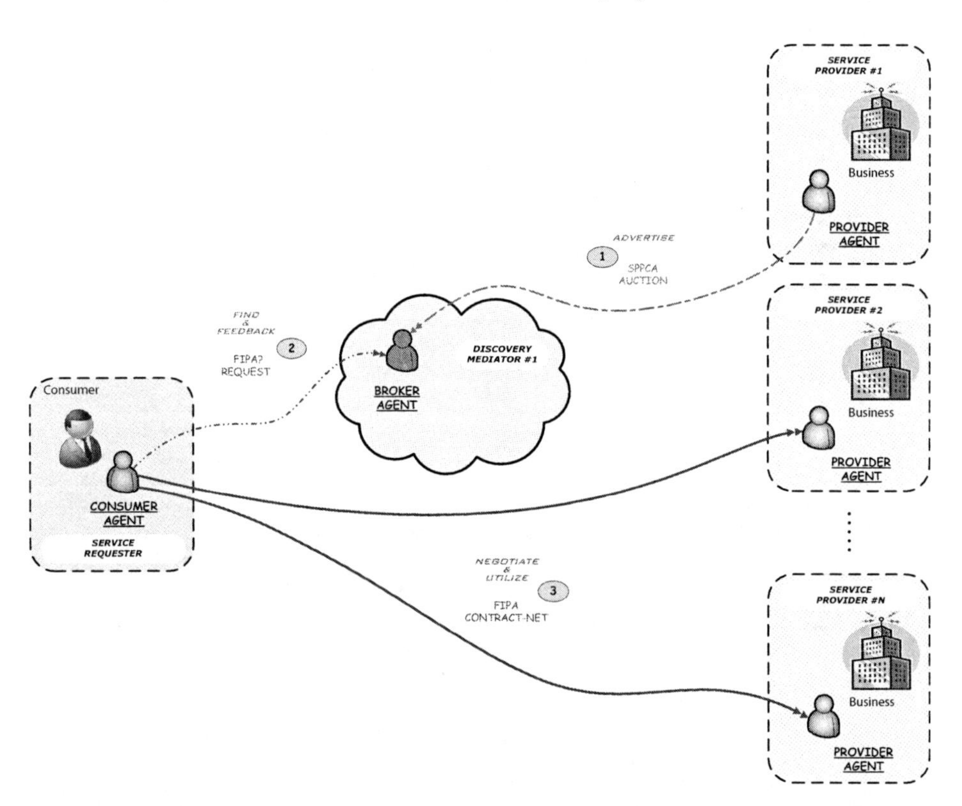

Fig. 4.11. An agent-mediated B2C e-market

which offers the best conditions (e.g., the lowest price) for utilizing the requested service. At the end of this conversation, the CA sends a message to the BA with information about its level of satisfaction regarding the proposed PAs.

4.4.2.2 The Discovery Model Used in the Implemented B2C e-Market

The discovery model for semantic-annotated services which is presented in this chapter uses a combination of AI mechanisms and CE concepts. CAs use BA's capability of two-level filtration of advertised service descriptions to efficiently discover adequate ones. First-level filtration is based on semantic matchmaking between descriptions of services requested by consumers (i.e., CAs) and those advertised by service providers (i.e., PAs). Services which pass the first level of filtration are then considered in the second filtration step. Second-level filtration combines information regarding the actual performance of businesses that act as service providers and the prices bid by PAs (in the SPPCA auction). The performance of a business (with respect to both price and quality) is calculated from the CAs' feedback ratings. Following filtration, a final ranked set of eligible services is chosen. This set is then recommended to the CAs in response to their requests.

Figure 4.12 shows the architecture of a discovery mediator entity, which acts as an intermediary in the modeled B2C e-market. Note that the BA serves as an interface agent between CAs/PAs and the discovery mediator entity. The *SPPCA Auction Agent (SAA)*, the *Matching Agent (MA)* and the *Discovery Agent (DA)* enable the necessary functionalities of the discovery mediator entity to be realized. These agents are allowed to make queries to the intermediary's databases. The SAA is in charge of conducting the SPPCA auction. Interaction **1.1** is used for registering/deregistering service providers in the auction, while the SAA uses interaction **1.2** to announce a new auction round. The MA facilitates semantic matchmaking which corresponds to the first level of filtration in the service discovery process. It receives semantic

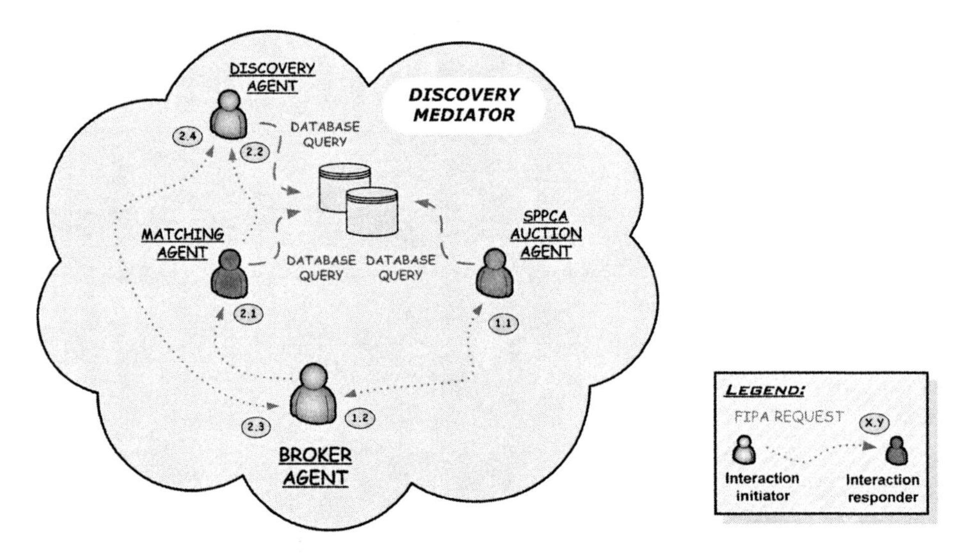

Fig. 4.12. The architecture of a discovery mediator

descriptions of requested service through interaction **2.1** and forwards a list of semantically suitable ones through interaction **2.2** to the DA which carries out second-level filtration and recommends top-ranked services (interaction **2.3**). Sometime later, the DA receives feedback information from the CA (through the BA) regarding the performance of the proposed services (interaction **2.4**).

Figure 4.13 describes communication between all three parties involved in the discovery process: consumers (i.e., CAs) as service requesters, discovery mediators (i.e., BAs) as intermediaries and businesses (i.e., PAs) as service providers. The presented interactions facilitate an efficient discovery process. The specific content of the exchanged messages is described in the following subsections to help clearly present the advertising concept, matchmaking mechanisms and performance evaluation techniques used for designing our service discovery model in the agent-mediated B2C e-market.

The SPPCA auction [46, 52] is divided into rounds which are defined by a fixed time duration (e.g., one SPPCA round lasts for 1 hour). To announce the beginning of a new auction round, the SAA broadcasts a CFB (*Call for Bid*) message to all the PAs which have registered their services for participation in the SPPCA auction. Every CFB message contains a status report (the `status_report(ir(OWL-`S_{adv}`))` parameter in Figure 4.13). In such a report, the SAA sends to the PA information regarding events related to its advertisement which occurred during the previous auction round. The most important information is that regarding how much of the PA's budget was spent (i.e., the advertisement bid price multiplied by the number of recommendations of its service to various CAs). The PA also receives information regarding the final ranking of its advertisement in discovery processes in which the respective advertisement passed first-level filtering (i.e., semantic matchmaking). On the basis of this information, the PA can conclude whether its bids or budget are too high or too low compared to its competitors. In response to a CFB message, a PA sends a BID message. In doing so, the PA assures that its service will be considered in the discovery processes which will occur during the next auction round. In addition to referencing the corresponding service description (the `ir(OWL-`S_{adv}`)` parameter in Figure 4.13), a BID message also contains information specifying the value of the bid (the `B` parameter in Figure 4.13) and information regarding the PA's budget (the `budget` parameter in Figure 4.13). Note that one business (i.e., a PA) can have multiple services simultaneously participating in the same SPPCA auction. If such is the case, all advertisements of the same business potentially have different bid values since a PA can advertise only one service per BID message, and yet all the services share the same budget. Thus, when multiple BID messages are received from a single PA for the same auction round, their budget values are cumulatively added to the budget balance unique for all advertisements originating from the same service provider. This way, PAs do not need to use complex optimization techniques to optimally distribute their budget among their multiple services, i.e., we have designed a SPPCA auction mechanism which makes such calculations unnecessary. Note that if the PA's budget is spent before the end of the round, all of its advertisements become inactive until the end of that round and are, therefore, not considered in any of the subsequent service discovery processes during that round.

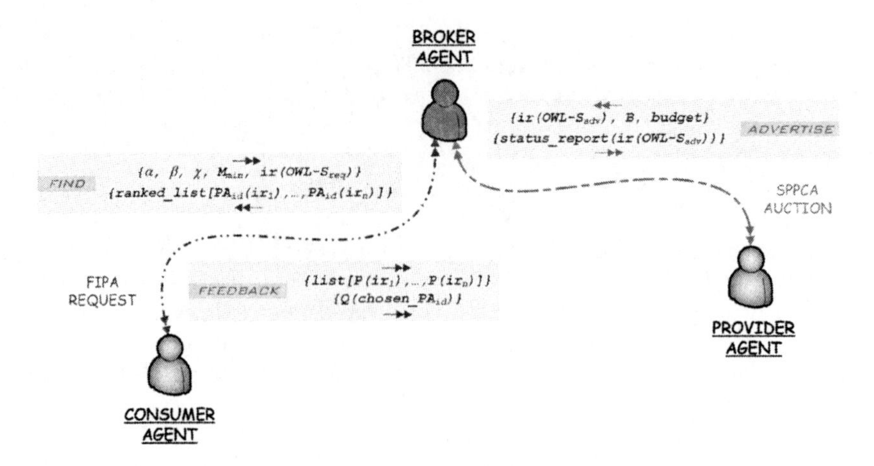

Fig. 4.13. Communication between the described Agents enabling efficient discovery

4.4.2.3 Semantic Matchmaking of Service Descriptions

Our Matching Agent (MA) uses OWLS-MX [53], a hybrid semantic matching tool which combines logic-based reasoning with approximate matching based on syntactic Information Retrieval (IR) similarity computations. As the notion of match rankings is very important, OWLS-MX enables computation of the degree of similarity between compared services, i.e., the comparison is assigned a *service correspondence factor (M)*, which we use as one of the parameters for calculation of a ranked final set of eligible services in our semantic-annotated information service discovery model. Such a similarity ranking is highly relevant since it is unlikely that there will always be a service available which offers the exact features requested. Namely, the OWLS-MX matchmaker takes as input the description of the CA's desired service (the $ir(OWL\text{-}S_{req})$ parameter in Figure. 4.13), and returns an ordered set of relevant services which match the query. Each relevant service is annotated with its individual degree of matching and syntactic similarity value. There are six possible levels of matching between services. The first level is a perfect match (also called an EXACT match) which is assigned factor M=5. Furthermore, we have four possible inexact matching levels which are as follows: a PLUG-IN match (M=4), a SUBSUMES match (M=3), a SUBSUMES-BY match (M=2) and a NEAREST-NEIGHBOUR match (M=1). If two services do not match according to any of the above mentioned criteria, they are assigned a matching level of FAIL (M=0). The EXACT, PLUG-IN and SUBSUMES criteria are logic-based only, whereas the SUBSUMES-BY and NEAREST-NEIGHBOUR matches are hybrid due to additional computation of syntactic similarity values. A CA specifies its desired matching degree threshold, i.e., the M_{min} parameter in Figure. 4.13, defining how relaxed the semantic matching can be. A discovery mediator, on the other hand, specifies its desired syntactic similarity threshold so as not to make the discovery process too complex for the CA (this is used only for finding SUBSUMES-BY and NEAREST-NEIGHBOUR matches, i.e., when M_{min} is set to values 2 or 1, respectively).

4.4.2.4 The Performance Model of Service Providers

A performance model tracks the past performance of businesses that act as service providers in the B2C e-market. This information can then be used to estimate its performance with respect to future requests [54]. Our model monitors two aspects of a service provider's performance – the reputation [55, 56, 57] of the business that acts as the service provider and the cost of utilizing the service it is offering. The reputation of a service provider reveals its former cooperative behavior and, thus, reduces the risk of financial loss for service requesters [58]. Additionally, the reputation of the business is a measure of the quality of services provided by that business. On the other hand, information regarding the cost of utilizing the offered service enables consumers to find the best-buy option and helps prevent them from spending their money where it is not necessary.

A CA does not provide the discovery mediator entity with feedback information regarding the quality and the price rating of a particular advertised service at the same time. Namely, a CA gives a BA feedback regarding a particular service's utilization cost (called the *price rating*) immediately after the discovery process ends (the `list[P(ir1),…,P(irn)]` parameter in Figure 4.13). On the other hand, the BA does not receive feedback regarding the service provider's reputation (called the *quality rating*) immediately after the discovery process, but sometime in the future after the CA's owner begins utilizing the chosen service (the `Q(chosen_PAid)` parameter in Figure 4.13). The rating value (both for quality and price) is a real number $r \in [0, 1]$. A rating of *0* is the worst (i.e., the service provider could not provide the service at all and/or utilizing the service is very expensive) while a rating of *1* is the best (i.e., the service provider provisions a service that perfectly corresponds to the consumer's needs and/or utilizing the service is very cheap).

The overall ratings of businesses and their services can be calculated a number of ways. In our approach, we use the EWMA (*Exponentially Weighted Moving Average*) method. An advantage of using EWMA is its adaptive nature, i.e., it can capture the trend of dynamic changes. Furthermore, it is computationally very simple since the new overall rating can be calculated from the previous overall rating and the current feedback rating (i.e., there is no need to store old ratings which is desirable due to scalability issues). EWMA is defined as follows:

$$\tilde{x}_t = \xi x_t + (1-\xi)\tilde{x}_{t-1} \qquad \text{for } t=1,2,\ldots \tag{4.1}$$

where $\tilde{x}_t$ is the new forecast value of x; x_t is the current observation value (in our case, the new feedback rating); $\tilde{x}_{t-1}$ is the previous forecast value; and $\xi \in [0, 1]$ is a factor that determines the depth of memory of the EWMA. As the value of ξ increases, more weight is given to the most recent values. Every discovery mediator entity sets this factor value according to its preferences.

4.4.2.5 Calculating a Recommended Ranked Set of Eligible Services

After a BA receives a discovery request message (the $\{\alpha, \beta, \chi, M_{min}, ir(OWL\text{-}S_{req})\}$ message in Figure 4.13) from a CA, the discovery mediator calculates a ranked set of the most suitable service advertisements. An ordered

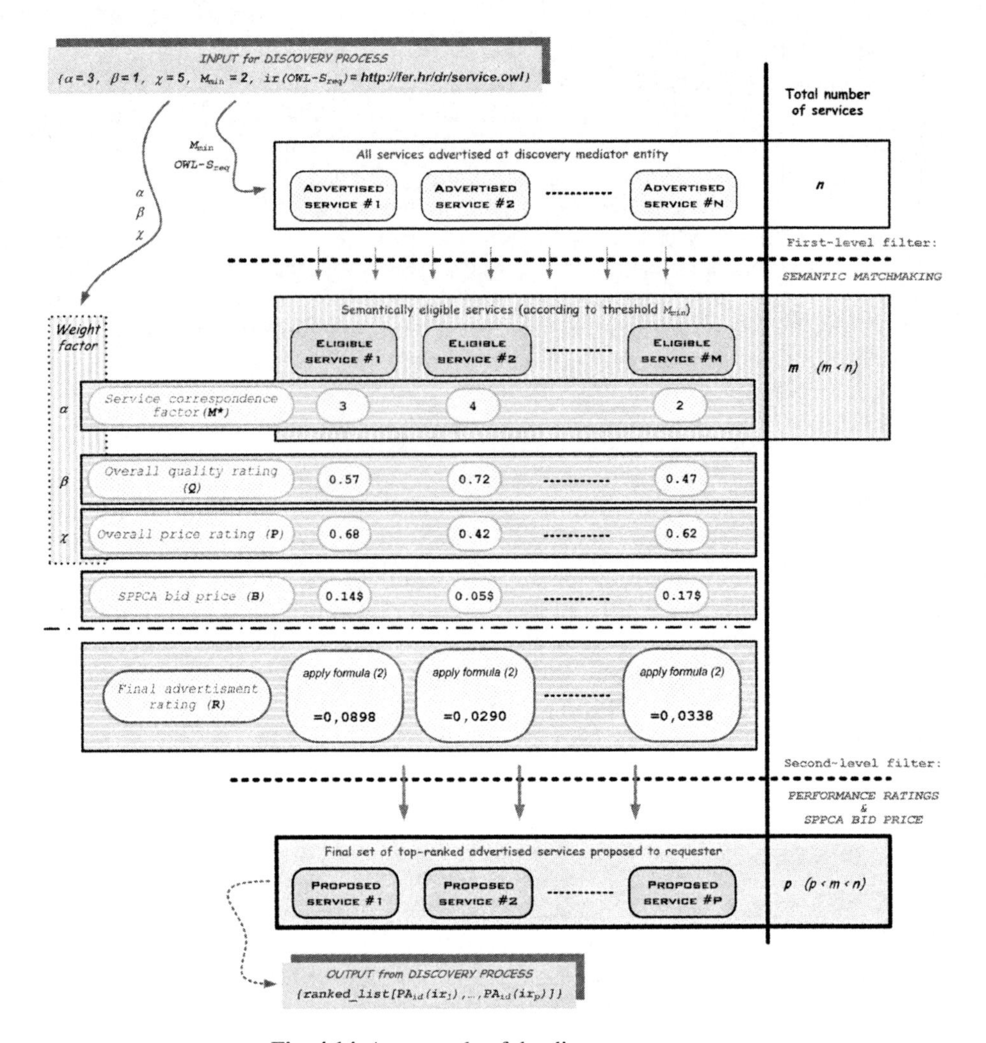

Fig. 4.14. An example of the discovery process

set of PAs which represent the businesses which provide the top-ranked services is then recommended to the CA in response to its request (the `{ranked_list[PA`$_{id}$`(ir`$_1$`),…,PA`$_{id}$`(ir`$_n$`)]}` message in Figure 4.13).

The final rating of a specific service advertisement at the end of the discovery process is calculated as follows:

$$R = \frac{\alpha \times \frac{M^*}{5} + \beta \times Q + \chi \times P}{\alpha + \beta + \chi} \times B \tag{4.2}$$

where R represents the final rating of the particular service advertisement after the discovery process has finished. A higher rating means that this particular service is more suitable for satisfying the consumer's needs; α , β and χ are weight factors which enable the consumer to profile his/her request according to his/her needs regarding the semantic similarity, quality and price of a service, respectively; M^* represents the *service correspondence factor (M)*, but only services with values of M higher than a threshold M_{min} are considered; Q and P represent the quality and price ratings of a particular service, respectively; finally, B is the bid value for advertising a particular service in the SPPCA auction.

Since our performance model monitors two aspects of the service provider's performance (i.e., its quality and price), the CA defines two weight factors which determine the significance of each of the two aspects in the process of calculating the final proposal (β represents a weight factor describing the quality of a service while χ represents a weight factor describing the price of a service). Furthermore, a consumer can specify whether information regarding the semantic similarity of a service is more important to him/her than information regarding a service provider's performance, and to what extent. Thus, the CA also defines parameter α which is a weight factor representing the semantic similarity aspect of a service. In our example (see Figure 4.14), α=3, β=1 and χ=5. This means that the requester is looking for a cheap service and (s)he is not very concerned with the service's quality.

It is obvious from Formula 4.2 that the final rating of a particular service advertisement at the end of the discovery process is calculated by multiplying two factors: a particular service conformance with the consumer needs and the bid value for advertising a particular service in the SPPCA auction. Such calculation enables honoring service low utilization cost and provider's high quality - businesses with better performance rating can put smaller bids and stay competitive. On the other hand, such calculation also enables punishing service high utilization cost and provider's low quality - businesses with worse performance rating must put higher bids to stay competitive.

4.5 Conclusion

In this chapter, we share some ideas on how to automate telecom processes. The advent of a novel telecom environment is elaborated upon and solutions for the automation of processes in such an environment, which combine economic and technical approaches, are proposed. Namely, we analyze the New Generation Network (NGN) environment and propose to organize the telecom market as a composition of Business-to-Business (B2B) and Business-to-Consumer (B2C) electronic markets (e-markets). Furthermore, we introduce intelligent software agents into these e-markets to enable automated interactions between stakeholders and define adequate auctions as efficient negotiating mechanisms. As a result, the suggested models and mechanisms are not only relevant for academia, but are also very amenable for application in real-world telecom environments.

The telecom market is a very complex system with a vast number of stakeholders and lots of interactions. Therefore, traditional mathematical-based methods are inappropriate for studying economic processes in such environment. The proposed implementation of the telecom e-market as a multi-agent system (MAS) enables application of the Agent-based Computational Economics (ACE) as the methodology for studying economic processes on the market. The main idea of the ACE approach is the experimental economic evaluation of market mechanisms and stakeholder behaviors characteristic for the telecom environment.

Acknowledgments

The work presented in this chapter was carried out within research projects 036-0362027-1639 "Content Delivery and Mobility of Users and Services in New Generation Networks", supported by the Ministry of Science, Education and Sports of the Republic of Croatia, and "Agent-based Service & Telecom Operations Management", supported by Ericsson Nikola Tesla, Croatia.

References

1. Leuf, B.: The Semantic Web: Crafting Infrastructure for Agency. Wiley, New York (2006)
2. Podobnik, V., Trzec, K., Jezic, G.: Context-Aware Service Provisioning in Next-Generation Networks: An Agent Approach. International Journal of Information Technology and Web Engineering 2(4), 41–62 (2007)
3. Podobnik, V., Jezic, G., Trzec, K.: A Multi-Agent System for Auction-Based Resource Discovery in Semantic-Aware B2C Mobile Commerce. International Transactions on Systems Science and Applications 3(2), 169–182 (2007)
4. Fasli, M.: Agent Technology for E-Commerce. Wiley & Sons, Chichester (2007)
5. Weiser, M., Brown, J.S.: The Coming Age of Calm Technology. In: Dening, P.J., Metcalfe, R.M., Burke, J. (eds.) Beyond Calculation: The Next Fifty Years of Computing. Springer, New York (1997)
6. Weiser, M.: The World is not a Desktop. ACM Interactions 1(1), 7–8 (1994)
7. Berners-Lee, T., Fischetti, M.: Weaving the Web, Harper San Francisco, New York (1999)
8. Berners-Lee, T., Hendler, J., Lassila, O.: The Semantic Web. Scientific American 284(5), 34–43 (2001)
9. Skorin-Kapov, L., Mosmondor, M., Dobrijevic, O., Matijasevic, M.: Application-Level QoS Negotiation and Signaling for Advanced Multimedia Services in IMS. IEEE Communications 45(7), 108–117 (2007)
10. Weiser, M.: The Computer for the 21st Century. Scientific American 265(3), 94–104 (1991)
11. Saha, D., Mukherjee, A.: Pervasive Computing: a Paradigm for the 21st Century. IEEE Computer 36(3), 25–31 (2003)
12. Ljubi, I., Podobnik, V., Jezic, G.: Cooperative Mobile Agents for Automation of Service Provisioning: A Telecom Innovation. In: Proc. of the 2nd IEEE International Conference on Digital Information Management (ICDIM 2007), pp. 817–822. IEEE Press, New York (2007)
13. Jurasovic, K., Kusek, M.: Verification of mobile agent network simulator. In: Nguyen, N.T., Grzech, A., Howlett, R.J., Jain, L.C., et al. (eds.) KES-AMSTA 2007. LNCS (LNAI), vol. 4496, pp. 520–529. Springer, Heidelberg (2007)

14. Foll, U., Fan, C., Carle, G., Dressler, F., Roshandel, M.: Service-oriented Accounting and Charging for 3G and B3G Mobile Environments. In: Proc. of the 9th IFIP/IEEE International Symposium on Integrated Network Management (IM 2005), pp. 1253–1256. IEEE Press, Los Alamitos (2005)
15. Fischer & Lorenz (European Telecommunications Consultants): Internet and the Future Policy Framework for Telecommunications. Report for the European Commission (2000)
16. Houssos, N., Gazis, E., Panagiotakis, S., Quesnel, S., Gessler, S., Schuelke, A.: Value Added Service Management in 3G Networks. In: Proc. the 9th IEEE/IFIP Network Operations and Management Symposium (NOMS 2002), pp. 529–544. IEEE Press, Florence (2002)
17. Yoon, J.-L.: Telco 2. 0: A New Role and Business Model. IEEE Communications 45(1), 10–12 (2007)
18. Jennings, N., Sycara, K., Wooldridge, M.: A Roadmap of Agent Research and Development. Journal of Autonomous Agents and Multi-Agent Systems 1(1), 7–36 (1998)
19. Bellavista, P., Corradi, A., Montanari, R., Tonin, A.: Context-Aware Semantic Discovery for Next Generation Mobile Systems. IEEE Communications 44(9), 62–71 (2006)
20. Cockayne, W.T., Zyda, M.: Mobile Agents. Manning Publications, Greenwich (1998)
21. Bradshaw, J.M.: Software Agents. MIT Press, Cambridge (1997)
22. Chorafas, D.N.: Agent Technology Handbook. McGraw-Hill, New York (1998)
23. Podobnik, V., Petric, A., Jezic, G.: An Agent-Based Solution for Dynamic Supply Chain Management. Journal of Universal Computer Science 14(7), 1080–1104 (2008)
24. Kusek, M., Lovrek, I., Sinkovic, V.: Agent team coordination in the mobile agent network. In: Khosla, R., Howlett, R.J., Jain, L.C. (eds.) KES 2005. LNCS (LNAI), vol. 3681, pp. 240–246. Springer, Heidelberg (2005)
25. Trzec, K., Lovrek, I.: Field-based coordination of mobile intelligent agents: An evolutionary game theoretic analysis. In: Apolloni, B., Howlett, R.J., Jain, L. (eds.) KES 2007, Part I. LNCS (LNAI), vol. 4692, pp. 198–205. Springer, Heidelberg (2007)
26. Hendler, J.: Agents and the Semantic Web. IEEE Intelligent Systems 16(2), 30–37 (2001)
27. Fensel, D.: Ontologies: A Silver Bullet for Knowledge Management and Electronic Commerce. Springer, Berlin (2004)
28. Davenport, T.H., Prusak, L.: Working Knowledge – How Organizations Manage What They Know. Harvard Business School Press, Boston (1998)
29. Antoniou, G., van Harmelen, F.: Semantic Web Primer. MIT Press, Cambridge (2004)
30. McAfee, R., McMillan, P.J.: Auctions and Bidding. Journal of Economic Literature 25, 699–738 (1987)
31. Courcobetis, C., Weber, R.: Pricing Communication Networks: Economics, Technology and Modeling. John Wiley & Sons, Chichester (2003)
32. Krishna, V.: Auction Theory. Elsevier Academic Press, San Diego (2002)
33. Amin, M., Ballard, D.: Defining New Markets for Intelligent Agents. IEEE IT Professional 2(4), 29–35 (2000)
34. Lim, W.S., Tang, C.S.: An Auction Model Arising from an Internet Search Service Provider. European Journal of Operational Research 172(3), 956–970 (2006)
35. Trzec, K., Lovrek, I., Mikac, B.: Agent behaviour in double auction electronic market for communication resources. In: Gabrys, B., Howlett, R.J., Jain, L.C. (eds.) KES 2006. LNCS (LNAI), vol. 4251, pp. 318–325. Springer, Heidelberg (2006)
36. Trzec, K., Lovrek, I.: Modelling Behaviour of Trading Agents in Electronic Market for Communication Resources. In: Proc. of the 2nd Conference on Networking and Electronic Commerce Research (NAEC 2006), pp. 171–186. ASTMA, Riva del Garda (2006)

37. He, M., Jennings, N.R., Leung, H.: On Agent-Mediated Electronic Commerce. IEEE Transactions on Knowledge and Data Engineering 15(4), 985–1003 (2003)
38. Do, V., Halatchev, M., Neumann, D.: A Context-based Approach to Support Virtual Enterprises. In: Proc. of the 33rd Hawaii International Conference on System Sciences (HICSS 2000), p. 6005. IEEE Computer Society, Island of Maui (2000)
39. Friedman, D., Rust, J.: The Double Auction Market: Institutions, Theories, and Evidence. Perseus Publishing, Cambridge (1993)
40. Ströbel, M.: Engineering Electronic Negotiations. Kluwer Academic/Plenum (2003)
41. Bichler, M., Kaukal, M., Segev, A.: Multi-Attribute Auctions for Electronic Procurement. In: Proc. of the 1st IBM IAC Workshop on Internet Based Negotiation Technologies (1999)
42. Bichler, M.: An Experimental Analysis of Multi-attribute Auctions. Decision Support Systems 29(3), 249–268 (2000)
43. Bichler, M., Kalagnanam, J.: Configurable Offers and Winner Determination in Multi-attribute Auctions. European Journal of Operational Research 16(2), 380–394 (2005)
44. Bichler, M., Werthner, H.: A Classification Framework of Multidimensional, Multi-unit Procurement Negotiations. In: Proc. ff the 11th International Workshop on Database and Expert Systems Applications (DEXA 2000), p. 1003. IEEE Computer Society, London (2000)
45. Guttman, R.H., Moukas, A.G., Maes, P.: Agent-Mediated Electronic Commerce: A Survey. Knowledge Engineering Review 13(2), 147–159 (1998)
46. Podobnik, V., Trzec, K., Jezic, G., Lovrek, I.: Agent-Based Discovery of Data Resources in Next-Generation Internet: An Auction Approach. In: Proc. of the 2007 Networking and Electronic Commerce Research Conference (NAEC 2007), pp. 28–51. ASTMA, Riva del Garda (2007)
47. Di Noia, T., Di Sciascio, E., Donini, F.M., Mong, M.: A System for Principled Matchmaking in an Electronic Marketplace. International Journal of Electronic Commerce 8(4), 9–37 (2004)
48. Jansen, B.J.: Paid Search. Computer 39(7), 88–90 (2006)
49. Aggarwal, G., Goel, A., Motwani, R.: Truthful Auctions for Pricing Search Keywords. In: Proc. of the 7th ACM Conference on Electronic Commerce (EC 2006), pp. 1–7. ACM, Ann Arbor (2006)
50. Kitts, B., LeBlanc, B.: Optimal Bidding on Keyword Auctions. Electronic Markets 14(3), 186–201 (2004)
51. Bellifemine, F.L., Caire, G., Greenwood, D.: Developing Multi-Agent Systems with JADE. John Wiley & Sons, Chichester (2007)
52. Podobnik, V., Trzec, K., Jezic, G.: An auction-based semantic service discovery model for E-commerce applications. In: Meersman, R., Tari, Z., Herrero, P. (eds.) OTM 2006 Workshops. LNCS, vol. 4277, pp. 97–106. Springer, Heidelberg (2006)
53. Klusch, M., Fries, B., Sycara, K.: Automated Semantic Web Service Discovery with OWLS-MX. In: Proc. of the 5th International Joint Conference on Autonomous Agents and Multiagent Systems (AAMAS 2006), pp. 915–922. ACM, Hakodate (2006)
54. Luan, X.: Adaptive Middle Agent for Service Matching in the Semantic Web: A Quantitive Approach. Thesis. Baltimore County, University of Maryland (2004)
55. Zhang, X., Zhang, Q.: Online Trust Forming Mechanism: Approaches and an Integrated Model. In: Proc. of the 7th International Conference on Electronic Commerce (ICEC 2005), pp. 201–209. ACM, Xi'an (2005)

56. Fan, M., Tan, Y., Whinston, A.B.: Evaluation and Design of Online Cooperative Feedback Mechanisms for Reputation Management. IEEE Transactions on Knowledge and Data Engineering 17(2), 244–254 (2005)
57. Rasmusson, L., Janson, S.: Agents, Self-Interest and Electronic Markets. The Knowledge Engineering Review 14(2), 143–150 (1999)
58. Padovan, B., Sackmann, S., Eymann, T., Pippow, I.A.: Prototype for an Agent-Based Secure Electronic Marketplace Including Reputation-Tracking Mechanisms. International Journal of Electronic Commerce 6(4), 93–113 (2002)

5

Multi-agent Systems and Paraconsistent Knowledge

Jair Minoro Abe[1,2] and Kazumi Nakamatsu[3]

[1] Graduate Program in Production Engineering, ICET - Paulista University
R. Dr. Bacelar, 1212, CEP 04026-002 São Paulo - SP – Brazil
[2] Institute For Advanced Studies – University of São Paulo – Brazil
jairabe@uol.com.br
[3] School of Human Science and Environment/H.S.E. – University of Hyogo – Japan
nakamatu@shse.u-hyogo.ac.jp

Abstract. Since impreciseness, inconsistency and paracompleteness are more and more common matter in distributed environments, we need more general and formal frameworks to deal to. In this work we discuss a logical framework for multi-agent reasoning based on Paraconsistent Annotated Systems, which can manipulate imprecise, inconsistent and paracomplete data.

Keywords: multi-agent, distributed system, paraconsistent logic, annotated logic.

5.1 Introduction

Conflicts, uncertainty and paracompleteness arise naturally in distributed environments in general. They originate for different reasons and are dealt with many different ways, depending on the domain where they are considered. Agents may have different opinions and beliefs on the same subject, may faced with uncertainty information, may have insufficient information to perform tasks, may have to share limited resources [20], [21].

Despite these concepts seem at first sight obstacles, it is interesting to study on their roles within a multiagent system, i.e. how this system may evolve thanks to, despite, or because of conflicts, uncertainty, and paracompleteness. Many of them can be avoided, kept, 'solved' or even created deliberated; for instance, a malicious individual can introduce conflicting information in a database corrupting it partially or entirely. Since more and more concern is attached to agents' teamwork and agents' dialogue, conflicts naturally arise as a key issue to be dealt with, not only with application dedicated techniques, but also with more formal and generic tools.

So, we discuss as central theme in this chapter, a logical system for reasoning with imprecise, inconsistent and paracomplete concepts in a non-trivial way in multi-agent systems. The desired aim, the system $M\tau$, is obtained by using ideas of [1] and [12] having the following characteristics:

a) The principle of contradiction, in the form $\neg(A \wedge \neg A)$, is not valid in general among hyper-literal formulas;
b) The law of excluded middle, in the form $A \vee \neg A$, is not valid in general among hyper-literal formulas;

L.C. Jain, N.T. Nguyen (Eds.): Knowl. Proc. & Dec. Mak. in Agent-Based Sys., SCI 170, pp. 101–121.
springerlink.com

c) From two contradictory hyper-literal formulas, A and $\neg A$, we cannot deduce any formula B whatsoever;
d) For complex formulas, it is valid all properties of the classical logic.
e) Mτ contains the most important schemes and rules of inference of the classical predicate calculus that are compatible with conditions a), b), and c).

Due these properties, this system can be the underlying logic for modeling uncertainty, conflicting and paracomplete common knowledge. The system can be adapted to cope conflicting beliefs and awareness. Due the existence of a logic programming based on annotated first-order predicate calculi, as well as logic programming for several modalities, we believe that a computational implementation for multi-agent systems is possible, although it is not accomplished yet. A short version of this work is to be found in [4].

We begin with basic concepts and definitions.

5.2 Paraconsistent, Paracomplete, and Non-alethic Logics

In what follows, we sketch the non-classical logics discussed in the paper, establishing some conventions and definitions.

Let T be a theory whose underlying logic is L. T is called inconsistent when it contains theorems of the form A and $\neg A$ (the negation of A). If T is not inconsistent, it is called *consistent.* T is said to be *trivial* if all formulas of the language of T are also theorems of T. Otherwise, T is called *non-trivial.* When L is classical logic (or one of several others, such as intuitionistic logic), T is inconsistent iff T is trivial. So, in trivial theories the extensions of the concepts of formula and theorem coincide. A *paraconsistent logic* is a logic that can be used as the basis for inconsistent but non-trivial theories. A *theory* is called *paraconsistent* if its underlying logic is a paraconsistent logic. Issues such as those described above have been appreciated by many logicians. In 1910, the Russian logician Nikolaj A. Vasil'év (1880-1940) and the Polish logician Jan Łukasiewicz (1878-1956) independently glimpsed the possibility of developing such logics. Nevertheless, the Polish logician Stanislaw Jaśkowski (1906-1965) was in 1948 effectively the first logician to develop a paraconsistent system, at the propositional level. His system is known as 'discussive propositional calculus'. Independently, some years later, the Brazilian logician Newton C.A. da Costa (1929-) constructed for the first time hierarchies of paraconsistent propositional calculi C_i, $1 \leq i \leq \omega$, of paraconsistent first-order predicate calculi (with and without equality), of paraconsistent description calculi, and paraconsistent higher-order logics (systems NF_i, $1 \leq i \leq \omega$). Also, independently of Da Costa [10], the American logician Nels David Nelson (1918-2003) [19] has considered a paraconsistent logic as a version of his known as constructive logics with strong negation.

Nowadays, paraconsistent logic has established a distinctive position in a variety of fields of knowledge.

Another important class of non-classical logics are the paracomplete logics. A logical system is called *paracomplete* if it can function as the underlying logic of theories in which there are formulas such that these formulas and their negations are simultaneously false. Intuitionistic logic and several systems of many-valued logics are paracomplete in this sense (and the dual of intuitionistic logic, Brouwerian logic, is therefore paraconsistent).

As a consequence, paraconsistent theories do not satisfy the principle of non-contradiction, which can be stated as follows: of two contradictory propositions, i.e., one of which is the negation of the other, one must be false. And, paracomplete theories do not satisfy the principle of the excluded middle, formulated in the following form: of two contradictory propositions, one must be true.

Finally, logics which are simultaneously paraconsistent and paracomplete are called *non-alethic logics.*

5.3 Paraconsistent Annotated Logics

Annotated logics are a family of non-classical logics initially used in logic programming by [24]. An extensive study of annotated logics was made in [1]. Some applications are summarized in [2]. In view of the applicability of annotated logics to differing formalisms in computer science, it has become essential to study these logics more carefully, mainly from the foundational point of view.

In general, annotated logics are a kind of paraconsistent, paracomplete, and non-alethic logic, The latter systems are among the most original and imaginative systems of non-classical logic developed in the past century. Also, annotated systems can viewed as many-valued logics [5], [6].

Annotated systems are surprisingly useful logics. Technically, they constitute 2-sorted logics, having a very clear syntax and semantics; thus, it is possible to investigate their metatheory. They admit an elegant Hilbertian axiomatics, which are sound and complete (when the lattice is finite) regarding their semantics [1]. They also have a natural deduction system, which make possible inferences more natural, as well as computability [7]. In fact, it was implemented a logic programming dubbed Paralog. Their high order logics can be established. Also annotated set theory was studied; these encompass *in totun* the Fuzzy set theory in different ways [1], [12]. Axiomatic version of certain class of Fuzzy systems was obtained via annotated logics [6]. Annotated logics also allow defeseable, deontic, non-monotonic, linear reasoning, among others in a different approaches than usual ones [18]. Regarding applications, some of them are summarized. It was built electronic logical gates Not, And Or which permit 'inconsistent' signals in a non-trivial manner; logical analyzer – Para-analyzer; logical controller – Para-control, logical simulator – Para-sim, all allowing manipulate inconsistent, imprecise and paracomplete signals [2]. Other logical controllers and integrated circuits were obtained. All these ideas were experimented in real-life robots or in simulators [3]. Underlying these studies, it was developed an automata theory based on annotated systems, close to weighted automata. Finally, Artificial Neural Networks were introduced based on annotated logics, with promising results [13].

Throughout this chapter, $A \cap B$ and $A \cup B$ indicate the set-theoretical intersection and union, respectively; and if $(A_i)_{i \in I}$ is a family of sets, its union is indicated by $\bigcup_{i \in I} A_i$.

$A \subseteq B$ means that A is a subset of B. $\#A$ indicates the cardinal number of A. $\mathbb{N}$ indicates the set of natural numbers and $\mathbb{N}^* = \mathbb{N} - \{0\}$. Some other usual conventions and notions of set theory are assumed without extensive comments.

5.4 The Paraconsistent Annotated First-Order Logics $Q\tau$

$Q\tau$ is a family of first-order logics, called annotated first-order predicate calculi. They are defined as follows: $\tau = \langle|\tau|, \leq, \sim\rangle$ will be some arbitrary, but fixed, finite lattice of truth-values. The least element of τ is denoted by $\perp$, while its greatest element by $\top$. We also assume that there is a fixed unary operator $\sim: |\tau| \rightarrow |\tau|$ which constitutes the "meaning" of our negation. $\vee$ and $\wedge$ denote, respectively, the least upper bound and the greatest lower bound operators (of τ) .

The language L of $Q\tau$ is a first-order language (with equality) whose primitive symbols are the following:

1. Individual variables: a denumerable infinite set of variable symbols.
2. Logical connectives: $\neg$, (negation), $\wedge$ (conjunction), $\vee$ (disjunction), and $\rightarrow$ (conditional).
3. For each n, zero or more n-ary function symbols (n is a natural number).
4. For each $n \neq 0$, zero or more n-ary predicate symbols.
5. Quantifiers: $\forall$ (for all) and $\exists$ (there exists).
6. The equality symbol: =.
7. Annotated constants: each member of τ is called an annotational constant.
8. Auxiliary symbols: parentheses and commas.

For each n, the number of n-ary function symbols may be zero or non-zero, finite or infinite. A 0-ary function symbol is called a *constant.* Also, for each n-1, the number of n-ary predicate symbol may be finite or infinite.

In the sequel, we suppose that $Q\tau$ possesses at least one predicate symbol.

We define the notion of *term* as usual. Given a predicate symbol p of arity n, an annotational constant λ and n terms $t_1, \ldots, t_n$, an *annotated atom* is an expression of the form $p_\lambda t_1 \ldots t_n$. In addition, if t_1 and t_2 are terms whatsoever, $t_1 = t_2$ is an *atomic formula.* We introduce the general concept of *formula* in the standard way. Among several intuitive readings, an annotated atom $p_\lambda t_1 \ldots t_n$ can be read is *it is believed that* $p_\lambda t_1 \ldots t_n$*'s truth-value is at least* λ.

In general, the syntactical notions, as well as the terminology, the notations, etc. are those of [1]. We will employ them without extensive comments.

Definition 5.1. Let A and B formulas of L. We put

$A \leftrightarrow B =_{\text{Def.}} (A \rightarrow B) \wedge (B \rightarrow A)$ and $\neg\!\!\!\neg\, A =_{\text{Def.}} A \rightarrow ((A \rightarrow A) \wedge (A \rightarrow A))$.

The symbol '$\leftrightarrow$' is called the *biconditional* and '$\neg\!\!\!\neg$ ' is called *strong negation.*

Let A be a formula. $\neg^0 A$ indicates A, $\neg^1 A$ indicates $\neg A$, and $\neg^n A$ indicates $\neg(\neg^{n-1} A)$, ($n \geq 1$). Also, if $\mu \in \tau$, $\sim^0\mu$ indicates μ, $\sim^1\mu$ indicates $\sim\mu$, and $\sim^n\mu$ indicates $\sim(\sim^{n-1}\mu)$, ($n \geq 1$).

Definition 5.2. Let $p_\lambda t_1 \ldots t_n$ be an annotated atom. A formula of the form $\neg^k p_\lambda t_1 \ldots t_n$ $(k \geq 0)$ is called a *hyper-literal.* A formula other than hyper-literal is called a *complex formula.*

We now introduce the concept of structures for L.

Definition 5.3. A *structure* S for L consists of the following objects:

1. A non-empty set $|S|$, called the *universe* of S. The elements of $|S|$ are called *individuals* of S.
2. For each n-ary function symbol f of L an n-ary function f_S: $|S|^n \rightarrow |S|$. (In particular, for each constant e of L, e_S is an individual of A.)
3. For each n-ary predicate symbol p of L an n-ary function p_S: $|S|^n \rightarrow |\tau|$.

Let A be a structure for L. The *diagram language* L_S is obtained as usual. If a is a free-variable term, we define the individual $S(a)$ of S. We use i and j as syntactical variables which vary over names.

We define a truth-value $S(A)$ for each closed formula A in L_S.

1. If A is $a = b$, $S(A) = 1$ iff $S(a) = S(b)$; otherwise $S(A) = 0$.
2. If A is $p_\lambda t_1 \ldots t_n$, $S(A) = 1$ iff $p_S(S(t_1) \ldots S(t_n) \geq \lambda$; $S(A) = 0$ iff it is not the case that $p_S(S(t_1) \ldots S(t_n) \geq \lambda$
3. If A is $B \wedge C$, or $B \vee C$, or $B \rightarrow C$, we let $S(B \wedge C) = 1$ iff $S(B) = S(C) = 1$; $S(B \vee C) = 1$ iff $S(B) = 1$ or $S(C) = 1$.
4. $S(B \rightarrow C) = 0$ iff $S(B) = 1$ and $S(C) = 0$.
5. If A is $\neg^k p_\lambda t_1 \ldots t_n$ $(k \geq 1)$, then $S(A) = S(\neg^{k-1} p_{\sim\lambda} t_1 \ldots t_n)$.
6. If A is a complex formula, then, $S(\neg A) = 1 - S(A)$.
7. If A is $\exists x B$, then $S(A) = 1$ iff $S(B_x[i]) = 1$ for some i in L_S.
8. If A is $\forall x B$, then $S(A) = 1$ iff $S(B_x[i]) = 1$ for all i in L_S.

A formula A of L is said to be *valid in S* if $S(A') = 1$ for every S-instance A' of A. A formula A is called *logically valid* if it is valid in every structure for L. In this case, we symbolize it by $\Vdash A$. If Γ is a set of formulas of L we say that A is a *semantic consequence* of Γ if for any structure S in what $S(B) = 1$ for all $B \in \Gamma$, it is the case that $S(A) = 1$. We symbolize this fact by $\Gamma \Vdash A$. Note that when $\Gamma = \varnothing$, $\Gamma \Vdash A$ iff $\Vdash A$.

Lemma 5.1. We have:

1. $\Vdash p_\perp t_1 \ldots t_n$
2. $\Vdash \neg^k p_\lambda t_1 \ldots t_n \leftrightarrow \neg^{k-1} p_{\sim\lambda} t_1 \ldots t_n)$ $(k \geq 1)$
3. $\Vdash p_\lambda t_1 \ldots t_n \rightarrow p_\mu t_1 \ldots t_n$, $\lambda \geq \mu$
4. $\Vdash p_{\lambda 1} t_1 \ldots t_n \wedge p_{\lambda 2} t_1 \ldots t_n \wedge \ldots \wedge p_{\lambda m} t_1 \ldots t_n \rightarrow p_\lambda t_1 \ldots t_n$, where $\lambda = \bigvee_{i=1}^{m} \lambda_i$

Now, we shall describe an axiomatic system which we call $A\tau$ whose underlying language is L: A, B, C are any formulas whatsoever, F, G are complex formulas, and $p_\lambda t_1 \ldots t_n$ an annotated atom. $A\tau$ consists of the following postulates (axiom schemes and primitive rules of inference), with the usual restrictions:

($\to_1$) $A \to (B \to A)$
($\to_2$) $(A \to (B \to C) \to ((A \to B) \to (A \to C))$
($\to_3$) $((A \to B) \to A \to A)$
($\to_4$) $\dfrac{A,\ A \to B}{B}$ (Modus Ponens)
($\wedge_1$) $A \wedge B \to A$
($\wedge_2$) $A \wedge B \to B$
($\wedge_3$) $A \to (B \to (A \wedge B))$
($\vee_1$) $A \to A \vee B$
($\vee_2$) $B \to A \vee B$
($\vee_3$) $(A \to C) \to ((B \to C) \to ((A \vee B) \to C))$
($\neg_1$) $(F \to G) \to ((F \to \neg G) \to \neg F)$
($\neg_2$) $F \to (\neg F \to A)$
($\neg_3$) $F \vee \neg F$
($\exists_1$) $A(t) \to \exists x A(x)$
($\exists_2$) $\dfrac{A(x) \to B}{\exists x A(x) \to B}$
($\forall_1$) $\forall x A(x) \to A(t)$
($\forall_2$) $\dfrac{B \to A(x)}{B \to \forall x A(x)}$
(τ_1) $p_{\perp} t_1 \ldots t_n$
(τ_2) $\neg^{k} p_{\lambda} t_1 \ldots t_n \to \neg^{k-1} p_{\sim\lambda} t_1 \ldots t_n,\ k \geq 1$
(τ_3) $p_{\lambda} t_1 \ldots t_n \to p_{\mu} t_1 \ldots t_n,\ \lambda \geq \mu$
(τ_4) $p_{\lambda 1} t_1 \ldots t_n \wedge p_{\lambda 2} t_1 \ldots t_n \wedge \ldots \wedge p_{\lambda m} t_1 \ldots t_n \to p_{\lambda} t_1 \ldots t_n$, where $\lambda = \bigvee_{i=1}^{m} \lambda_i$
($=_1$) $x = x$
($=_2$) $x = y \to (A[x] \leftrightarrow A[y])$

Theorem 5.1. In $Q\tau$, the operator $\urcorner\!\!\urcorner$ has all properties of the classical negation. For instance, we have:

1. $\vdash A \vee \urcorner\!\!\urcorner A$
2. $\vdash \urcorner\!\!\urcorner (A \wedge \urcorner\!\!\urcorner A)$
3. $\vdash (A \to B) \to ((A \to \urcorner\!\!\urcorner B) \to \urcorner\!\!\urcorner A)$
4. $\vdash A \to \urcorner\!\!\urcorner \urcorner\!\!\urcorner A$
5. $\vdash \urcorner\!\!\urcorner A \to (A \to B)$
6. $\vdash (A \to \urcorner\!\!\urcorner A) \to B$

among others, where *A*, *B* are any formulas whatsoever.

Corollary 5.1. In $Q\tau$ the connectives $\urcorner\!\!\urcorner$, $\wedge$, $\vee$, and $\to$ together with the quantifiers $\forall$ and $\exists$ have all properties of the classical negation, conjunction, disjunction, conditional and the universal and existential quantifiers, respectively. If *A*, *B*, *C* are formulas whatsoever, we have, for instance:

1. $\vdash (A \wedge B) \leftrightarrow \text{╥}(\text{╥}A \vee \text{╥}B)$
2. $\vdash \text{╥}\forall A \leftrightarrow \exists x \text{╥}A$
3. $\vdash \exists xB \vee C \leftrightarrow \exists x(B \vee C)$
4. $\vdash B \vee \exists xC \leftrightarrow \exists x(B \vee C)$

Theorem 5.2. If A is a complex formula, then $\vdash \neg A \leftrightarrow \text{╥}A$.

Definition 5.4. We say that a *structure* S is *non-trivial* if there is a closed annotated atom $p_\lambda t_1 \ldots t_n$ such that $S(p_\lambda t_1 \ldots t_n) = 0$.

Hence a structure S is non-trivial iff there is some closed annotated atom that is not valid in S.

Definition 5.5. We say that a *structure* A is *inconsistent* if there is a closed annotated atom $p_\lambda t_1 \ldots t_n$ such that $S(p_\lambda t_1 \ldots t_n) = 1 = S(p_\lambda t_1 \ldots t_n)$.

So, a structure S is inconsistent iff there is some closed annotated atom such that it and its negation are both valid in S.

Definition 5.6. A *structure* S is called *paraconsistent* if S is both inconsistent and non-trivial. The *system* $Q\tau$ is said to be *paraconsistent* if there is a structure S for $Q\tau$ such that S is paraconsistent.

Definition 5.7. A *structure* S is called *paracomplete* if there is a closed annotated atom $p_\lambda t_1 \ldots t_n$ such that $S(p_\lambda t_1 \ldots t_n) = 0 = S(p_\lambda t_1 \ldots t_n)$.

The *system* $Q\tau$ is said to be *paracomplete* if there is a structure S such that S is paracomplete.

Theorem 5.3. $Q\tau$ is paraconsistent iff $\#|\tau| \geq 2$.

Proof. Suppose that $\#|\tau| \geq 2$. There is at least one predicate symbol p. Let τ be a non-empty set which $\#|\tau| \geq 2$. Let us define $p_S{:}|S|^n \rightarrow |\tau|$ setting $p_S(a_1, \ldots, a_n) = \perp$ and $p_S(b_1, \ldots, b_n) = \top$ where $(a_1, \ldots, a_n) \neq (b_1, \ldots, b_n)$.

Then, $S(p_\perp i_1 \ldots i_n) = 1$, where i_j the name of b_j, $j = 1, \ldots, n$, and $S(\neg p_\perp i_1 \ldots i_n) = 1$. Likewise, $S(p_\top j_1 \ldots j_n) = 0$, where j_i is the name of a_i, $i = 1, \ldots, n$. So, $Q\tau$ is paraconsistent. The converse is immediate.

Theorem 5.4. For all τ there are systems $Q\tau$ that are paracomplete; and also systems that are not paracomplete. If $Q\tau$ is paracomplete, then $\#|\tau| \geq 2$.

Proof. Similar to the proof of the preceding theorem.

Definition 5.8. A *structure* S is called *non-alethic* if S is both paraconsistent and paracomplete. The *system* $Q\tau$ is said to be *non-alethic* if there is a structure S for $Q\tau$ such that S is non-alethic.

Theorem 5.5. For all $\#|\tau| \geq 2$ there are systems $Q\tau$ that are non-alethic; and also systems that are not non-alethic. If $Q\tau$ is non-alethic, then $\#|\tau| \geq 2$.

Given a structure S, we can define the theory Th(S) associated with S to be the set Th(S) = $C_n(\Gamma)$, where Γ is the set of all annotated atoms which are valid in S. $C_n(\Gamma)$ indicates the set of all semantic consequences of elements of Γ.

Theorem 5.6. Given a structure S for $Q\tau$, we have:

1. Th(S) is a paraconsistent theory iff S is a paraconsistent structure.
2. Th(S) is a paracomplete theory iff S is a paracomplete structure.
3. Th(S) is a non-alethic theory iff S is a non-alethic structure.

In view of the preceding theorem, $Q\tau$ is, in general, a paraconsistent, paracomplete and non-alethic logic.

Next, we present another deduction method for the logics Qτ, by natural deduction [7].

5.4.1 Natural Deduction System NQτ

The natural deduction system NQτ proposed in [7] is composed by those classical predicate calculi plus the rules below.

(Axiom)

$$(\tau_1)\ \frac{}{p_{\perp}(t_1,\ldots,t_n)}$$

(Rules)

$$(\wedge_I)\ \frac{A\quad B}{A\wedge B}\qquad (\wedge_E)\ \frac{A\wedge B}{A}\quad \frac{A\wedge B}{B}\qquad (\vee_I)\ \frac{A}{A\vee B}\quad \frac{B}{A\vee B}\qquad (\vee_E)\ \frac{A\vee B\quad \begin{matrix}[A]\\ C\end{matrix}\quad \begin{matrix}[B]\\ C\end{matrix}}{C}$$

$$(\rightarrow_I)\ \frac{\begin{matrix}[A]\\ B\end{matrix}}{A\rightarrow B}\qquad (\rightarrow_E)\ \frac{A,\ A\rightarrow B}{B}\qquad (\mathrm{P})\ \frac{\begin{matrix}[A\rightarrow B]\\ A\end{matrix}}{A}\qquad (\neg\mathrm{I})\ \frac{\begin{matrix}[F]\\ G\wedge\neg G\end{matrix}}{\neg G}$$

$$(\neg\neg\mathrm{E})\ \frac{\neg\neg F}{F}\qquad (\neg\mathrm{E})\ \frac{F\quad \neg F}{A}$$

$$(\forall\mathrm{I})\ \frac{A(a)}{\forall x A(x)}\qquad (\forall\mathrm{E})\ \frac{\forall x A(x)}{A(t)}\qquad (\exists\mathrm{I})\ \frac{A(t)}{\exists x A(x)}\qquad (\exists\mathrm{E})\ \frac{\exists x A(x)\quad \begin{matrix}[A(a)]\\ B\end{matrix}}{B}$$

$$(\tau_2)\ \frac{\neg^k p_{\lambda}(t_1,\ldots,t_n)}{p^{k-1}{}_{\bar\lambda}(t_1,\ldots,t_n)}\quad \frac{p^{k-1}{}_{\bar\lambda}(t_1,\ldots,t_n)}{\neg^k p_{\lambda}(t_1,\ldots,t_n)}\qquad (\tau_3)\ \frac{p_{\lambda}(t_1,\ldots,t_n)\ \ \lambda\geq\mu}{p_{\mu}(t_1,\ldots,t_n)}$$

$$(\tau_4)\ \frac{p_{\lambda_1}(t_1,\ldots,t_n)\ \ldots\ p_{\lambda_m}(t_1,\ldots,t_n)\ \ \lambda=\bigvee_{i=1}^{m}\lambda_i}{p_{\lambda}(t_1,\ldots,t_n)}$$

As it is known, natural deduction systems are very useful for logic programming; so annotated systems are suitable for programming.

5.4.2 Completeness

We give a Henkin-type proof of the completeness theorem for the logics $Q\tau$.

Definition 5.9. A theory T based on $Q\tau$ is said to be *complete* if for each closed formula A we have $\vdash_T A$ or $\vdash_T \urcorner A$.

Lemma 5.2. Let $\lambda_0 = \vee\{\lambda \in |\tau|: \vdash_T p_\lambda t_1 \ldots t_n\}$. Then $\vdash_T p_{\lambda 0} t_1 \ldots t_n$.

Now let T be a non-trivial theory containing at least one constant. We shall define a structure S that we call the *canonical structure* for T. If a and b are variable-free terms of T, then we define aRb to mean $\vdash_T a = b$. It is easy to check that R is an equivalence relation. We let $|S|$ be the quotient set F/R, where F indicates the set of all formulas of L. The equivalence class determined by a is designed by a^o. We complete the definition of S by setting

$f_S(a_1^0, \ldots, a_m^0) = (f_S(a_1, \ldots, a_m))^0$ and $p_S(a_1^0, \ldots, a_n^0) = \vee\{\lambda \in |\tau|: \vdash_T p_\lambda a_1 \ldots a_n\}$.

It is straightforward to check the formal correctness of the above definitions.

Theorem 5.7. If $p_\lambda a_1 \ldots a_n$ is a variable-free annotated atom, then

$S(p_\lambda a_1 \ldots a_n) = 1$ iff $\vdash_T p_\lambda a_1 \ldots a_n$.

Proof. Let us suppose that $S(p_\lambda a_1 \ldots a_n) = 1$. Then $p_S(a_1^0, \ldots, a_n^0) \geq \lambda$. But $p_S(a_1^0, \ldots, a_n^0) = \vee\{\lambda \in |\tau|: \vdash_T p_\lambda a_1 \ldots a_n\}$; so, $\vdash_T p_{\lambda 0} a_1 \ldots a_n$ by the preceding lemma. As $\lambda_0 \geq \lambda$, it follows that $\vdash_T p_\lambda a_1 \ldots a_n$ by axiom (τ_3). Conversely, let us suppose that $\vdash_T p_\lambda a_1 \ldots a_n$. Then $\vee\{\mu \in |\tau|: \vdash_T p_\mu a_1 \ldots a_n\}$. Let $\lambda_0 = \vee\{\mu \in |\tau|: \vdash_T p_\mu a_1 \ldots a_n\}$;). Then it follows that $\lambda_0 \geq \lambda$. But $\lambda_0 = p_S(a_1^0, \ldots, a_n^0)$, and so $p_S(a_1^0, \ldots, a_n^0) \geq \lambda$; hence $S(p_\lambda a_1 \ldots a_n) = 1$.

A formula A is called *variable-free* if A does not contain free variables.

Theorem 5.8. Let $a = b$ be a variable-free formula. Then $S(a = b) = 1$ iff $\vdash_T a = b$.

We define the Henkin theory as in the classical case. Now, suppose that T is a Henkin theory and S the canonical structure for T.

Theorem 5.9. Let $\neg^k p_\lambda a_1 \ldots a_n$ be a variable-free hyper-literal. Then

$S(\neg^k p_\lambda a_1 \ldots a_n) = 1$ iff $\vdash_T \neg^k p_\lambda a_1 \ldots a_n$

Proof. By induction on k taking into account the axiom (τ_2) and Lemma 1.

Theorem 5.10. Let T be a complete Henkin theory, S the canonical structure for T, and A a closed formula. Then, $S(A) = 1$ iff A.

Proof. By induction on the length of A.

Corollary 5.2. Under the conditions of the Theorem 10, the canonical structure for T is a model of T.

We construct Henkin theories as in the classical case.

Theorem 5.11. (*Lindenbaum's theorem*). If T is a non-trivial theory, then T has a complete simple extension.

Theorem 5.12. (*Completeness theorem*). A theory (consistent or not) is non-trivial iff it has a model.

Theorem 5.13. Let Γ be a set of formulas. Then $\Gamma \vdash A$ iff $\Gamma \Vdash A$.

Theorem 5.14. A formula A of a theory T is a theorem of T iff it is valid in T.

The basic study of model theory was made in [1]. Abe has shown that almost all classical results in classical model theory can be adapted to annotated systems. In particular, Łós theorem is valid for annotated logics, showing that annotated model theory in a certain precise sense generalize the classical model theory.

5.5 Programming with Paraconsistent Annotated Systems

One particular annotated system is obtained when we consider the lattice $\tau = [0, 1]^2 = [0, 1] \times [0, 1]$ (where [0, 1] is the unitary real number interval). The order relation on $[0, 1]^2$ is defined as follows: $(\mu_1, \lambda_1) \leq (\mu_2, \lambda_2) \Leftrightarrow \mu_1 \leq \mu_2$ and $\lambda_1 \leq \lambda_2$. The corresponding annotated logic is called Paraconsistent Annotated Evidential Logic $E\tau$. The atomic formulas of the logic $E\tau$ are of the type $p_{(\mu, \lambda)}$, where $(\mu, \lambda) \in [0, 1]^2$ and [0, 1] is the real unitary interval (p denotes a propositional variable). $p_{(\mu, \lambda)}$ can be intuitively read: "It is assumed that p's favorable evidence is μ and contrary evidence is λ." Thus,

- $p_{(1.0, 0.0)}$ can be read as a true proposition.
- $p_{(0.0, 1.0)}$ can be read as a false proposition.
- $p_{(1.0, 1.0)}$ can be read as an inconsistent proposition.
- $p_{(0.0, 0.0)}$ can be read as a paracomplete (unknown) proposition.
- $p_{(0.5, 0.5)}$ can be read as an indefinite proposition.

Table 5.1. Extreme and Non-extreme states

Extreme States	Symbol	Non-extreme states	Symbol
True	V	Quasi-true tending to Inconsistent	$QV \to T$
False	F	Quasi-true tending to Paracomplete	$QV \to \bot$
Inconsistent	T	Quasi-false tending to Inconsistent	$QF \to T$
Paracomplete	$\bot$	Quasi-false tending to Paracomplete	$QF \to \bot$
		Quasi-inconsistent tending to True	$QT \to V$
		Quasi-inconsistent tending to False	$QT \to F$
		Quasi-paracomplete tending to True	$Q\bot \to V$
		Quasi-paracomplete tending to False	$Q\bot \to F$

We introduce the following concepts: Uncertainty Degree: $G_{un}(\mu, \lambda) = \mu + \lambda - 1$; Certainty Degree: $G_{ce}(\mu, \lambda) = \mu - \lambda$ ($0 \leq \mu, \lambda \leq 1$). With the uncertainty and certainty degrees we can get the following 12 output states: extreme state and non-extreme states, showed in the Table 5.1.

All states are represented in the next figure.

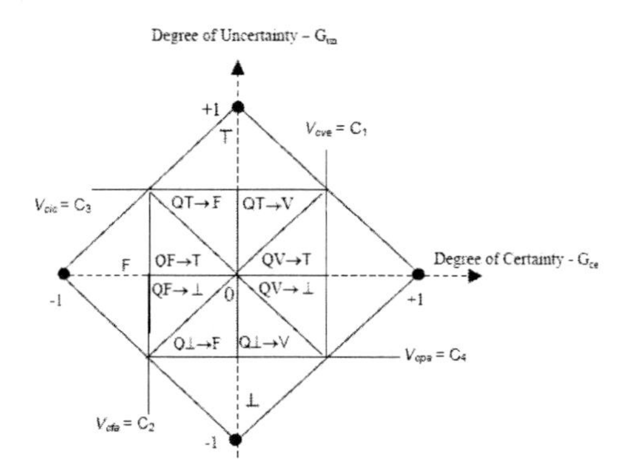

Fig. 5.1. Extreme and non-extreme states with certainty and uncertainty degrees

Some additional control values are:

- V_{cic} = maximum value of uncertainty control = C_3
- V_{cve} = maximum value of certainty control = C_1
- V_{cpa} = minimum value of uncertainty control = C_4
- V_{cfa} = minimum value of certainty control = C_2

In [11] it was developed a paraconsistent logic programming language – Paralog. As it is well known, the development of computationally efficient programs in it should exploit the following two aspects in its language:

1. The declarative aspect that describes the logic structure of the problem, and
2. The procedural aspect that describes how the computer solves the problem.

However, it is not always an easy task to conciliate both aspects. Therefore, programs to be implemented in Paralog should be well defined to evidence both the declarative aspect and the procedural aspect of the language.

It must be pointed out that programs in Paralog, like programs in standard Prolog, may be easily understood or reduced - when well defined - by means of addition or elimination of clauses, respectively.

A small knowledge base in the domain of Medicine is presented as a Paralog program. The development of this small knowledge base was subsidized by the information provided by three experts in Medicine. The first two specialists - clinicians - provided six[1] diagnosis rules for two diseases: disease1 and disease2. The

[1] The first four diagnosis rules were supplied by the first expert clinician and the two remaining diagnosis rules were provided by the second expert clinician.

last specialist - a pathologist - provided information on four symptoms: symptom1, symptom2, symptom3 and symptom4.

Example 5.1. A small knowledge base in Medicine implemented in Paralog

```
disease1 (X): [1.0, 0.0] <--
symptom1(X): [1.0,0.0] &
symptom2(X): [1.0, 0.0]
disease2(X): [1.0, 0.0] <--
symptom1(X): [1.0, 0.0] &
symptom3(X): [1.0, 0.0]
disease1(X): [0.0, 1.0] <--
disease2(X): [1.0, 0.0].
disease2(X): [0.0, 1.0] <--
disease1(X): [1.0, 0.0].
disease1(X): [1.0, 0.0] <--
symptom1(X): [1.0, 0.0] &
symptom4(X): [1.0, 0.0] .
disease2(X): [1.0, 0.0] <--
symptom1(X): [0.0, 1.0] &
symptom3(X): [1.0, 0.0] .
symptom1(john) : [1.0, 0.0].
symptom1(bill): [0.0,1.0].
symptom2(john) : [0.0, 1.0].
symptom2(bill): [0.0, 1.0].
symptom3(john): [1.0, 0.0].
symptom3(bill): [1.0, 0.0].
symptom4(john): [1.0, 0.0].
symptom4(bill): [0.0, 1.0].
```

Table 5.2. Query and answer forms in Paralog

Item	Query and answer form		Meaning
1	Query	D1(bill):[1.0, 0.0]	Does Bill have disease 1 ?
	Evidence	[0.0, 0.0]	The information on Bill's disease1 is unknown
2	Query	D2(bill):[1.0, 0.0]	Does Bill have disease 2 ?
	Evidence	[1.0, 0.0]	Bill has disease2
3	Query	D1(john):[1.0, 0.0]	Does John have disease 1 ?
	Evidence	[1.0, 1.0]	The information on John's disease1 is inconsistent
4	Query	D2(john):[1.0, 0.0]	Does John have disease 2 ?
	Evidence	[1.0, 1.0]	The information on John's disease2 is inconsistent
5	Query	D1(bob):[1.0, 0.0]	Does Bob have disease 1 ?
	Evidence	[0.0, 0.0]	The information on Bob's disease1 is unknown

In this example, several types of queries can be performed. Table 5.2 below shows some query types, the evidences provided as answers by the Paralog inference engine and their respective meaning.

The knowledge base implemented in the Example 5.1 may also be implemented in standard Prolog, as shown in the Example 5.2.

Example 5.2. Knowledge base of Example 1 implemented in standard Prolog

```
disease1(X) :-
symptom1(X),
symptom2(X).
disease2(X):-
symptom1(X),
symptom3(X).
disease1 (X):-
not disease2(X).
disease2(X) :-
not disease1(X).
disease1(X) :-
 symptom1(X),
 symptom4(X).
disease2(X) :-
 not symptom1(X),
 symptom3(X).
 symptom1 (john).
 symptom3(john).
 symptom3(bill).
 symptom4(john).
```

In this example, several types of queries can be performed as well. Table 5.3 shows some query types provided as answers by the standard Prolog and their respective meaning.

Table 5.3. Query and answer forms in standard Prolog

Item	Query and answer form		Meaning
1	Query	D1(bill)	Does Bill have disease 1 ?
	Answer	Loop	System enters into an infinite loop
2	Query	D2(bill)	Does Bill have disease 2 ?
	Answer	Loop	System enters into an infinite loop
3	Query	D1(john)	Does John have disease 1 ?
	Answer	Yes	John has disease1
4	Query	D2(john)	Does John have disease 2 ?
	Answer	Yes	John has disease2
5	Query	D1(bob)	Does Bob have disease 1 ?
	Answer	No	Bob does not have disease1

Starting from Examples 5.1 and 5.2 it can be seen that there are different characteristics between implementing and consulting in Paralog and standard Prolog. Among these characteristics, the most important are:

1. The semantic characteristic; and
2. The execution control characteristic.

The first characteristic may be intuitively observed when the program codes in Examples 5.1 and 5.2 are placed side by side. That is, when compared to Paralog, the standard Prolog representation causes loss of semantic information on facts and rules. This is due to the fact that standard Prolog cannot directly represent the negation of facts and rules.

In Example 5.1, Paralog program presents a four-valued evidence representation. However, the information loss may be greater for a standard Prolog program, when the lattice is τ. This last characteristic may be observed in Tables 5.2 and 5.3. These two tables show five queries and answers, presented and obtained both in Paralog and standard Prolog program.

The answers obtained from the two approaches present major differences. That is, to the first query: "Does Bill have diseasel?", Paralog answers that the information on Bill's diseasel is unknown, while the standard Prolog enters into a loop. This happens because the standard Prolog inference engine depends on the ordination of facts and rules to reach deductions. This, for standard Prolog to be able to deduct an answer similar to Paralog, the facts and rules in Example 5.2 should be reordered. On the other hand, as the Paralog inference engine does not depend on reordering facts and rules, such reordering becomes unnecessary.

In the second query: "Does Bill have disease2?", Paralog answers that "Bill has disease2", while the standard Prolog enters into a loop. This happens for the same reasons explained in the foregoing item.

In the third query: "Does John have diseasel?", Paralog answers that the information on John's diseasel is inconsistent, while the standard Prolog answers that "John has diseasel". This happens because the standard Prolog inference engine, after reaching the conclusion that "John has diseasel" does not check whether there are other conclusions leading to a contraction. On the other hand, Paralog performs such check, leading to more appropriate conclusions.

In the fourth query: "Does John have disease2?", Paralog answers that the information on John's disease2 is inconsistent, while the standard Prolog answers that "John has disease2". This happens for the same reasons explained in the foregoing item.

In the last query: "Does Bob have diseasel", Paralog e answers that the information on Bob's diseasel is unknown, while the standard Prolog answers that "Bob does not have diseasel". This happens because the standard Prolog inference engine does not distinguish the two possible interpretations for the answer not. On the other hand, the Paralog inference engine, being based on an infinitely valued paraconsistent evidential logic, allows the distinction to be made.

In view of the above, it is shown that the use of the Paralog language may handle several Computer Science questions more naturally.

5.6 The Paraconsistent Annotated Multimodal Logics Mτ

We present, in this section, the multimodal predicate calculi Mτ, extension of the logics Qτ studied in the section 5.4.

The language of Mτ has the following primitive symbols:

1. Individual variables: a denumerable infinite set of variable symbols: x_1, x_2, ...
2. Logical connectives: $\neg$ (negation), $\wedge$ (conjunction), $\vee$ (disjunction), and $\rightarrow$ (implication).
3. For each n, zero or more n-ary function symbols (n is a natural number).
4. For each $n \neq 0$, n-ary predicate symbols.
5. The equality symbol: =
6. Annotational constants: each member of τ is called an annotational constant.
7. Modal operators: $[]_1$, $[]_2$, ... , $[]_n$, ($n \geq 1$), $[]_G$, $[]_G^C$, $[]_G^D$ (for every nonempty subset G of $\{1, ... , n\}$).
8. Quantifiers: $\forall$ (for all) and $\exists$ (there exists).
9. Auxiliary symbols: parentheses and comma.

We introduce the general concept of (*annotated*) *formula* in the standard way. For instance, if A is a formula, then $[]_1A$, $[]_2A$, ... , $[]_nA$, $[]_GA$, $[]_G^C A$, and $[]_G^D A$ are also formulas.

Definition 5.10. Let A and B be formulas. The formulas $A \leftrightarrow B$ (biconditional) and $\urcorner\!\!\urcorner A$ (strong negation) are as in Definition 1. Also, $\neg^k A$ indicates $\neg(\neg^{k-1}A)$ and if $\mu \in \tau$, $\sim^k\mu$ indicates $\sim(\sim^{k-1}\mu)$.

The postulates (axiom schemata and primitive rules of inference) of Mτ are the same of the logics Qτ plus the following listed below, where A, B, and C are any formulas whatsoever, $p(t_1, \ldots, t_n)$ is a basic formula, and λ, μ, μ_j are annotational constants.

M1) $[]_i(A \to B) \to ([]_iA \to []_iB)$, $i = 1, 2, \ldots, n$
M2) $[]_iA \to []_i[]_iA$, $i = 1, 2, \ldots, n$
M3) $\urcorner\!\!\urcorner []_iA \to []_i \urcorner\!\!\urcorner []_iA$, $i = 1,2, \ldots, n$
M4) $[]_iA \to A$, $i = 1, 2, \ldots, n$
M5) $\dfrac{A}{[]_i A}$, $i = 1, 2, \ldots, n$
M6) $[]_G A \leftrightarrow \wedge_{i\in G}[]_iA$
M7) $[]_G^C A \to []_G(A \wedge []_G^C A)$
M8) $[]_{\{i\}}^D A \leftrightarrow []_iA$, $i = 1, 2, \ldots, n$
M9) $[]_G^D A \to []_{G'}^D A$ if $G' \subseteq G$
M10) $\dfrac{A \to []_G (B \wedge A)}{A \to []_G^B B}$
M11) $\forall x[]_iA \to []_i\forall xA$, $i = 1, 2, \ldots, n$
M12) $\urcorner\!\!\urcorner(x = y) \to []_i\urcorner\!\!\urcorner(x = y)$, $i = 1, 2, \ldots, n$

with the usual restrictions.

Mτ is an extension of the logic Qτ. As Qτ contains classical predicate logic, Mτ contains classical modal logic S5, as well as the multimodal system studied in [14] in at least two directions. So, usual all valid schemes and rules of classical positive propositional logic are true. In particular, the deduction theorem is valid in Mτ ant it contains intuitionistic positive logic.

Theorem 5.15. Mτ is non-trivial.

6.6.1 Semantical Analysis: Kripke Models

Definition 5.11. A Kripke model for Mτ (or Mτ structure) is a set theoretical structure $K = [W, R_1, R_2, \ldots, R_n, I]$ where
W is a nonempty set of elements called 'worlds'
R_i ($i = 1, 2, \ldots, n$) is a binary relation on W such that it is an equivalence relation.

I is an interpretation function with the usual properties with the exception that for each n-ary predicate symbol p we associate a function $p_I: W^n \rightarrow |\tau|$.

Given a Kripke model K for the language L of Mτ, the *diagram* language $L(K)$ is obtained as usual. Given a free variable term a of $L(K)$ we define, as usual, the individual $K(a)$ of K. We use i and j as meta-variables for names.

Definition 5.12. If A is a closed formula of Mτ, and $w \in W$, we define the relation $K,w \Vdash A$ (K,w *force* A) by recursion on A:

1. If A is atomic of the form $p_\lambda(t_1, \ldots, t_n)$, then
   $K,w \Vdash A$ iff $p_I(K(t_1), \ldots, K(t_n)) \geq \lambda$.
2. If A is of the form $\neg^k p_\lambda(t_1, \ldots, t_n)$ $(k \geq 1)$, $K,w \Vdash A$ iff $K,w \Vdash \neg^{k-1} p_{\sim\lambda}(t_1, \ldots, t_n)$.
3. Let A and B formulas. Then, $K,w \Vdash (A \wedge B)$ iff $K,w \Vdash A$ and $K,w \Vdash B$; $K,w \Vdash (A \vee B)$ iff $K,w \Vdash A$ or $K,w \Vdash B$; $K,w \Vdash (A \rightarrow B)$ iff it is not the case that $K,w \Vdash A$ or $K,w \Vdash B$;
4. If F is a complex formula, then $K,w \Vdash (\neg F)$ iff it is not the case that $K,w \Vdash F$.
5. If A is of the form $(\exists x)B$, then $K,w \Vdash A$ iff $K,w \Vdash B_x[i]$ for some i in $L(K)$.
6. If A is of the form $(\forall x)B$, then $K,w \Vdash A$ iff $K,w \Vdash B_x[i]$ for all i in $L(K)$.
7. If A is of the form $[]_i B$ then $K,w \Vdash A$ iff $K,w' \Vdash B$ for each $w' \in W$ such that $wR_i w'$, $i = 1, 2, \ldots, n$

Definition 5.13. Let $K = [W, R_1, R_2, \ldots, R_n, I]$ be a Kripke structure for Mτ. The Kripke structure K *forces* a formula A (in symbols, $K \Vdash A$), if $K,w \Vdash A$ for each $w \in W$. A formula A is called Mτ-*valid* if for any Mτ-structure K, $K \Vdash A$. A formula A is called *valid* if it is Mτ-valid for all Mτ structure. We symbolize this fact by $\Vdash A$.

Theorem 5.16. Let $K = [W, R_1, R_2, \ldots, R_n, I]$ be a Kripke structure for Mτ. For all formulas A, B, then

1. If A is an instance of a propositional tautology then, $K \Vdash A$
2. If $K \Vdash A$ and $K \Vdash A \rightarrow B$, then $K \Vdash B$
3. $K \Vdash []_i(A \rightarrow B) \rightarrow ([]_i A \rightarrow []_i B)$, $i = 1, 2, \ldots, n$
4. $K \Vdash []_i A \rightarrow []_i[]_i A$, $i = 1, 2, \ldots, n$
5. $K \Vdash []_i A \rightarrow A$, $i = 1, 2, \ldots, n$
6. If $K \Vdash A$ then $K \Vdash []_i A$, $i = 1, 2, \ldots, n$

Theorem 5.17. Let K be a Kripke model for Mτ and F a complex formula. Then we have not simultaneously $K,w \Vdash \neg F$ and $K,w \Vdash F$.

Theorem 5.18. Let $p(t_1, \ldots, t_n)$ be a basic formula and $\lambda, \mu, \rho \in |\tau|$. We have

1. $\Vdash p_\perp(t_1, \ldots, t_n)$
2. $\Vdash p_\lambda(t_1, \ldots, t_n) \rightarrow p_\mu(t_1, \ldots, t_n)$, if $\lambda \geq \mu$
3. $\Vdash p_\lambda(t_1, \ldots, t_n) \wedge p_\mu(t_1, \ldots, t_n) \rightarrow p_\rho(t_1, \ldots, t_n)$, where $\rho = \lambda \vee \mu$

Proof. 1. For any Kripke model K, we have $p_I(K(t_1), \ldots, K(t_n)) \geq \bot$, for all $w \in K$. So, $K \Vdash p_\bot(t_1, \ldots, t_n)$ for every K, and therefore $\Vdash p_\bot(t_1, \ldots, t_n)$.
2. Let us suppose that there exists a K such that it is not the case that $K \Vdash p_\lambda(t_1, \ldots, t_n) \rightarrow p_\mu(t_1, \ldots, t_n)$, that is $K \Vdash p_\lambda(t_1, \ldots, t_n)$ and it is not the case that $K \Vdash p_\mu(t_1, \ldots, t_n)$, for some $w \in K$. So, $p_I(K(t_1), \ldots, K(t_n)) \geq \lambda$. As $\lambda \geq \mu$, we have $p_I(K(t_1), \ldots, K(t_n)) \geq \mu$, which contradicts the hypothesis. Therefore, we have $\Vdash p_\lambda(t_1, \ldots, t_n) \rightarrow p_\mu(t_1, \ldots, t_n)$, if $\lambda \geq \mu$.
3. Similar to the preceding item.

Theorem 5.19. Let A and B be arbitrary formulas and F a complex formula. Then:

1. $\Vdash ((A \rightarrow B) \rightarrow ((A \rightarrow \urcorner B) \rightarrow \urcorner A))$
2. $\Vdash (A \rightarrow (\urcorner A \rightarrow B))$
3. $\Vdash (A \vee \urcorner A)$
4. $\Vdash (\neg F \leftrightarrow \urcorner F)$
5. $\Vdash A \leftrightarrow \urcorner\urcorner A$
6. $\Vdash \forall x A \leftrightarrow \exists x \urcorner A$
7. $\Vdash (A \wedge B) \leftrightarrow \urcorner(\urcorner A \vee \urcorner B)$
8. $\Vdash \forall A \leftrightarrow \exists x \urcorner A$
9. $\Vdash \forall x A \vee B \leftrightarrow \exists x(A \vee B)$
10. $\Vdash A \vee \exists x B \leftrightarrow \exists x(A \vee B)$

Corollary 5.3. In the same conditions of the preceding theorem, we have not simultaneously $K \Vdash \urcorner A$ and $K \Vdash A$. The set of all formulas together with the connectives $\wedge$, $\vee$, $\rightarrow$, and $\urcorner$ has all properties of the classical logic.

Theorem 5.20. There are Kripke models K such that for some hyper-literals A and B and some worlds w and $w' \in W$, we have $K,w \Vdash \neg A$ and $K,w \Vdash A$ and it is not the case that $K,w' \Vdash B$.

Proof. Let $W = \{\{a\}\}$ and $R = \{(\{a\},\{a\})\}$ (that is $w = \{a\}$) and $p(t_1, \ldots, t_n)$ and $q(t'_1, \ldots, t'_n)$ basic (closed) formulas such that $I(p) \equiv \top$ and $I(q) \equiv \bot$ (so $K = [W, R, I]$). As $\top \geq \top$, it follows that $p_I(K(t_1), \ldots, K(t_n)) \geq \top$. Also, $\top \geq {\sim}\top$. So, $p_I(K(t_1), \ldots, K(t_n)) \geq {\sim}\top$. Therefore, $K,w \Vdash p_\top(t_1, \ldots, t_n)$ and $K,w \Vdash p_{\sim\top}(t_1, \ldots, t_n)$. It follows that $K,w \Vdash \neg p_\top(t_1, \ldots, t_n)$. On the other hand, as it is false that $\bot \geq \top$; it follows that it is not the case that $q_I(K(t'_1), \ldots, K(t'_n)) \geq \top$, and so, it is not the case that $K,w \Vdash q_\bot(t'_1, \ldots, t'_n)$.

Theorem 5.21. For some systems $M\tau$ there are Kripke models K such that for some hyper-literal formula A and some world $w \in W$, we don't have $K,w \Vdash A$ nor $K,w \Vdash \neg A$.

Proof. Let us define the operator ${\sim}: |\tau| \rightarrow |\tau|$ by setting ${\sim}\top = \top$. Then, let I be the interpretation such that $I(p) \equiv \bot$. So, it is no the case that $I(p) \geq \top$ and also, it is not the case that $I(p) \geq \sim\top$ (or, equivalently, not $K,w \Vdash p_{\top}(t_1, \ldots, t_n)$ and not $K,w \Vdash \neg p_{\top}(t_1, \ldots, t_n)$).

Corollary 5.4. For some systems Mτ there are Kripke models *K* such that for some hyper-literal formulas *A* and *B*, and some worlds $w, w' \in W$, we have $K,w \Vdash \neg A$ and $K,w \Vdash A$ and we don't have $K,w \Vdash B$ nor $K,w \Vdash \neg B$.

Proof. Consequence of the Theorems 5.20 and 5.21.

The earlier results show us that there are systems Mτ such that we have "inconsistent" worlds, "paracomplete" worlds, or both.

Now we present a strong version these results linking with paraconsistent, paracomplete, and non-alethic logics.

Definition 5.14. A Kripke model *K* is called *paraconsistent* if there are basic formulas $p(t_1, \ldots, t_n)$, $q(t_1, \ldots, t_n)$, and annotational constants $\lambda, \mu \in |\tau|$ such that $K,w \Vdash p_\lambda(t_1, \ldots, t_n)$, $K,w \Vdash \neg p_\lambda(t_1, \ldots, t_n)$, and it is not the case that $K,w \Vdash q_\mu(t_1, \ldots, t_n)$.

Definition 5.15. A system Mτ is called *paraconsistent* if there is a Kripke model *K* for Mτ such that *K* is paraconsistent.

Theorem 5.22. Mτ is a paraconsistent system iff $\#|\tau| \geq 2$.

Proof. Define a structure $K = [\{w\}, \{(w, w)\}, I]$ such that $\begin{cases} q_I = \bot \\ p_I = \top \end{cases}$

It is clear that $p_I \geq \top$, and so $K \Vdash p_{\top}(t_1, \ldots, t_n)$. Also, $p_I \geq \sim\top$, and, so $K \Vdash p_{\sim\top}(t_1, \ldots, t_n)$, or $K \Vdash \neg p_{\top}(t_1, \ldots, t_n)$. Also, it is not the case that $q_I(t_1, \ldots, t_n) \geq \bot$, so it is not the case that $K,w \Vdash q_\bot(t_1, \ldots, t_n)$.

Definition 5.16. A Kripke model *K* is called *paracomplete* if there are a basic formula $p(t_1, \ldots, t_n)$ and an annotational constant $\lambda \in |\tau|$ such that it is false that $K,w \Vdash p_\lambda(t_1, \ldots, t_n)$ and it is false that $K,w \Vdash \neg p_\lambda(t_1, \ldots, t_n)$. A system Mτ is called *paracomplete* if there is a Kripke models *K* for Mτ such that *K* is paracomplete.

Definition 5.17. A Kripke model *K* is called *non-alethic* if *K* are both paraconsistent and paracomplete. A system Mτ is called *non-alethic* if there is a Kripke model *K* for Mτ such that *K* is non-alethic.

Theorem 5.23. If $\#|\tau| \geq 2$, then there are systems Mτ which are paracomplete and systems Mτ' that are not paracomplete, $\#|\tau'| \geq 2$.

Proof. Similar to the preceding theorem.

Corollary 5.5. If $\#|\tau| \geq 2$, then there are systems Mτ which are non-alethic and systems Mτ' that are not non-alethic, $\#|\tau'| \geq 2$.

5.6.2 Soundness and Completeness

Theorem 5.24. Let U be a maximal non-trivial maximal (with respect to inclusion of sets) subset of the set of all formulas. Let A and B formulas whatsoever. Then

If A is an axiom of Mτ, then $A \in U$

1. $A \wedge B \in U$ iff $A \in U$ and $B \in U$.
2. $A \vee B \in U$ iff $A \in U$ or $B \in U$.
3. $A \rightarrow B \in U$ iff $A \notin U$ or $B \in U$.
4. If $p_\mu(t_1, \dots, t_n) \in U$ and $p_\lambda(t_1, \dots, t_n) \in U$, then $p_\rho(t_1, \dots, t_n) \in U$, where $\rho = \mu \vee \lambda$
5. $\neg^k p_\lambda(t_1, \dots, t_n) \in U$ iff $\neg^{k-1} p_{\sim\lambda}(t_1, \dots, t_n) \in U$.
6. If A and $A \rightarrow B \in U$, then $B \in U$.
7. $A \in U$ iff $\urcorner\!\urcorner A \notin U$. Moreover $A \in U$ or $\urcorner\!\urcorner A \in U$
8. If A is a complex formula, $A \in U$ iff $\neg A \notin U$. Moreover $A \in U$ or $\neg A \in U$.
9. If $A \in U$, then $[]_i A \in U$.

Proof. Let us show only 3. In fact, if $p_\mu(t_1, \dots, t_n) \in U$ and $p_\lambda(t_1, \dots, t_n) \in U$, then $p_\mu(t_1, \dots, t_n) \wedge p_\lambda(t_1, \dots, t_n) \in U$ by 2. But it is an axiom $p_\mu(t_1, \dots, t_n) \wedge p_\lambda(t_1, \dots, t_n) \rightarrow p_\rho(t_1, \dots, t_n)$, where $\rho = \mu \vee \lambda$. It follows that $p_\mu(t_1, \dots, t_n) \wedge p_\lambda(t_1, \dots, t_n) \rightarrow p_\rho(t_1, \dots, t_n) \in U$, and so $p_\rho(t_1, \dots, t_n) \in U$, by 6.

We give a Henkin-type proof of the completeness theorem for the logics Mτ. For this we define relations R_i on the set of all free-variable terms of Mτ as usual and we indicate by $\overset{\circ}{t}$ the equivalence class determined by t. Also, we will consider the quotient set F/R_i, where F indicates the set of all formulas.

Given a set U of formulas, define $U/[]_i = \{A \mid []_i A \in U\}$, $i = 1, 2, \dots, n$. Let us consider the canonical structure $K = [W, R_i, I]$ where $W = \{U \mid U$ is a maximal non-trivial set$\}$ and the interpretation function is as usual with the exception that given a n-ary predicate symbol p we associate the function $p_I : W^n \rightarrow |\tau|$ defined by $p_I(\overset{\circ}{t}_1, \dots, \overset{\circ}{t}_n) =_{def.} \vee\{\mu \in |\tau| \mid p_\mu(t_1, \dots, t_n) \in U\}$ (such function is well defined, so $p_\perp(t_1, \dots, t_n) \in U$). Moreover, define $R_i =_{Def.} \{(U, U') \mid U/[]_i \subseteq U'\}$.

Lemma 5.3. For all propositional variable p and if U is a maximal non-trivial set of formulas, we have $p_{pI(t^\circ 1, \dots, t^\circ n)}(t_1, \dots, t_n) \in U$.

Proof. It is a simple consequence of the previous theorem, item 5.

Theorem 5.25. For any formula A and for any non-trivial maximal set U, we have $(K, U) \Vdash A$ iff $A \in U$.

Proof. Let us suppose that A is $p_\lambda(t_1, \dots, t_n)$ and $(K, U) \Vdash p_\lambda(t_1, \dots, t_n)$. It is clear by previous lemma that $p_{pI(t^\circ 1, \dots, t^\circ n)}(t_1, \dots, t_n) \in U$. It follows also that $p_I(\overset{\circ}{t}_1, \dots, \overset{\circ}{t}_n) \geq \lambda$. It is an axiom that $p_{pI(t^\circ 1, \dots, t^\circ n)}(t_1, \dots, t_n) \rightarrow p_\lambda(t_1, \dots, t_n)$. Thus, $p_\lambda(t_1, \dots, t_n) \in U$. Now, let us suppose that $p_\lambda(t_1, \dots, t_n) \in U$. By previous lemma, $p_{pI(t^\circ 1, \dots, t^\circ n)}(t_1, \dots, t_n) \in U$. It follows that $p_I(\overset{\circ}{t}_1, \dots, \overset{\circ}{t}_n) \geq \lambda$. Thus, by definition, $(K, U) \Vdash p_\lambda(t_1, \dots, t_n)$. By Theorem 5.20, $\neg^k p_\lambda(t_1, \dots, t_n) \in U$ iff $\neg^{k-1} p_{\sim\lambda}(t_1, \dots, t_n) \in U$. Thus, by Definition 5.12, $(K, U) \Vdash$

$\neg^k p_\lambda(t_1, \ldots, t_n)$ iff $(K, U) \Vdash \neg^{k-1} p_{\sim\lambda}(t_1, \ldots, t_n)$. So, by induction on k the assertion is true for hyper-literals.

The other cases, the proof is as in the classical case.

Corollary 5.6. A is a provable formula of Mτ iff $\Vdash A$.

5.7 Concluding Remarks

In this work we have presented a class of paraconsistent annotated multimodal syste ms. It has shown that there is a logic in which it is possible to deal with imprecise, inconsistent and paracomplete knowledge, conflicting beliefs and awareness. Our main point is that imprecise, inconsistency and paracomplete concepts are natural phenomena in distributed systems. Disagreements among agents, inconsistent, paracomplete information, in doing a task is a common matter. Thus it is clear that formal methods are needed to handle this problem. The system proposed has provided a means to reasoning about inconsistent system of knowledge.

Acknowledgments. The authors are grateful to the anonymous referees providing useful comments to improving this version of the work. Also, the first author is grateful to Dr. Ngoc Thanh Nguyen for his kind invitation.

References

1. Abe, J.M.: Fundamentos da Lógica Anotada (Foundations of Annotated Logics), Ph.D. thesis, University of São Paulo, São Paulo (1992) (in Portuguese)
2. Abe, J.M.: Some Aspects of Paraconsistent Systems and Applications. Logique et Analyse 15, 83–96 (1997)
3. Abe, J.M., Da Silva Filho, J.I.: Manipulating Conflicts and Uncertainties in Robotics. Multiple-Valued Logic and Soft Computing 9, 147–169 (2003)
4. Abe, J.M., Nakamatsu, K.: Manipulating paraconsistent knowledge in multiagent systems. In: Nguyen, N.T., Grzech, A., Howlett, R.J., Jain, L.C. (eds.) KES-AMSTA 2007. LNCS (LNAI), vol. 4496, pp. 159–168. Springer, Heidelberg (2007)
5. Akama, S., Abe, J.M.: Many-valued and annotated modal logics. In: IEEE International Symposium on Multiple-Valued Logic Proceedings, pp. 114–119 (1998)
6. Akama, S., Abe, J.M., Murai, T.: On the relation of fuzzy and annotated logics. In: Proc. of the 7th International Conference Artificial Intelligence and Soft Computing, pp. 46–51 (2003)
7. Akama, S., Nakamatsu, K., Abe, J.M.: A natural deduction system for annotated predicate logic. In: Apolloni, B., Howlett, R.J., Jain, L. (eds.) KES 2007, Part II. LNCS (LNAI), vol. 4693, pp. 861–868. Springer, Heidelberg (2007)
8. Cresswell, M.J.: Intensional logics and logical truth. Journal of Philosophical Logic 1, 2–15 (1972)
9. Cresswell, M.J.: Logics and Languages. Methuen and Co., London (1973)
10. Da Costa, N.C.A.: On the theory of inconsistent formal systems. Notre Dame J. of Formal Logic 15, 497–510 (1974)

11. Da Costa, N.C.A., Prado, J.P.A., Abe, J.M., Ávila, B.C., Rillo, M.: Paralog: Um Prolog Paraconsistente baseado em Lógica Anotada, Institute For Advanced Studies, University of São Paulo TR-18 (1995)
12. Da Costa, N.C.A., Abe, J.M., Subrahmanian, V.S.: Remarks on annotated logic. Zeitschr. f. Math. Logik und Grundlagen d. Math. 37, 561–570 (1991)
13. Da Silva Filho, J.I., Abe, J.M.: Fundamentos das Redes Neurais Paraconsistentes – Destacando Aplicações em Neurocomputação, Arte & Ciência (2001) (in Portuguese)
14. Fagin, R., Halpern, J.Y., Moses, Y., Vardi, M.Y.: Reasoning about knowledge. The MIT Press, London (1995)
15. Fischer, M.J., Immerman, N.: Foundation of knowledge for distributed systems. In: Halpern, J.Y. (ed.) Theoretical Aspects of Reasoning about Knowledge: Proc. Fifth Conference, pp. 171–186. Morgan Kaufmann, San Francisco (1986)
16. Lipman, B.L.: Decision theory with impossible possible worlds, TR, Queen's University (1992)
17. McCarthy, J.: Ascribing mental qualities to machines. TR-CRL 90/10, DEC-CRL (1979)
18. Nakamatsu, K., Mita, Y., Shibata, T., Abe, J.M.: Defesiable deontic action control based on paraconsistent logic program and its implementation. In: Mohammadia, M. (ed.) International Conference on Computational Intelligence for Modeling, Contrl & Automation, Proceedings, pp. 233–246 (2003)
19. Nelson, D.: Negation and separation of concepts in constructive systems. In: Heyting, A. (ed.) Constructivity in Mathematics, pp. 208–225. North-Holland, Amsterdam (1959)
20. Nguyen, N.T.: Consensus System for Solving Conflicts in Distributed Systems. Journal of Information Sciences 147, 91–122 (2002)
21. Nguyen, N.T., Malowiecki, M.: Deriving consensus for conflict situations with respect to its susceptibility. In: Negoita, M.G., Howlett, R.J., Jain, L.C. (eds.) KES 2004. LNCS (LNAI), vol. 3214, pp. 1179–1186. Springer, Heidelberg (2004)
22. Parikh, R., Ramamujan, R.: Distributed processing and the logic of knowledge. In: Parikh, R. (ed.) Proc. Workshop on Logics of Program, pp. 256–268 (1985)
23. Rosenschein, S.J.: Formal theories of AI in knowledge and robotics. New Generation Computing 3, 345–357 (1985)
24. Subrahamanian, V.S.: On the Semantics of Quantitative Logic Programs. In: Proc. 4th IEEE Symposium on Logic Programming, pp. 173–182. IEEE Computer Society Press, Washington (1987)
25. Sylvan, R., Abe, J.M.: On general annotated logics, with an introduction to full accounting logics. Bulletin of Symbolic Logic 2, 118–119 (1996)

6

An Agent-Based Negotiation Platform for Collaborative Decision-Making in Construction Supply Chain

Xiaolong Xue[1,2] and Zhaomin Ren[3]

[1] Department of Construction and Real Estate, School of Management, Harbin Institute of Technology, Harbin 150001, China
[2] National Center of Technology, Policy and Management, Science Park, Harbin Institute of Technology, No.2, Yikuang Street, Harbin 150001, China
xlxue@hit.edu.cn
[3] Division of Built Environment, Faculty of Advanced Technology, University of Glamorgan, UK
zren@glam.ac.uk

Abstract. Negotiation is an effective and popular decision-making and coordination behavior in inter-organization systems, especially in construction supply chain (CSC) which is characterized by fragmentation, low efficiency and multiple partners. This chapter defines the concepts of CSC and CSC management, especially regarding CSC management as the coordination of inter-organization decision-making in CSC and the integration of key construction business processes and key members involved in CSC. Much research and practice indicates that there are still many problems in construction, most of which are supply chain problems. The research analyses the problems in CSC. In order to resolve these problems and improve the performance of construction, this research proposes a multi-agent based negotiation platform for improving the effectiveness and efficiency of collaborative decision-making in CSC by adopting agent technology and regarding CSC as a typical multi-agent system. The general structure of the agent-based negotiation platform (ANeP) is designed, which includes two kinds of agent group: specialty agents and service agents. Since different members in CSC have different preferences on the decision attributes (such as cost, time, quality, safety and environment), a multi-attribute negotiation model is established by designing a negotiation protocol and describing the negotiation process. Considering the negotiation failure or man-made termination of negotiation for increasing the efficiency of negotiation, this research presents a relative entropy method for improving agent-based multi-attribute negotiation efficiency (REAMNE) in ANeP. This method aggregates the preference information of negotiators in CSC twice. Firstly, compromise group preference order is ascertained by using preference information and compromise preference information. Secondly, group preferences are aggregated by using a relative entropy model, which is established based on entropy theory whilst, meanwhile, considering the multiple attributes in CSC negotiation. The prototype of ANeP is developed based on the ZEUS agent development toolkit. The trial run reveals the feasibility of implementing ANeP for improving collaborative working in CSC.

Keywords: Agent, Multi-attribute negotiation, Negotiation efficiency, Collaborative decision-making, Construction supply chain, Collaborative working.

6.1 Introduction

The construction industry is characterized by high fragmentation, which results in complexity of the construction supply chain (CSC). For example, the separation of design

L.C. Jain, N.T. Nguyen (Eds.): Knowl. Proc. & Dec. Mak. in Agent-Based Sys., SCI 170, pp. 123–145.
springerlink.com

and construction, the lack of coordination and integration among organizations in the supply chain, poor communication, uncertain production conditions, etc., are the important impact factors causing performance-related problems, such as low productivity, cost and time overruns, conflicts and disputes Love et al. [1] According to Vrijhoef et al. [2], the major problems originate at the interfaces of different participants or stages involved in CSC. These problems are caused by myopic and independent decision-making in CSC operation. With economic globalization and demands for improving construction performance and quickly responding to requirements of owners, collaborative working has become the core strategy in CSC management.

The application of supply chain management (SCM) philosophy to the construction industry has been widely investigated as an effective and efficient management measure and strategy for improving the performance of construction since the middle of the 1990s [3], [4], and to address the adversarial inter-organizational relationships of organizations by increasing the number of construction organizations and researchers [1] [3], [5], [6]. SCM, in a sense, can be considered as the coordination and cooperation of distributed decision-making of organizations or participants on material flow, information flow, human flow and cash flow in a supply chain from systems perspective.

Negotiation is defined as one kind of decision-making process where two or more decision makers jointly search a space for a solution based on a goal-achieving consensus. It is an effective mechanism for supply chain coordination and cooperation.

Multi-agent systems (MAS) technology offers new means and tools for SCM [7]. The main advantage of MAS is that responsibilities for acting as the various components of an engineering process or participants of a business process which is delegated to a number of agents. MAS is suitable for domains that involve interactions between different organizations with different objectives and proprietary information [8]. The SCM system is a typical MAS, where the participants are delegated to different agents. Furthermore, agent-based supply chain cooperation has been proved to be an effective mechanism to improve the performance of SCM [9]. The core principles of SCM and agent technology provide new perspectives for collaborative working in CSC.

Whereas some researchers have addressed a number of key issues and have applied agent technology in construction management [10], [11], little research has been conducted to investigate the application of intelligent agenta to support collaborative working in CSC operation from chain-wide perspective. This chapter introduces an agent-based negotiation platform for collaborative working in CSC (ANeP) by adopting agent technology and multi-attribute negotiation (MAN) theory. In the following sections CSC is regarded as a typical multi-agent system.

6.2 Concepts and Problems of CSC

6.2.1 The Concepts of CSC

CSC consists of all construction processes, from the initial demands by the client/owner, through design and construction, to maintenance, replacement and eventual demolition of the projects. It also consists of organizations involved in the construction process, such as client/owner, designer, GC, subcontractor, and suppliers. CSC is not only a chain of construction businesses with business-to-business relationships but also a network of multiple organizations and relationships, which includes the flow of information, the flow of materials, services or products and the flow of funds between owner, designer, GC, subcontractors and suppliers [12], [13], [2], [14], [15]. According to Muya et al. [16], there are three types of CSC: the primary supply chain, which delivers the materials that are incorporated into the final construction products; the support chain, which provides equipment and materials that facilitate construction, and the human resource supply chain, which involves the supply of labor. In this chapter, CSC is considered from the stage of owner demands to the stages of design, construction, and handover. A typical model of CSC is shown in Fig. 6.1. In the model of CSC, GC is the core of the CSC with the owner and designer as the other two main partners in CSC. Excepting the direct suppliers of GC, subcontractors are also regarded as the suppliers of GC, meanwhile, subcontractors have their own suppliers.

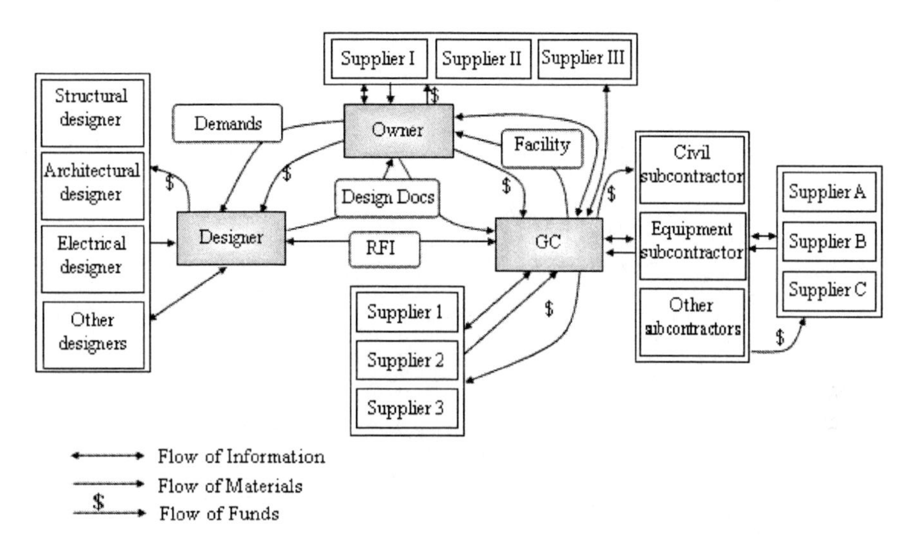

Fig. 6.1. Model of the construction supply chain

Although a number of researchers have provided definitions for CSC management [17], [1], for consistency, this research defines CSC management as follows: CSC management is the coordination of inter-organization decision-making in CSC and the integration of key construction business processes and key members involved in CSC, including client/owner, designer, GC, subcontractors, suppliers, etc. CSC management focuses on how firms utilize their suppliers' processes, technology and capability to enhance their own competitive advantage. It is a management philosophy that extends traditional intra-enterprise activities by bringing partners together with the common goal of optimization and efficiency. CSC management emphasizes long-term, win-win, and cooperative relationships between stakeholders from a systemic perspective. Its ultimate goal is to improve construction performance and add client value at less cost.

6.2.2 Problems in CSC

Although there have been many changes in the construction industry as a result of the development of technology and culture over the last decades, CSCs do not seem to have changed much. Many problems still exist in CSC. According to Vrijhoef et al. [13], the major problems originate at the interfaces of different participants or stages involved in the CSC, as shown in Fig. 6.2. The problems are caused by myopic and independent control of the CSC.

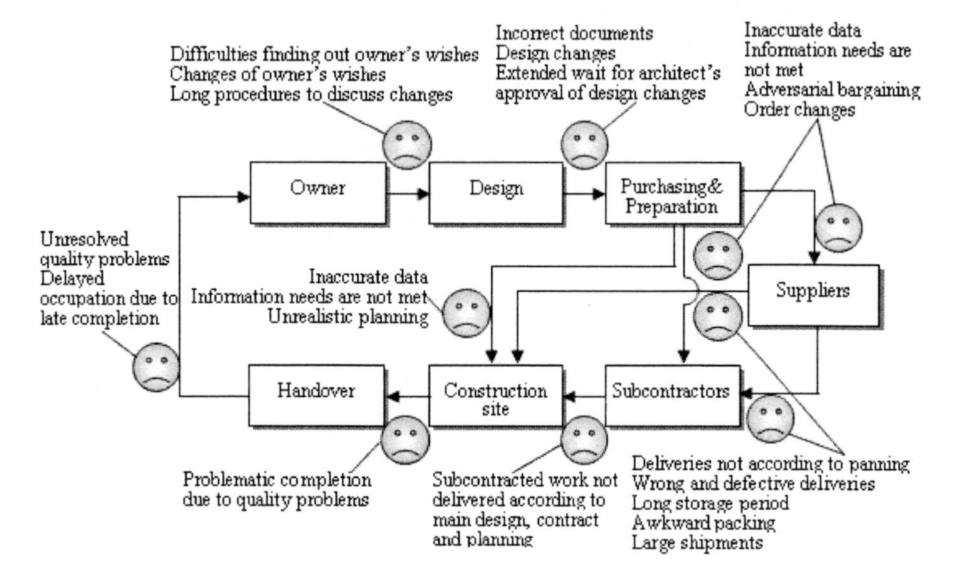

Fig. 6.2. Problems in CSC (adapted from Vrijhoef et al., [13])

Love et al. [1] and Mohamed [18] noted the highly fragmented characteristics of the construction industry. For example, the separation of design and construction, lack of coordination and integration between various functional disciplines, poor communication, etc., are the important impact factors causing performance-related problems, such as low productivity, cost and time overruns, conflicts and disputes. Palaneeswaran et al. [19] revealed the weak links in CSC as follows:

- Adversarial relationships between clients and contractors;
- Inadequate recognition of the sharing of risks and benefits;
- Fragmented approaches;
- Narrow-minded "win-lose" attitudes and short-term focus;
- Power domination and frequent contractual non-commitments resulting in adverse performance track records with poor quality, conflicts, disputes and claims;
- Prime focus on bid prices (with inadequate focus on life-cycle costs and ultimate value);
- Low transparency coupled with inadequate information exchanges and limited communications;
- Minimal or no direct interactions that foster sustainable long-term relationships.

In order to overcome these shortcomings (weak links) and resolve the problems in CSC, and to further improve the performance of the whole CSC, this research presents a solution that integrates agent technology and multi-attribute negotiation technology. This solution will be explained in detail in the following sections.

6.3 General Structure Design of ANeP

CSC is a typical MAS, which involves multiple agents that delegate the organizations to autonomously perform tasks through exchanging information. CSC involves increasing participants along with the magnifying scale of more and more construction projects. So CSC also is a complex system, and collaborative working becomes a challenging and significant feature that is vital to improve the performance of CSC and add value for the client. This requires an agent-based negotiation platform that has the ability to efficiently communicate, to perfect structure and to an effective

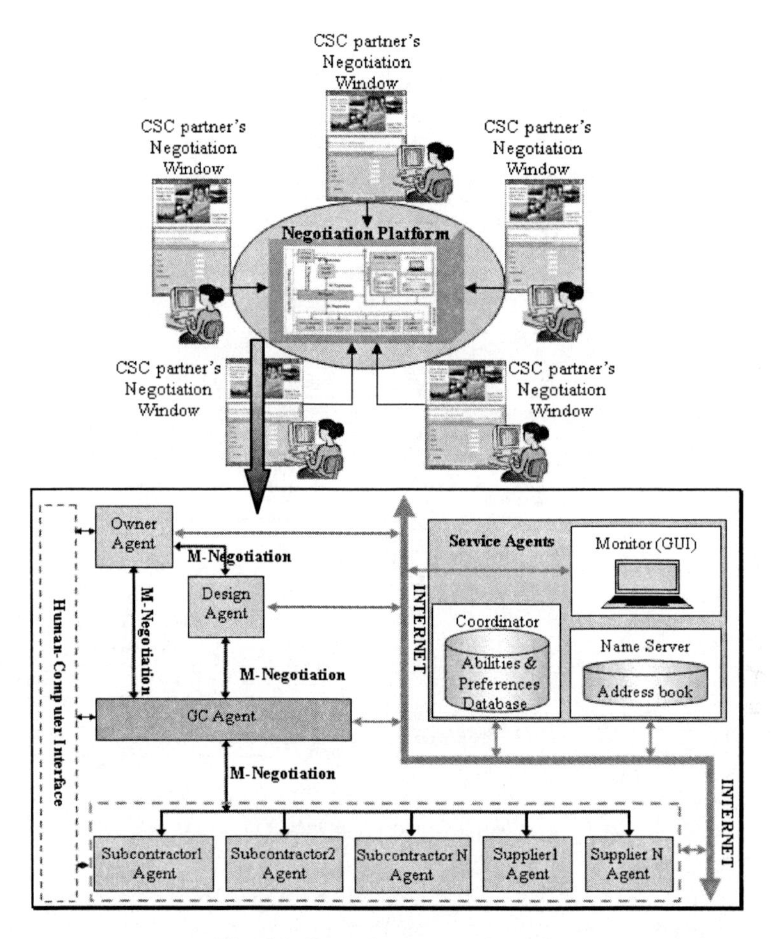

Fig. 6.3. General structure of ANeP

coordination mechanism, stability, and flexibility. Considering the above factors, we designed the general structure of ANeP as shown in Fig. 6.3.

In ANeP, the domain agents include both 'service' agents: coordinator agent, monitor agent, and name server agent, and 'specialty' agents: owner agent, design agent, general contractor (GC) agent, subcontractor agents, and supplier agents. The collaborative working process in CSC is supported through MAN between specialty agents. We hypothesize that all materials and human resources are organized by GC or subcontractors and do not consider the owners' suppliers in this structure. All agents communicate and cooperate through the Internet. Decision makers can control the negotiation process through a Human-Computer Interface (Negotiation Window). Fig. 6.4 provides a detailed insight into how these agents function. The interactions shown are time-ordered, with those at the top occurring before those further down.

6.4 Multi-attribute Negotiation Model in ANeP

This research adopts the MAN technology to coordinate the collaborative working in CSC. Since a number of factors such as cost, time, quality, safety, environment, must be considered in the decision-making process of CSC management. The above factors are seen as the attributes involved in CSC decision-making. MAN technology is developed based on the multi-attribute utility theory (MAUT), which is an analytical tool for making decisions involving multiple interdependent objectives based on uncertainty and utility analyses.

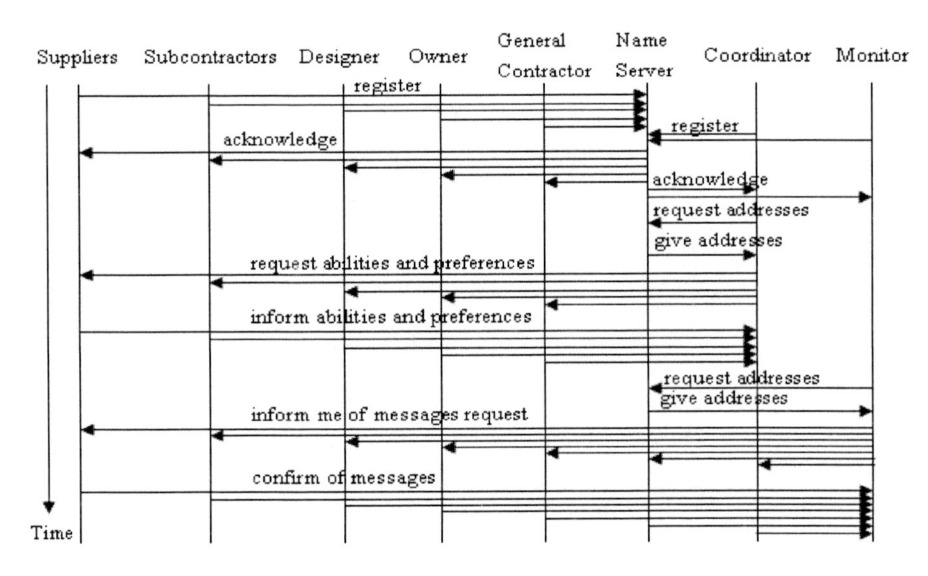

Fig. 6.4. Interactions of agents in ANeP

This research presents an agent-based MAN model for collaborative working in CSC, which creatively extends the general negotiation model for A/E/C [10] from SCM and utility theory perspectives and integrates the compositional MAN model

[20], as shown in Fig. 6.5. The model consists of four elements: the negotiation protocol, CSC participants, the MAN process, and the outcome. In the following we will focus on the negotiation protocol and negotiation process.

6.4.1 Multi-attribute Negotiation Protocol in ANeP

Negotiation protocol (NP) controls the interactions between agents by constraining the way the agents interact [11]. It also specifies the kinds of deals that the agents can make, as well as the sequence of offers and counter-offers that are allowed. In ANeP, multiple attributes are involved in the process of negotiation between agents. For example, the GC agent needs to negotiate with subcontractor agents, supplier agents, designer agent, and owner agent with considering the different attributes of decision-making, such as time, cost, safety, and quality based on the overall utility. So, in ANeP, NP is called 'multi-attribute NP'. The multi-attribute NP is shown in Fig. 6.6 (adapted from Barbuceanu and Lo [21]).

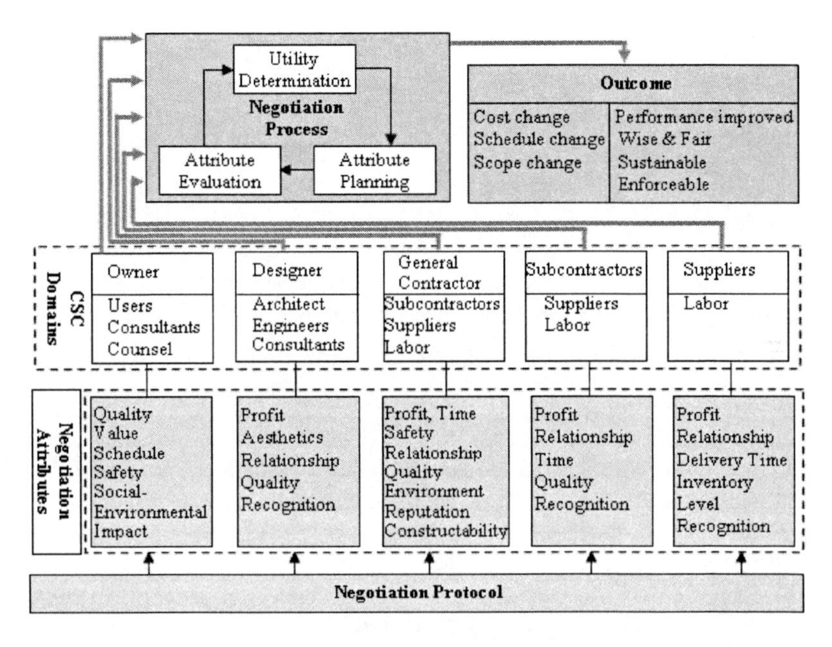

Fig. 6.5. Multi-attribute negotiation model in ANeP

Each agent offers its current best solution that is saved locally. Then the solution is sent to the other agents. The process waits for a message from other agents. If the message is an acceptance, it indicates that the sent solution is consistent with the other agents' solutions, and the negotiation is successful. If the message is 'NoMoreSolution', the other agent has run out of solutions to generate. If the same is true for this agent as well, then the negotiation ends unsuccessfully. Otherwise the agent will continue to generate a new solution. If the message includes a changed solution from the other agent, this solution is checked for compatibility with any of the past solutions

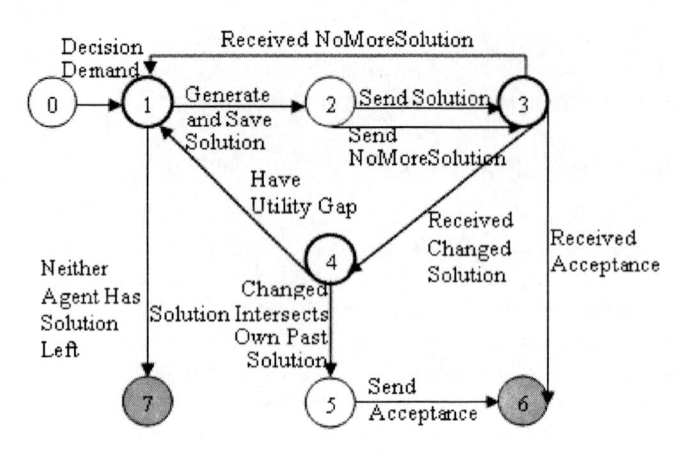

Fig. 6.6. Multi-attribute negotiation protocol in ANeP

generated by this agent. If an intersection is found, it presents a mutually acceptable solution, which determines the negotiation successfully. Otherwise this indicates that utility gap remains between the own agent and the other agent. The top loop is repeated. The most advantage of the protocol is that it guarantees the discovery of the Pareto optimum [21].

6.4.2 Multi-attribute Negotiation Process in ANeP

In this chapter, a five-step MAN process, as described by Jonker and Treur [20] is predigested to three processes: attributes evaluation, utility determination and attribute planning.

6.4.2.1 Attributes Evaluation

Many attributes are involved in the process of SCC decision-making in construction, as shown in part of the 'Representative Negotiation Attributes', see Fig. 6.5. Attribute evaluation evaluates the value of the attributes based on the preferences of the participants in LCPSC. All attributes in general are classified into two categories: quantitative attributes and qualitative attributes. For the quantitative attributes, such as cost and time, their value can be directly calculated according to relative principles. For the qualitative attributes, such as quality, safety and environment, it is necessary to construct a scale representing the levels of these attributes. In this paper, a scale, from 0 (worst) to 10 (best), serves as the measure of evaluation.

6.4.2.2 Utility Determination

In this process, target utility (TU) is determined. TU is given by

$$TU = U_{BOW} + CS \tag{6.1}$$

where U_{BOW} is the utility of the own decision-making, and the concession step (CS) is determined by

$$CS = \beta(1 - \mu/U_{BOW})(U_{BOT} - U_{BOW}) \tag{6.2}$$

where U_{BOT} is the utility of the other participant's decision-making. The factor $(1\text{-}\mu/U_{BOW})$ expresses that CS will decrease to 0 if the U_{BOW} approximates the minimal utility μ and $(U_{BOT}\text{-}U_{BOW})$ expresses the current utility gap. β stands for the negotiation speed.

The utility of ith participant's decision-making (U_i) is given by

$$U_i = \sum_{j=1}^{n} w_j y_{ij} \tag{6.3}$$

where the w_j is the weight of the jth attribute. y_{ij} is give by

$$y_{ij} = x_{ij} \Big/ \sqrt{\sum_{i=1}^{m} x_{ij}^2} \tag{6.4}$$

where x_{ij} is the value of jth attribute evaluated by the ith participant in a decision-making process.

6.4.2.3 Attribute Planning

The attribute planning process refers to target evaluation and configuration determination. Target evaluation of the jth attribute, TE_j, is given by

$$TE_j = (1-\tau)BTE_j + \tau E_{BOT,j} \tag{6.5}$$

where BTE_j is the basic target evaluation of jth attribute, which is determined in such a away that $\sum w_j BTE_j = TU$. $E_{BOT,j}$ is the jth attribute evaluation value of other agents. τ stands for the configuration tolerance.

The configuration determination for the next decision-making includes three steps. Firstly, attribute values are determined with an evaluation that is as close as possible to the target evaluation value. Then a partial configuration (excepting the quantitative attributes, such as cost and time) is selected from the closest value of the attribute. The final step is to reevaluate the quantitative attributes [20].

6.5 Method for Improving Negotiation Efficiency in ANeP

Considering the negotiation failure or man-made termination of negotiation for increasing efficiency of negotiation, this research presents a relative entropy method for improving agent-based multi-attribute negotiation efficiency (REAMNE) in ANeP. This method aggregates preference information of negotiators in CSC twice. Firstly, compromise group preference order is ascertained by using preference information, which is newly provided by negotiators, and compromise preference information (CPI), which is formed previously in the agent-based auto-negotiation process and calculated by using a compromise preference model. Secondly, group preferences are aggregated by using a relative entropy model, which is established based on entropy theory whilst, meanwhile, considering the multiple attributes in CSC negotiation.

6.5.1 Compromise Group Preference Order

6.5.1.1 Definitions

Suppose that there are m substitutable negotiation solutions S_1, S_2, ..., S_m with attributes (criteria) P_1, P_2, ..., P_n, which are given and shared by negotiators when negotiation fails or is terminated. The assessment matrix is given by

	P_1	P_2	...	P_n
S_1	a_{11}	a_{12}	...	a_{1n}
S_2	a_{21}	a_{22}		a_{2n}
$\vdots$	$\vdots$	$\vdots$		$\vdots$
S_m	a_{m1}	a_{m2}	...	a_{mn}

where $0 \le a_{ij} \le 1$, is the consequence with an assessment numerical value of solution i in respect of the criterion j, for i=1, ..., m; j=1, ..., n. Note that the requirement of $a_{ij} \le 1$ is only for notation convenience, and it can always be obtained through a normalization.

Definition 6.1. [22] If there is a $\beta \in E^1$ and weight

$\omega = (\omega_1, \omega_2, ..., \omega_n)^T \in W \bigcap \{\omega \mid e^T \omega = 1, \omega \ge 0\}$

satisfying

$$\sum_{k=1}^{n} \omega_k a_{ik} \ge \sum_{k=1}^{n} \omega_k a_{jk} + \beta \tag{6.6}$$

we say that there is a preference order between S_i and S_j,

$$S_i \succ S_j \tag{6.7}$$

where $1 \le i,\ j \le m,\ i \ne j,\ e = (1,1,...,1)^T \in E^n$, β represents the amount of value that makes solution *i* preferable to solution *j*, and *W* represents the relative importance of attributes, i.e. a kind of preference of an attribute's weight which is accepted by all negotiators.

Definition 6.2. [23] Let $\overline{\beta} \ge 0, \overline{\omega} = (\overline{\omega}_1, \overline{\omega}_2, ..., \overline{\omega}_n)^T \in W$, and $e^T \overline{\omega} = 1,\ \overline{\omega} \ge 0$, such that

$$a_s^r \overline{\omega} \ge \overline{\beta}, (r,s) \in \bigcup_{k=1}^{l} I_k \tag{6.8}$$

Then $\overline{\omega}$ is called a set of 'compromise weights' for the group decision-making, and $\overline{\beta}$ is called the 'compromise index'. The meanings of symbols in the equation will be described in the following section.

6.5.1.2 Method to Calculate Compromise Weight in CSC Negotiation

Referring to the method to calculate compromise weight described by Wei and Yan [24], the method to calculate compromise the weight in CSC multi-attribute negotiation decision is given as follows:

Step 1: According to the negotiation process, the preference orders of l negotiators are listed by

$$S_{k(1)} \succ S_{k'(1)},\ S_{k(2)} \succ S_{k'(2)},\dots,\ S_{k(t_k)} \succ S_{k'(t_k)} \tag{6.9}$$

where t_k represents the number of the preference order of the kth negotiator on negotiation solutions, and $k(t), k'(t) \in \{1,2,\dots,m\}, k(t) \neq k'(t), t = 1,2,\dots,t_k, t_k \geq 1$.

Denote:

$$I_k = \{(k(t),k'(t)) \mid S_{k(t)} \succ S_{k'(t)}, t = 1,2,\dots,t_k\};$$

$$a_s^r = (a_{r1} - a_{s1}, a_{r2} - a_{s2}, \dots, a_{rn} - a_{sn}),\ \text{where}\ (r,s) \in \bigcup_{k=1}^{l} I_k .$$

Step 2: Set up a maximization objective function for LP (Linear Programming) problem (LP1):

$$\begin{aligned}
&\max \beta \\
&s.t.\quad a_r^s \omega \geq \beta \qquad (r,s) \in \bigcup_{k=1}^{l} I_k \\
\text{(LP1)}\quad &e^T \omega = 1 \qquad \omega \geq 0, \omega \in W
\end{aligned}$$

Let $\hat{\beta}$ and $\hat{\omega}$ be the optimal solution of above LP problem. If $\hat{\beta} \geq 0$, then the all-preference order of l negotiators $S_r \succ S_s, (r,s) \in \bigcup_{k=1}^{l} I_k$ is coordinated, and $\hat{\omega}$ is the compromise weight of all negotiators. If $\hat{\beta} < 0$, then turn to the following, step 3.

Step 3: Denote:

$$\hat{I}_k = \{(r,s) \mid a_s^r \hat{\omega} = \hat{\beta}, (r,s) \in I_k\} \mid k = 1,2,\dots,l .$$

For $\forall (r',s') \in \bigcup_{k=1}^{l} \hat{I}_k$, we set up the following LP problem (LP2):

$$\begin{aligned}
&\max a_{s'}^{r'} \omega \\
\text{(LP2)}\quad &s.t.\quad a_r^s \omega \geq \hat{\beta} \qquad (r,s) \in \bigcup_{k=1}^{l} I_k \\
&e^T \omega = 1 \qquad \omega \geq 0, \omega \in W
\end{aligned}$$

Suppose that $\omega_{s'}^{r'}, (r',s') \in \bigcup_{k=1}^{l} \hat{I}_k$ is the optimal solution.

Denote:

$$\hat{\hat{I}}_k = \{(r',s') \mid a_{s'}^{r'} \omega_{s'}^{r'} = \hat{\beta}, (r',s') \in \hat{I}_k\}, k = 1,2,...,l.$$

Step 4: Set up the following LP problem (LP3):

$$\begin{aligned} &\max \beta \\ \text{(LP3)} \quad & s.t. \quad a_r^s \omega \geq \beta \qquad (r,s) \in \bigcup_{k=1}^{l} I_k \setminus \bigcup_{k=1}^{l} \hat{\hat{I}}_k \\ & e^T \omega = 1 \qquad \omega \geq 0, \omega \in W \end{aligned}$$

Let $\overline{\beta}$ and $\overline{\omega}$ be the optimal solution of the above LP problem. If $\overline{\beta} \geq 0$, then $\overline{\omega}$ is the compromise weight of all negotiators. If $\overline{\beta} < 0$, then turn to step 3 and re-coordinate by replacing $\hat{\beta}$ with $\overline{\beta}$.

After confirming the compromise weight following the above steps, the priority of negotiation solutions can be judged by calculating utility value using the model in section 6.4. Then, we can obtain a compromise preference order of all negotiators on former negotiation solutions. This compromise preference order is seen as the preference of the virtual third-party negotiation coordinator. Note that compromise preference order can be only used as a preference relationship and cannot be seen as the real utility value of the negotiation solution since compromise weight does not stand for the real evaluation weight of a negotiator.

6.5.2 Transformation of Compromise Group Preference Order to Utility Value

Suppose that $O^k = \{o^k(1),...,o^k(n)\}$ is the compromise preference ordering of a set of negotiation solutions *S*, where $o^k(j) \in N = \{1,2,...,n\}$ is a permutation function over the index set $\{1,2,...,n\}$ for the negotiator *k* [25], [26]. Generally, $o^k(j)$ more to less, corresponding solution more tp better, and vice versa [27]. For example, there are three solutions in a set of solution $S = \{S_1, S_2, S_3\}$, where the compromise preference order is $O^k = \{2,1,3\}$, then the solution S_2 is the best and S_3 is the worst.

Suppose that $u^k(j)$ is the utility value of a solution S_j after transformation. Since the less the $o^k(j)$, the bigger the corresponding utility value, so we can establish the following equation to transform preference ordering into utility value [28]:

$$u^k(j) = \frac{n - o^k(j)}{n-1} \tag{6.10}$$

6.5.3 Aggregating Group Preference of CSC Negotiators Using Relative Entropy Method

The choice of solutions after negotiation failure or man-made termination is a typical group decision-making. The nature of group decision-making is a process of preference aggregation, whose goal is to maximize consensus of group preference, i.e. to find a solution with which the gap between the preference utility value of the group and an individual is minimum. So the group decision-making problem is one kind of optimization problem as well. Entropy-based optimization theory as an effective tool has great success in the decision analysis area [29]. The definition of relative entropy is given as follows:

Definition 6.3. Suppose that $x_i, y_i \geq 0, i = 1,2,...,n$ and $1 = \sum_{i=1}^{n} x_i \geq \sum_{i=1}^{n} y_i$, we say that $h(X,Y) = \sum x_i \log \frac{x_i}{y_i} \geq 0$ is the relative entropy of X relative to Y, where $X = (x_1, x_2,..., x_n)^T$, $Y = (y_1, y_2,..., y_n)^T$. The main properties of relative entropy are given by [30], [31], [32]

(1) $\sum_i x_i \log \frac{x_i}{y_i} \geq 0$;

(2) $\sum_i x_i \log \frac{x_i}{y_i} = 0$, only when $x_i = y_i$, $\forall i$.

According to the above properties, minimum relative entropy is obtained when two discrete variables X and Y have the same distribution. So we can use relative entropy to measure the degree of consensus.

The value of the normalization of utility values, including utility value transformed from compromise preference order relations, can be seen as a discrete probability measurement of the negotiators' preference. The probability distribution of the set of solutions $S = \{S_1, S_2,..., S_n\}$ is formed by the discrete probability measurement of each negotiator on *S*. Here, suppose that the probability distribution of each negotiation is independent.

Let $U_g = \{u_{gj}\}, j = 1,2,...,n$, be the group preference vector of utility value given by every negotiator, where u_{gj} is the group utility value of the jth negotiation solution. Negotiators can use this vector to choose the best solution according to the value of u_{gj}.

Let $W = \{\omega_1, \omega_2, \omega_3\}$ is the weight vector of the negotiators, where $\omega_1 + \omega_2 + \omega_3 = 1$ and ω_i is the weight of negotiator *i*. Note that *i*=3 since the virtual negotiation environment includes three parties in this paper. According to the properties of relative entropy, minimizing relative entropy of the group preference vector of the utility value contrasts with the vector of utility value of negotiation solution of each negotiator and group consensus preference vector, U_g, can be obtained. This is a process to find the optimal solution of the following optimization problem:

$$\min Q(U_g) = \sum_{i=1}^{3} \omega_i \sum_{j=1}^{n} \left(\log \frac{u_{gj}}{u_{ij}} \right) u_{gj}$$
$$\text{s.t.} \sum_{j=1}^{n} u_{gj} = 1, \quad u_{gj} > 0 \tag{6.11}$$

where u_{ij} is the normalized utility value of negotiator i to negotiation solution j. Note that the utility value of the negotiation solution regenerated by the real two-party negotiators is the value in the above model; the utility value of the virtual third-party negotiator is obtained by transforming his preference ordering into a utility value.

Qiu [29] has proved that the optimization problem of (11) has an optimal solution in whole scope when i=1,2,…,m, which is given by $U_g^* = \{u_{gj}^*\}, j = 1,2,...,n$, where

$$u_{gj}^* = \frac{\prod_{i=1}^{3} (u_{ij})^{\omega_i}}{\sum_{j=1}^{n} \prod_{i=1}^{3} (u_{ij})^{\omega_i}} \tag{6.12}$$

Then negotiators can get the satisfied negotiation solution by all negotiators according to the value of u_{gj}^*.

6.6 Prototype Development of ANeP

The prototype of ANeP is developed by using the ZEUS agent building toolkit. ZEUS is an advanced development toolkit for constructing distributed multi-agent applications. ZEUS is the culmination of a careful synthesis of established agent to provide an integrated and visual environment for the rapid software engineering of collaborative agent applications [33]. The developing process consists of two steps: role modeling and application design.

6.6.1 Role Modeling in ANeP

The ZEUS agent building toolkit adopts role modeling to address the specification, analysis, design, implementation and maintenance of agents. Role models formalize the definition of an agent role and provide a readily comprehensible means of analyzing the problem in question. The role models are grouped into domains. The domains provide a context that enables developers to compare their planned system with

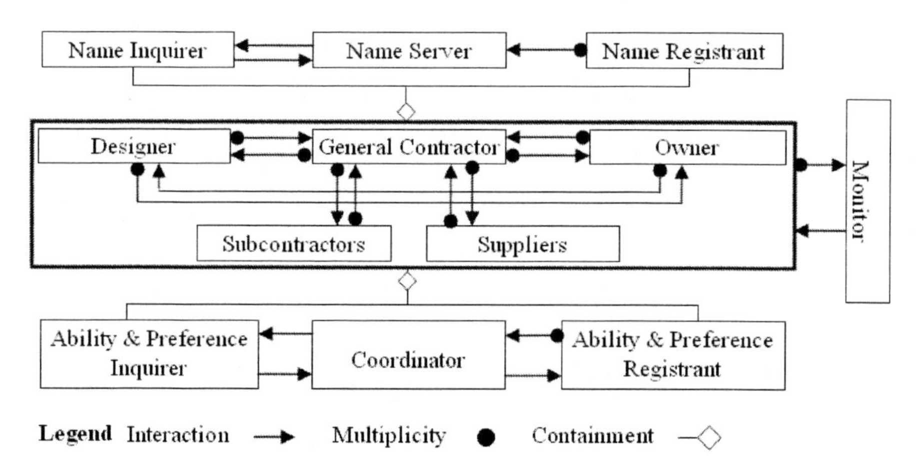

Fig. 6.7. Roles represented in ANeP

existing applications. Role models, which describe the dynamic interaction between roles, are architectural patterns that depict the high-level similarities between related systems, i.e. the problems inherent to each domain, but not how they were solved. The role models of ANeP are illustrated in Fig. 6.7. The interactions between these roles are shown in Fig. 6.8. Table 6.1 summarizes the collaboration relationships.

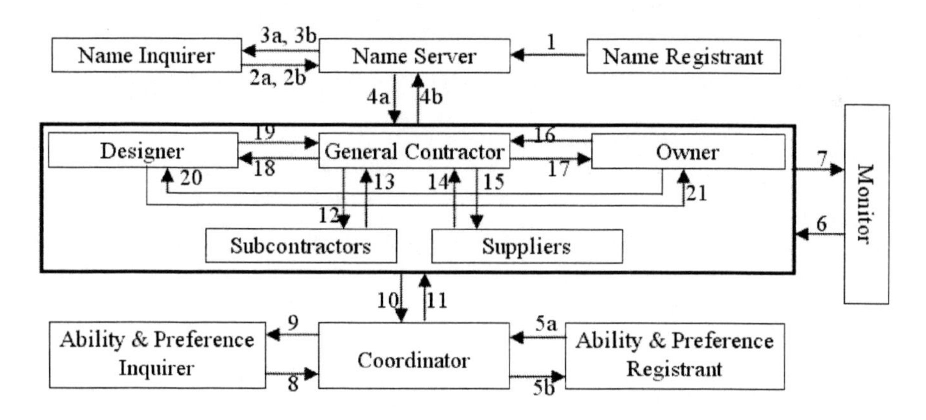

Fig. 6.8. Interactions of agents in ANeP

6.6.2 Application Design

This process is illustrated based on a virtual CSC, which involves the following participants: owner, designer, GC, Groundwork subcontractor, Civil and structure subcontractor, building services subcontractor, finishing works subcontractor, concrete supplier and finishing materials supplier. Each participant is delegated to corresponding agent.

Table 6.1. Summary of Interactions

	Collaboration	Explanation		Collaboration	Explanation
1	Registration	Agents notify the NS of their presence	10	Inquire	A agent inquire from facilitator about the other agent's abilities and preference
2a	Resolve query	A request for the network location of a named agent	11	Answer	The facilitator answers the inquires from agents
2b	List query	A request for all agents of a particular type	12	Negotiate with Subcontractors	GC negotiates the relative construction solutions with Subcontractors
3a	Location response	The location of the agent previously in question	13	Sent solution (decision-making)	Subcontractors send their decision-making to GC
3b	List response	The list of agents previously in question	14	Send demands information	GC sends its demands plan to suppliers
4a	Answer agent name	NS answers the inquiries about the agent name	15	Response GC demands	Suppliers sends their supply plan to GC
4b	Inquire agent name	Agent inquire from NS about the other agent name it needs to communicate with them	16	Inquire project information and send change	The owner inquires from GC about the project information and sends demands change
5a	Ability & preference response	Information about an agent's current abilities & preference	17	Report and response change	GC reports the project information to Owner and response the demands change
5b	Ability & preference request	Ask for information about recipient's abilities & preference	18	Require	GC requires for relative problems of the design drawings
6	Information request	Asks all activity be forwarded	16	Explain	Designer explains the design drawings
7	Activity notification	A copy of ant message sent	20	Put forward new demands	The owner puts forward his new demands, i.e. design change
8	Find request	Asks for agents with particular abilities and preference	21	Offer design drawings	The designer offers drawings to meet the owner's demands
6	Find response	A list of agents matching the desired criteria			

6.6.2.1 Ontology Creation

Ontology is a set of declarative knowledge representing every significant concept within a particular application domain. The significance of a concept is easily assessed. If meaningful interaction cannot occur between agents without both parties being aware of it, then the concept is significant and must be modeled. Ontology contains the key concepts within the specific application domain, the attributes of each concept, the types of each attribute and any restrictions on the attributes. In ZEUS an individual domain concept is described by using the term 'fact'. ZEUS provides two kinds of fact: abstract and entity. In ANeP, all the concepts refer to the entity, such as drawing, concrete, rebar, finishing materials, designer, various workers, supplier, subcontractors, etc. The ontology created in ANeP is shown in Fig. 6.9.

6.6.2.2 Agent Creation

In ZEUS, agent creation includes three steps: agent definition, agent organization and agent coordination. Agent definition determines the planning parameters, the task and the initial resources allocation. Agent organization illustrates the relationship between

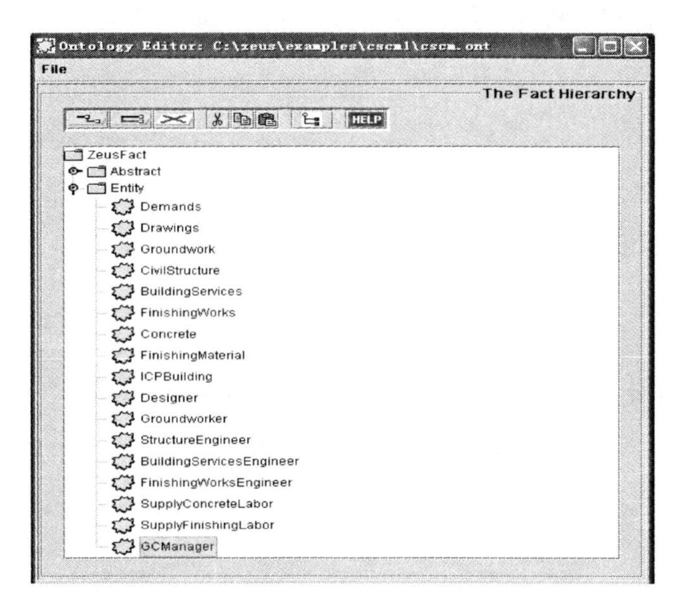

Fig. 6.9. Ontology in ANeP

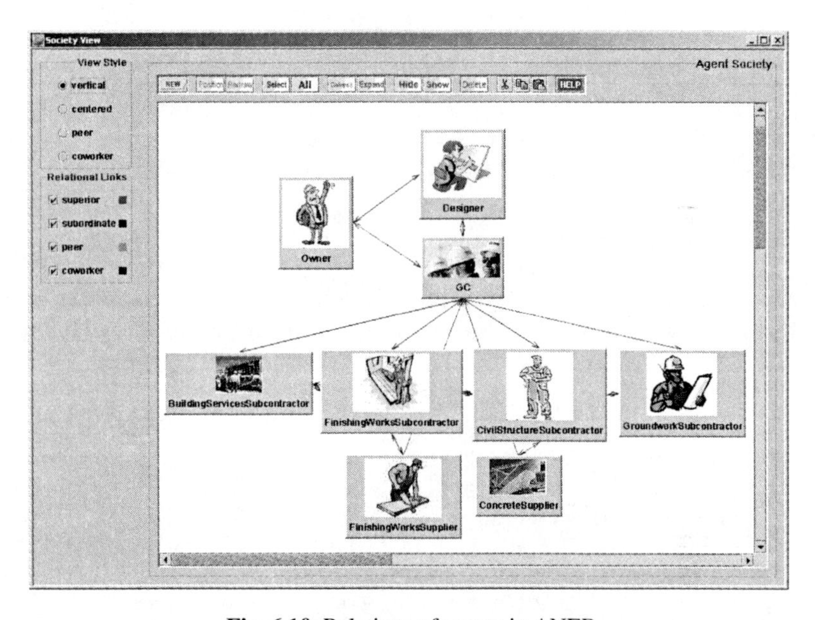

Fig. 6.10. Relations of agents in ANEP

the own agent and other agents and acquaintance abilities from other agents. Agent coordination defines the coordination protocols and strategies between the own agent and other agents. As previously mentioned, multi-attribute NP is integrated into the prototype of ANeP as the coordination protocol in ZEUS. Table 6.2 lists the agents in the prototype of ANeP.

The GC agent, as the CSC head, possesses the core position compared with other agents. GC is the superior of subcontractors and suppliers. The owner supervises the activities of GC and designer. The relationships of GC and owner, subcontractor and subcontractor are cooperative partnering. The relations between agents in ANeP are shown in Fig. 6.10.

Table 6.2. Agent created in the ANeP

Agent Name		Roles Played
Specialty Agents	GC agent	CSC head (Negotiation initiator, Manager and constructor)
	Owner agent	Client
	Designer agent	CSC participant (Negotiation partner, supplier, designer)
	Groundwork subcontractor agent	CSC participant (Negotiation partner, supplier, constructor)
	Civil and structure subcontractor agent	CSC participant (Negotiation partner, supplier, constructor, consumer)
	Building services subcontractor agent	CSC participant (Negotiation partner, supplier, constructor)
	Finishing works subcontractor agent	CSC participant (Negotiation partner, supplier, constructor, consumer)
	Concrete supplier agent	CSC participant (Negotiation partner, supplier)
	Finishing material supplier agent	CSC participant (Negotiation partner, supplier)
Service Agents	ANS agent	Agent name server
	Monitor agent	Monitor (to view, analyse or debug societies of agents)
	Construction coordinator agent	Coordinator (receive and respond to queries from agents about the abilities and preferences of other agents)

6.6.2.3 External Program

ZEUS allows users to link an external java class (program) to execute a ZEUS agent program. Once linked to the agent program, the external program can utilize the agent's public methods to query or modify the agent's internal state. In the prototype of ANeP, each specialty agent is linked to an external program. Each external program provides a user interface (negotiation window) through which the users, for example GC, subcontractor, supplier, etc., input their preferences in collaborative working of CSC, as shown in Fig. 6.11.

6.6.3 Trial Run and Discussions

The commands for running the prototype of ANeP can be generated through the 'Code Generator' panel in ZEUS. Fig. 6.12 gives a screen shot of the prototype system. When running the prototype, the negotiation windows will display on the screen waiting for the decision makers input the preferences and relative weightings of attributes. In this prototype system, we only consider five negotiation attributes: cost, time, quality, safety and environment, which present the decision maker's different preferences and utility, as shown in Fig. 6.11. The above information will be sent to

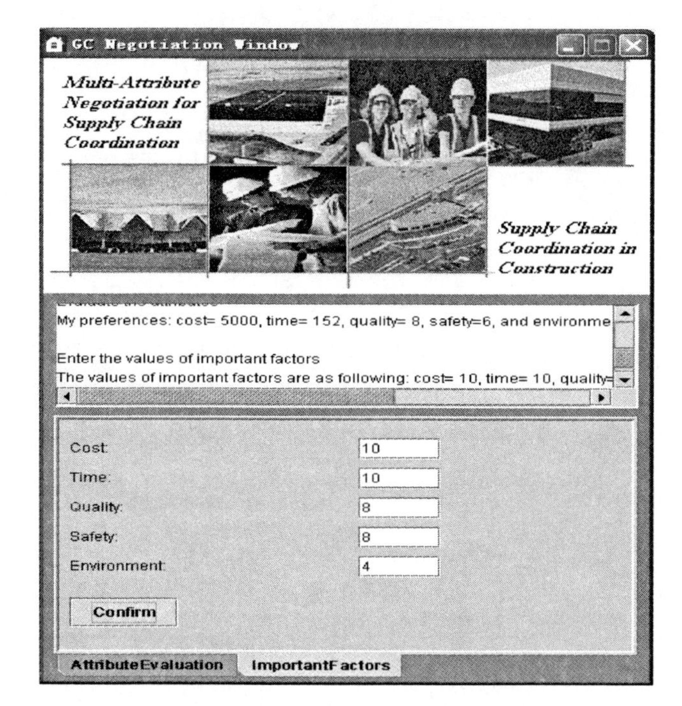

Fig. 6.11. Negotiation window of ANeP

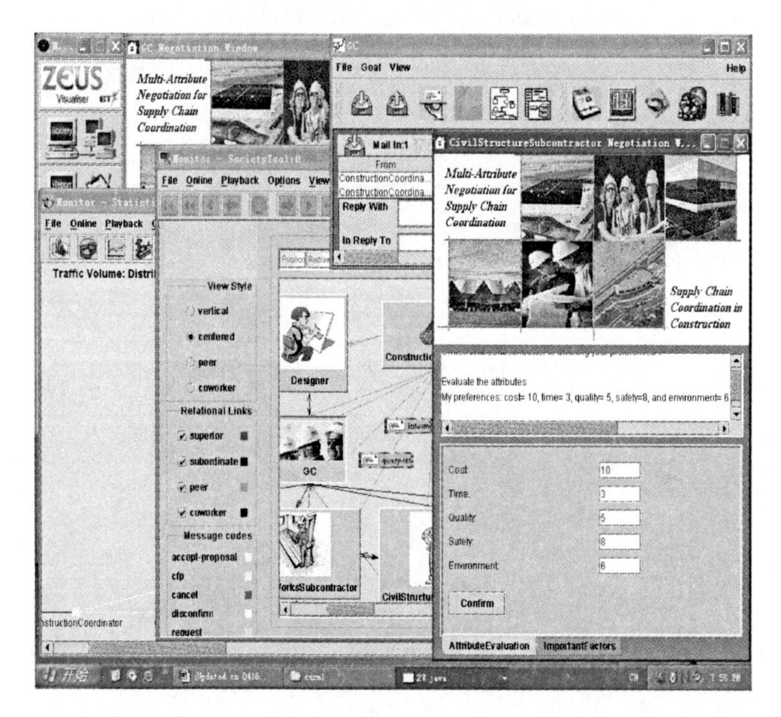

Fig. 6.12. A screen shot when the prototype system is running

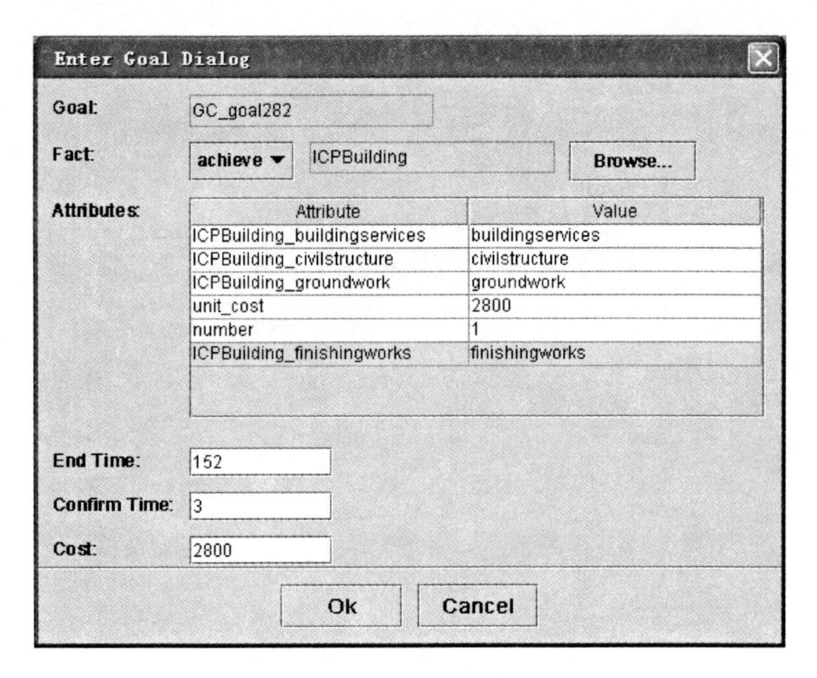

Fig. 6.13. Dialogue for new goal of agents

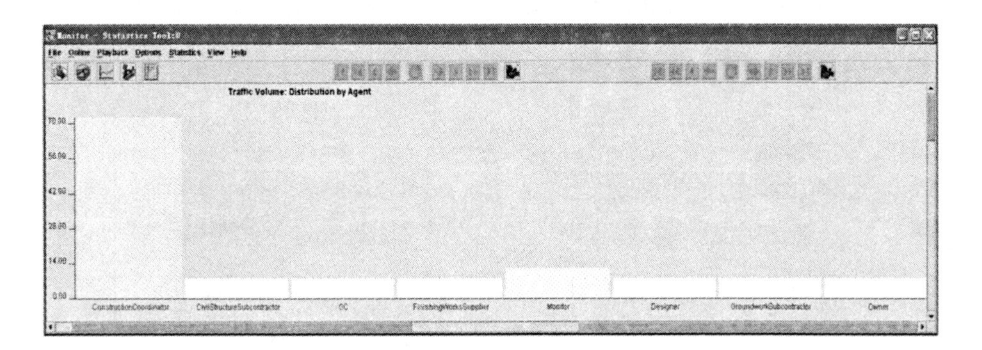

Fig. 6.14. Traffic volume of each agent

the coordinator agent and corresponding negotiation agent. When a new round of negotiation starts, the new goal of the agent can be created through the 'Enter Goal Dialog' (Fig. 6.13).

The prototype of ANeP provides a tool (monitor agent) to dynamically monitor the traffic volume of each agent and inter-agent traffic volume, as shown in Fig. 6.14. The user can analyze and evaluate the performance and activities of agents. From Fig.6.14, it can be concluded that the running process is stable. Although the frequency of communication of the construction coordinator much higher than for other agents, this does not present a problem. The phenomenon rightly reflects the hub role of coordinator agent in ANeP.

6.7 Conclusions

There are increasing demands for collaborative working in the construction industry. This chapter defines the concepts of CSC and especially CSC management as the coordination of inter-organization decision-making in CSC and the integration of key construction business processes and key members involved in CSC. The research finds that there still are many problems in CSC.

Negotiation is an effective and popular coordination behavior in a collaborative working process. This research designs an agent-based negotiation platform for improving the efficiency of collaborative working in CSC by regarding CSC as a typical multi-agent system. Since different members in CSC have different preferences on the decision attributes, a multi-attribute negotiation model is established by designing a negotiation protocol and describing the negotiation process.

The relative entropy method REAMNE presented in this chapter provides an effective approach to solve the group decision-making problem when agent-based automated negotiation fails or is terminated by a decision-maker in real world when they cannot efficiently reach an optimal solution for a negotiation issue. REAMNE mainly includes three steps: calculating a compromise weight, transforming compromise preference orderings into a utility value, and aggregating group preference relations.

The development of a prototype system provides an effective platform to simulate the collaborative working in CSC. It is helpful to reduce the duration of inter-organization decision-making and to enhance the quick response ability to meet the change of demands in competitive marketing environment.

Since CSC, like other economic sectors, involves various participants and activities, so how to use agents to efficiently support collaborative working presents a real challenge. Furthermore, how agents can efficiently elicit participant's preferences and the relevant utility functions will be the significant research direction in the future.

Acknowledgement. The research was supported by the National Natural Science Foundation of China (NSFC) (Grant No. 70801023). The work described in this chapter was also funded by the Development Program for Outstanding Young Teachers in Harbin Institute of Technology Grant No. HITQNJS.2007.027, and the Research Grants Council of the Hong Kong Special Administrative Region, China (PolyU 5114/03E and 5264/06E).

References

1. Love, P.E.D., Irani, Z., Edwards, D.F.: A Seamless Supply Chain Model for Construction. Supply Chain Management: An International Journal 9(1), 43–56 (2004)
2. Vrijhoef, R., Koskela, L., Voordijk, H.: Understanding Construction Supply Chains: A Multiple Theoretical Approach to Inter-organizational Relationships in Construction. In: Proceedings of International Group of Lean Construction 11th Annual Conference, Virginia, USA (2003)
3. Briscoe, G.H., Dainty, A.R.J., Millett, S.J., Neale, R.H.: Client-led Strategies for Construction Supply Chain Improvement. Construction Management and Economics 2(2), 193–201 (2004)

4. Xue, X.L., Li, X.D., Shen, Q.P., Wang, Y.W.: An agent-based framework for supply chain coordination in construction. Automation in Construction 14(3), 413–430 (2005)
5. London, K.A., Kenley, R.: An Industrial Organization Economic Supply Chain Approach for the Construction Industry: A Review. Construction Management and Economics 19, 777–788 (2001)
6. Arbulu, R.J., Tommelein, I.D.: Contributors to Lead Time in Construction Supply Chains: Case of Pipe Supports Used in Power Plants. In: Proceedings of Winter Simulation Conference: Exploring New Frontiers, pp. 1745–1751 (2002)
7. Frey, D., Stockheim, T., Woelk, P., Zimmermann, R.: Integrated Multi-agent-based Supply Chain Management. In: Proceedings of the Twelfth IEEE International Workshops on Enabling Technologies: Infrastructure for Collaborative Enterprises (2003)
8. Ren, Z., Anumba, C.J.: Multi-agent Systems in Construction-State of the Art and Prospects. Automation in Construction 13, 421–434 (2004)
9. Lou, P., Zhou, Z.D., Chen, Y.P., Wu, A.: Study on the Multi-agent-based Agile Supply Chain Management. International Journal of Advanced Manufacturing Technology 23, 197–203 (2004)
10. Pena-Mora, F., Wang, C.Y.: Computer-supported Collaborative Negotiation Methodology. Journal of Computing in Civil Engineering 12(2), 64–81 (1998)
11. Kim, K., Paulson, B.C.: An Agent-based Compensatory Negotiation Methodology to Facilitate Distributed Coordination of Project Schedule Changes. Journal of Computing in Civil Engineering 17(1), 10–18 (2003)
12. Edum-Fotwe, F.T., Thorpe, A., McCaffer, R.: Organizational relationships within the construction supply-chain. Proceedings of a Joint CIB Triennial Symposium, Cape Town 1, 186–194 (1999)
13. Vrijhoef, R., Koskela, L., Howell, G.: Understanding construction chains: an alternative interpretation. In: Proceedings of 9th Annual Conference International Group for Lean Construction, Singapore (2001)
14. O'Brien, W.J., London, K., Vrijhoef, R.: Construction supply chain modeling: a research review and interdisciplinary research agenda. In: Proceedings of 10th Annual Conference International Group for Lean Construction, Brazil (2002)
15. Saad, M., Jones, M., James, P.: A review of the progess towards the adoption of supply chain management (SCM) relationships in construction. European Journal of Purchasing & Supply Management 8, 173–183 (2002)
16. Muya, M., Price, A.D.F., Thrope, A.: Contractors' supplier management. In: Proceedings of a Joint CIB Triennial Symposium, Cape Town, vol. 2, pp. 632–640 (1999)
17. Fisher, N., Morledge, R.: Supply chain management. Best Value in Construction. Blackwell Science Ltd., RICS Foundation (2002)
18. Mohamed, S.: Web-based technology in support of construction supply chain networks. Work Study 52(1), 13–19 (2003)
19. Palaneeswaran, E., Kumaraswamy, M., Ng, S.T.: Formulating a framework for relationally integrated construction supply chains. Journal of Construction Research 4(2), 189–205 (2003)
20. Jonker, C.M., Treur, J.: An Agent Architecture for Multi-attribute Negotiation. In: Proceedings of the International Joint Conferences on Artificial Intelligence, Washington, USA (2001)
21. Barbuceanu, M., Lo, W.K.: A Multi-attribute Utility Theoretic Negotiation Architecture for Electronic Commerce. In: Proceedings of the Fourth International Conference on Autonomous Agents (2000)

22. Wei, Q.L., Yan, H., Ma, J., Fan, Z.: A compromise weight for multi-criteria group making with individual preference. Journal of Operational Research Society 51, 625–634 (2000)
23. Yan, H., Wei, Q.L.: Determining compromise weights for group decision-making. Journal of Operational Research Society 53, 680–687 (2002)
24. Wei, Q.L., Yan, H.: Generalized optimization theory and model, pp. 310–320. Science and Technology Press, Beijing (2003)
25. Chiclana, F., Herrera, F., Herrera, V.E., Potatos, M.C.: A classification method of alternatives for multiple preference ordering criteria based on fuzzy majority. Journal of Fuzzy Mathematics (4), 118–135 (1996)
26. Seo, F., Sakawa, M.: Fuzzy mutiattribute utility analysis for collective choice. IEEE Tansaction on Systems Man and Cybernetics (15), 45–53 (1985)
27. Chiclana, F., Herrera, F., Herrera, V.E.: Integrating three representation models in fuzzy multipurpose decision-making based on fuzzy preference relations. Fuzzy Sets and Systems 97(1), 33–48 (1998)
28. Zhou, H.A., Liu, S.Y.: Method Based on OWGA Operators for Aggregating Preference Information and Its Application in Group Decision-making Problems. Operation Research and Management Science 14(6), 29–32 (2005)
29. Qiu, W.H.: Management Decision and Entropy Theory, pp. 290–293. China Machine Press, Beijing (2002)
30. Berger, J.O.: Statistical Decision Theory, pp. 235–269. Springer, New York (1980)
31. Guiasu, S.: Information Theory with Application, pp. 125–148. McGraw-Hill, New York (1977)
32. Wei, C.P., Qiu, W.H., Yang, J.P.: Minimum Relative Entropy Aggregation Model on Group Decision-making. System Engineering-Theory and Practice 19(8), 38–41 (1999)
33. Nwana, H.S., Ndumu, D.T., Lee, L.C., Collis, J.C.: ZEUS: A Toolkit for Building Distributed Multi-Agent Systems (accessed May 19, 2005), `http://agent.aitia.ai/download`

7

An Event-Driven Algorithm for Agents on the Web

Anne Håkansson

Department of Information Science, Computer Science, Uppsala University,
P.O. Box 513,
S-751 20 Uppsala, Sweden
Anne.Hakansson@dis.uu.se

Abstract. This chapter describes how meta-agents in a multi-agent system can be used to effectively search for services in networks with an event-driven algorithm. These services can be attained in a range of different ways, including a simultaneous combination of several services in order to optimize costs and time. A challenge with a network is finding and extracting an optimal combination of the different services to implement a complex requested service. To solve this problem, it is possible to develop multi-agent systems in which the task steers the agents. While finding information on the web, the search path becomes an event-driven algorithm. The algorithm acts as a search method to extract information from several different services. Once built up, the algorithm can guide future search and optimize the searching.

7.1 Introduction

In networks or graphs, multi-agent systems are used to search and retrieve information, but also to combine information collected from distributed elements. In these multi-agent systems, finding solutions for tasks implies deciding a searching strategy and applying this with the agents. A challenge with network problems is finding and extracting information in the network within an acceptable time bound. In particular, finding paths between nodes based on properties of the graph elements given at the time. This is especially difficult when information must be extracted and combined from several different services. Reducing time and making the agents work together requires a plan or an effective algorithm.

Several different effective graph search algorithms have been developed and applied, e.g., breadth-first, depth-first search [10], best-first search, heuristics [22; 52], A* [54], Dijkstra's algorithm [13], the nearest neighbour algorithm, Prim's algorithm [48], Kruskal's algorithm [31] and minmax and alpha-beta techniques [54]. Some of these algorithms have been applied in agent systems for searching information. However, searching in enormous networks still remains a challenge, with many of the interesting problems being NP-hard, i.e., problems with finding and verifying a solution which can also solve all NP-problems [50]. NP-hard problems can be search problems, decision problems and optimization problems where examples for graphs include Hamiltonian cycle, travelling salesman (TSP), Clique, Vertex cover, Independent set, Graph partition, Edge cover, and Graph isomorphism [52].

On the web, agents have been searching for information using corresponding key word strategy as text-based search strings. These strategies are basic and only looked

L.C. Jain, N.T. Nguyen (Eds.): Knowl. Proc. & Dec. Mak. in Agent-Based Sys., SCI 170, pp. 147–174.
springerlink.com

for lexical similarities to the query. Other used search strategies for agents, based on hypertext links, have proven to be more useful for semantic search, such as page ranking algorithm [38]. The technique use the page's linkage to other pages to estimates its relevance. However, the ranking algorithm use indexing and only the pages that are indexed can be retrieved [38].

Commonly, the service request is sent to a web server, which responds by serving up the requested web page. A challenge with networks, such as the web, is finding information from several services corresponding to the intension of the users when searching for particular information but also combining these services to constitute the required and valuable information. Moreover, the time aspect becomes more challenging when it requires extracting an optimal combination of different services to implement a complex requested service. Services can be attained in many different ways, especially, when a service is a combination of several services. The combination must be valid and at the same time optimize costs and time. The cost is the message passing through the network and the time is the response time including the combination of several services. A valid combination is, of course, due to the predefined constraints given to the agents.

In distributed settings, the agents cannot be assumed to follow an algorithm [42]. This is especially true for agents that work at the web. A solution used by these agents is that they can manipulate the algorithm themselves out of self-interest [42]. Thus, besides applying the strategies by the agents to solve a problem, it is possible to develop a multi-agent system in which the task steers the agents, i.e., generate a strategy directly from the executing multi-agent system [26]. More precisely, the task of the agents can create the strategy and, by using this strategy, manoeuvre the agents. These agents operate at the code level, called object-level agents, while fetching the relevant information [23; 24]. The object-level agents can also operate to find relevant services for information logistics processes [1]. Note, in our work, the agents are moving between the nodes like the web crawlers, one agent per link, which is different from common use of agents on the web. The web crawlers or spiders look for pages while the agents look for the next server, router or other components that either contain service or hold a connection. While moving between nodes on the web, the agents are gathering information using an event-driven algorithm. The algorithm works as event loop dispatches available events, which in turn, drives the agents' action in the network.

The number of agents dispatched grows with the number of connections from a node, which quickly leads to exponential growth and requires a supervisor agent, a meta-agent. The meta-agent can reflect or introspect the object-level agents [11] and act using information about properties of the object-level [23]. The meta-agents possess the characteristics of meta-reasoning agents, which allow them to perform reasoning, plan actions and model individual agents but also classify conflicts and resolve these [8]. Simultaneously as object-level, ground-level, agents work, the meta-level agents monitor the ground-level agents working with the same task and, thereby, can extract the event-driven algorithm.

The algorithm decides the actions for the agents, which means that these agents do not, necessarily, have to be intelligent. Nonetheless, these agents can become intelligent agents when performing tasks and, thereby, learn the environment while moving in the network and following the connections in the network. At the meta-level, the meta-level agents can also be intelligent and adapt to the changes of the object-level agents, which include adapt to a changed algorithm and execute the algorithm by applying it to the intelligent agents.

Depending on the task assigned to the system, the meta-agents can comprise everything from one agent to a couple of hundreds or more agents, which will be used to create the event-driven algorithm. There will be a lot of different algorithms gained from performing a particular task but these paths will be tuned into an optimal search algorithm, which will be stored by a meta-agent and reused for future service searching. This chapter focuses on the intelligent agents, meta-agents and event-driven algorithm for searching and combining services. Using this algorithm, the meta-agents drive and control the behaviour of the ground-level agent moving between states in a network, as well as, inspect the agents when they reach a result and perceive the reason for that result. The benefit of using meta-agents for a multi-agent system is the ability to provide the optimal path between nodes, under given circumstances, and to handle a vast number of nodes in the graph and the network.

In this chapter, we present an event-driven algorithm that acts as a search method for extracting information from several services on the web, wherefrom the algorithm can guide future search and optimize the searching for services. The chapter is structured as follows: In Section 7.2, a brief description of the most common graph search algorithms used today is provided. This section also includes a couple of algorithms that have been applied on the web. In Section 7.3, the event-driven algorithm is presented by briefly describing event-driven programming, the intelligent ground-level agents and the meta-agents. Also the search technique for the algorithm is presented in the section followed by a formal description of the algorithm. In section 7.4, an explanation of how the algorithm can be applied to services at the web is given and in section 7.5 the conclusions and discussion is presented.

7.2 Graph Search Algorithms

Graph algorithms solve problems related to graph theory. A graph search algorithm is used to systematically go through all the nodes in a graph and find solutions. Due to the graph's formation, these algorithms can traverse the vertices of the graph, learning the structure properties as the algorithms go through the graph [51].

The storage of the graph depends on the data structure and the algorithm used to manipulate the graph. Numerous effective graph search algorithms have been developed, where many of the algorithms are used for sequential implementation [5]. For this work, there are two approaches for consideration: a distributed search and multiple local searches that can recombine results [10]. Distributed search algorithms can allow collectively searching and locating a number of targets. They are designed to support parallel computation and the results from the searching are combined to form the final solution. When using a distributed search algorithm, the reassembly of results can take several forms. For simple searching the reassembly can be a simple concatenation of results with duplicate removal. On the more complex end of the spectrum, the reassembly must also look at combining partial results from different partitions of the search process into final forms [20].

A multiple local search algorithm creates multiple search processes, which are sent off in parallel. This approach requires a reassembly process to combine the results after the searches complete, which can be a complex design problem. One dimension on the design is whether the search space is partitioned or the searches are allowed to overlap the search space. The partition is cleaner, but will require a more complex reassembly process. Allowing some degree of overlap will make sure the search locates all the possible results.

The most basic search algorithms for graphs are brute force, or exhaustive search techniques, that solve problems with limited size or problem-specific, including the bread-first search and, the backtrack search technique, depth-first search [10]. These techniques have a very strong property of finding solutions and, if a solution exists, these techniques will find it. However, they have a runtime cost that can be exponential.

Many of the effective algorithms for graph search use these search techniques as a base. Some are greedy algorithms, which means that they never reconsider a decision once it is made [10]. Some of greedy algorithms are very effective for simpler problems that are polynomial runtime algorithms, like Prim's [48] and Kruskal's [31]. Both these algorithms extract a tree from a given graph in an iterative fashion, and in weighted graphs they can select minimum weight of the edges. Another example of greedy algorithms is the nearest neighbour algorithm applied for problems in discrete or combinatorial optimization. However, there remain a lot of problems that do not yield a computationally tractable solution, which are the ones of interest for the multi-agent system working on the web.

More challenging problems require the search algorithm to reach solutions faster and/or can handle large search spaces. There are several extensions that can improve performance and a variety of techniques have been applied such as the heuristic search and the best-first search algorithm [22; 52]. Heuristic searches are used to extend the basic depth-first search with applying heuristics [52]. The challenge is always to find an effective heuristic function where the heuristic is usually problem specific. Likewise, heuristics can have problems with search spaces that have features described as foothills, plateaus and ridges. Heuristics can be used to only follow the "best path" or be employed to order the choices in depth-first search.

By a similar extension, best-first searches can be optimised by beam search, A* search algorithm and Dijkstra's algorithm [13]. In the beam search, paths expand at each level of the breadth-first search tree, where the number of paths is limited by applying a heuristic function. The breadth-first search, on the other hand, evaluates the merit of each alternative path and then limits the list of paths to include only the best. The limitation is typically accomplished by selecting some maximum number of paths to expand, like in best-first search.

Other searching techniques often referred to as optimum techniques, such as branch and bound algorithm [10]. This algorithm is an extension of the depth-first search. The extension is to always expand the lowest cost path at each step, which is similar to best-first algorithm without the heuristic function. In many cases, the current best path is highly likely to extend to a good solution. Branch and bound keeps all the paths, so it is guaranteed to find a solution. The sorting of the paths, to the current lowest cost, tends to make the search more efficient and, thereby, the algorithm lends itself for parallel and distributed searching [17].

Branch and bound can be extended with a heuristic to get A* search algorithm [54]. The A* combines an estimation of the cost, to reach a solution from the current state, with the cost encountered so far. Again a serious issue is finding an estimate function. If the estimation overestimates the cost, there is no guarantee that a solution is found. A* algorithm is a generalization of Dijkstra's algorithm which, in turn, is similar to the Prim's algorithm. Dijkstra's algorithm can be used find a path of minimum total length between two given nodes [13]. This algorithm can also be applied to cases where the length of a branch depends on the direction in which the graph is traversed. Since Dijkstra's is using an internal loop [49], the event-driven algorithm, in our work, is similar to Dijkstra's with some differences. For example, the event-driven algorithm is built up by traversing the graph and uses of a function that looks ahead for links to explore.

The minmax and alpha-beta techniques are commonly used in game searches [54]. Minmax assumes that alternative levels of the search tree should be selected to give advantage to the alternating players in a game. Thus, from the view of either one of the players, one level is selected to maximize their gain and the next level is selected to minimize the other player gain. Alpha-beta uses bounding to prevent expanding all the alternatives and does not explore the branches of the search tree that the analysis finds, at a given point, to be uninterested. Hence, alpha limits every other level of the tree and beta is used to limit the other levels.

While using parallel approaches will not reduce the problems from the class NP, it is an interesting approach to explore. NP-hard problems can be search problems, decision problems and optimization problems and since solving these problems is a key factor for handling the vast number of information on the web, it also initialized the event-driven algorithm.

Some of the examples of NP-hard problems for graphs include Hamiltonian cycle, travelling salesman (TSP), Clique, Vertex cover, Independent set, Graph partition, Edge cover, and Graph isomorphism [52]. In the class of problems of class NP in computational complexity theory, the Hamiltonian cycle and Travelling salesman problem (TSP) are listed. Hamiltonian cycle visits each vertex exactly once whereas the TSP find the least-cost round-trip, i.e., the Hamiltonian cycle with the least weight.

The problems with finding sets that are connected in a graph are independent set, clique, vertex cover and edge cover. Independent set seeks the maximum independent set of vertices in the graph where no two vertices are connected by an edge, called the maximum independent set in the graph. Similar to independent set is clique. Clique, or complete graph, computes the number of nodes in undirected graphs largest complete sub-graph. Another search technique for undirected graphs is Vertex cover, which works with finding the minimum size set in the graph. Closely, to vertex cover is edge cover. Edge cover search looks for the smallest set of edges where each vertex is included in one of the edges.

A similar problem is graph partition problem, which divide the graph into pieces that are about the same size and few connections between the pieces. The graph partition can be applied to load balancing for parallel computing. The vertices correspond to the computational tasks and the edges correspond to data dependencies. Another class of problems is mapping between graphs, such as graph isomorphism. Isomorphism of graphs implies there is a bijection between two graphs, i.e., one-to-one correspondence between the sets in the graphs. The presumption for the web is the edges are the transport connections between vertices for the message passing including the data about the delivery and the vertices are computational tasks or information sources.

7.2.1 Search Algorithms on the Web

The web is a complex environment in which the intelligent agents are used to perform tasks such as searching and retrieving information from sources but also accessing services. Some of the agents can work efficiently with matching text strings, usually offering a list of web pages on some criteria. Other agents can take actions on behalf of the users, such as, assembling important information from several services. Using intelligent agents when looking for information and services at the web is beneficial, since they can search information more intelligently than common search engines [21]. Additionally, the individual agents can create their own knowledge base and communicate and cooperate with other agents. These agents can also support users finding various services, perform tasks continually and in parallel, search for information based on context and adjust to preferences of individual users.

On the web, agent technology is applied in several application areas [21]. The agents can enhance the system and network management software by filtering information and taking actions, but also by detecting and reacting to patterns in system behaviour. The agents can manage and search information and also categorize, prioritize, disseminate, annotate and share information and documents. Moreover, agents can assist in electronic commerce by supporting, and negotiating, purchasing products for individuals and companies [36].

Commonly, agents on the web have been searching for services with corresponding keywords. In multi-agent systems, finding solutions for a service often implies letting the users deciding the search query, applying the query to the agents and then looking for a match for the service. These agents only work with one service at the time, which excludes searching for best solution while incorporating several constraints.

Agents provide an interesting mechanism to support distributed processing for search algorithms in graphs. One example of the search algorithms is the extended real-time search algorithm for autonomous agents [27]. The real-time search algorithm adaptively controls the search process by allowing suboptimal solutions with a number of errors and another search mechanism, which balances the trade-offs between exploration and exploitation. The search is used in uncertain situations where the goal may change during the course of search.

There are distributed algorithms that are based on the event concept. One approach is building distributed algorithms on an event driven base [14]. The authors take the view that recasting an algorithm into event driven requires complex state machines to handle the events, called Protothreads [14]. This is not always a problematic issue, since it will be dependent of the algorithm being implemented. However, their approach is useful since they develop a structure, which give a sequential programming model by "hiding" the event loop below the user-programming layer. The event layer is efficient and provides the synchronization calls. which requires writing the algorithms in an easy to work with style. The one deficiently with the algorithm is handling automatic variables across thread calls.

A different approach is an implementation model for event driven systems, which separates the programming task into two parts [47]. One part is single threaded message receiver that handles incoming messages and the other part is execution modules that make computations based on incoming messages. The execution modules are not interruptible and the single threaded approach does restrict the class of application that can be used. In a multi-agent system, this means the design needs to use a modularization strategy by designing smaller short execution time components. There is an extension to this architecture, which is using multiple channels to allow simpler dispatching loops. While the authors have not used the algorithm as such, it could allow multi-threading.

There are also a number of search algorithms used by search engines on the web. Some of the algorithms use web crawlers to browse the web and provide up-to-date data. The web crawlers are software agents that copy all visited pages, index the pages and process those pages by a search engine. Some of the problems that confront the web crawlers are large volumes and fast changing information. An example of using web crawlers for larger volumes of information is distributed web crawlers, which spread out the required resources of computation and bandwidth to many computers and networks [9]. Another measure is to assign importance to the web pages and reduce indexing of duplicated pages. In order to avoid crawling the same resource more than once, web crawlers can perform URL normalization. The URL normalization is the process that transforms URL into a normalized form and determines whether the links have been visited or pages have been cached. Other crawlers are used to find only the relevant web pages, such as, the goal-directed focused crawler [7] and topic-driven crawlers that attempt to download web pages based on predefined topics [37]. The crawlers focus on predicting the probability of a link's relevance using anchor text, the complete content in the page, or reinforced learning and evolutionary adaptation.

Other algorithms such as the link analysis algorithms, HITS algorithm and PageRank, are based on the linkage of the documents [30; 43]. HITS algorithm rates web pages for authority based on relationship between s set of relevant authoritative pages and the set of hub pages that join those [30]. The PageRank is a method for ranking Web pages objectively [43]. Both of the approaches are based in the event driven concept, i.e., distributed algorithms on an event driven basis and implementation model, provide interesting models for event driven agents. The event-driven approach with an event layer together with synchronization calls is most attractive. We also use an event layer, so called meta-level layer, which synchronizes calls. Also the separation of incoming messages and execution modules that execute computations is of interest since finding services can require searching for services in the intranet and the extranet. Also communication between results from the net is needed for information combination. Searching for our agents finds relevant web pages as the search algorithms do for the web but also to return values found in the web page.

Our design uses the Java Agent Development Framework JADE framework [28], which is an infrastructure that simplifies implementation of a multi-agent system through a middle-ware layer. The event models can be constructed in the Jade agents as a set of classes that support the event communication and handling. This allows the search as an extension layer of the event handling. The actual problem solving can then be implemented as execution modules invoked by the search. The execution module approach even allows us to mix in functional or logic programming approaches in the agents. The choice of infrastructure between the Pfeifer *et al.* [47], separating the programming task into two parts, or the Dunkels *et al.*, [14] giving a sequential programming model by "hiding" the event loop, is chosen based on the execution requirements.

7.3 Event-Driven Algorithm

The event-driven algorithm is based on the notion of event-driven programming. When examining event-driven programming for agents, the prototypical example of event-driven programming, namely GUI's, is not the interesting or a useful model for the agents finding services. Instead, we focus on the characteristics of the events that are beneficial for finding and combining services at the web, such as launching events, performing operations and store values in variables while utilising parallel computation.

Event-driven programming is a programming paradigm in which the control flow of the program is determined by messages, from other programs, or by user actions [15]. Thus, in event-driven systems, the program itself does not fully determine the order of execution, rather the order, or event delivery, controls the flow of execution. The events are directing the computation to the next actions to be taken in the program. The events launch manuscripts that can either contain a single event or a script including several events, functions and operations. These scripts only execute when the event dispatches them. In the meantime, the script is idle, waiting for an event.

The event-driven algorithm uses messages as events where message passing is the communication method that processes the event and acts as a launching function. Thus, the events constitute link between objects sending messages. From the sender object, i.e., the event generator, the event is directed towards the next destination for the message. The receiving object, i.e., the event handler, picks up the message and carries out the commands in the message. Hence, the event handler depends on the content of message but also how this message was sent. With the messages, parameters and variables can be sent, which needs to be picked up by the handler. The parameters are either events or functions whereas the variables contain values.

To illustrate the similarities between the architecture of web and the execution of the events, we use the archetype diagram of event-driven programming, presented in Ferg's paper [15] as the data flow graph of a typical transaction center of an application see Figure 7.1.

The transactions are the events and the transaction centre is a dispatcher for the events [15]. Commonly, the dispatcher processes a stream of input events, which is stored as an event queue in the dispatcher. The event loop receives an event, dispatches it and loops back again to obtain and process next event in the stream. The transactions or events illustrate a graph of events. This event graph resembles the web with links, where the data flow corresponds to the links. On the web, a client-server architecture can include a dispatcher, or teller. This dispatcher is a server that runs in an infinite loop finding pages according to the URL:s [15]. However, in our work the dispatcher is not utilized as one unit sending all events. Instead, the agents in clude event scripts that dispatch agents at the web finding information similar to

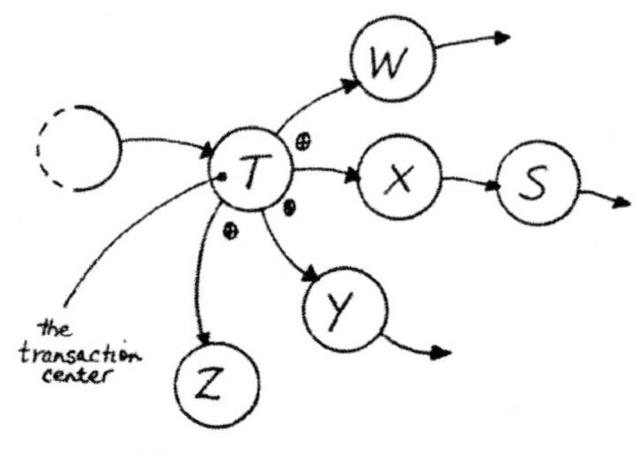

Figure 11.1. Data flow graph of a typical transaction center of an application.

Fig. 7.1. Data flow graph for event-driven programming. In: Ferg, page 9 [15].

topic-driven distributed web crawlers. The scripts follow the message and are utilized during the message route through the web. Thus, the event launches following agent but must also search for the required service and find the destination where the information is accessible. The message continues to execute until the script has reached the end of the event script. Together with the script, parameters and variables are adopted. The parameters can hold composite information about the services with the variables containing corresponding values. The facility is useful in an information logistics process especially when several pieces of information are needed from a single page with certain values or within given intervals of values.

7.3.1 The Different Kinds of Agents

The event-driven algorithm is applied in multi-agent systems and accomplishes tasks by handling agents at different levels, i.e., object-level agents and meta-level agents. The object-level agents are the event handlers, which are pre-programmed with event loops that look for message to process. These object-level agents are also event generators since they dispatch the other agents on the web, while the agents are looking for services. When the meta-level agents are developed, the meta-level agents, or meta-agents, become the event generators for a certain service and dispatch all the object-level agents connected to each meta- agents. Thus, these meta-agents are built up from the object-level agents and comprise the object-level agents that have successfully accomplished their tasks. Successfully accomplish has two different meanings; one is successfully reaching the receiver without finding the web page that is asked for and the other is successfully reaching the receiver finding the web page that satisfy the constraints. All these object-agents are present in the meta-agent and due to the set of object-agents in the meta-agents the contents of the meta-agents are different. Conclusively, the meta-agents correspond to an event queue that, from the beginning, holds all the unprocessed events to be dispatched.

The object-level agents, also called ground-level agents, collect information about the environment, while moving on the web. The events drive the action of the agents, which can collect information at the different levels of the protocol. Simultaneously, meta-level agents, also called meta-agents, monitor the intelligent agents and, thereby, extract the performance of the event-driven algorithm. There will be a lot of different paths explored when performing a particular task. All these paths will be tuned into an optimal search algorithm. The search algorithm for achieved task is stored by a meta-agent and reused for future information searching.

The event-driven algorithm can work in connected and directed graphs. For purposes of this work, we consider only connected graphs. This represents a useful set without much of a restriction. Additionally, we can consider either directed and di-graphs, or undirected graphs since, in some of our prior work, directed graphs were found to be useful for road networks. In those networks, the costs can differ depend on the direction, making the directions important for the calculation. Even though directed graphs are more commonly used subtype they are not considered at this time since the goal of this work is to consider graph algorithms in general.

Another interesting property to consider is the graph representation. For some graphs, a single representation can work well. The single representation can be distributed in whole to the agents. However, for complex graphs, the representation is distributed over multiple machines. The graph itself might be large and needs to be spread out to conserve memory in a single processor or node. An interesting reason for using multiple representation is that a single authority or source does not produce the graph. Thus, the graph is produced by multiple sources and, hence, the representation of the graph is logically a distributed representation. Moreover, when the graph is a part of a real world dynamic system, the graph is a representation of a part of the world and contains data sourced from geographically distributed systems. Examples of a geographic distributed system are the web and the highway system with sensors. Such systems clearly call for a distributed agent architecture.

The most useful search technique for the event-driven algorithm is to examine the assigned task. Each task has its best path through the network and there is no single search algorithm that covers all the cases for tasks. The only way to find the path is applying the task on the system and finding solutions using a dispatching function on the agents. Thus, the task is the event generator. Several agents perform the same task and, in the meantime, the agents are checked for performance and the successful agents are kept as an event in the meta-agents.

7.3.2 The Multi-agent System Environment

The multi-agents systems, in which our event-driven algorithm has been applied, contain agents in graphs and networks [25; 24]. Some networks, such as the web, are usually complex and, thus, require several collaborating agents. On the web, the agents execute the assignments between the nodes and, in particular, there is only one agent per arc. The agents do not request any information about the assignment to perform the task. Instead, the only external influence is that the agent is dispatched by an event as a message from another agent.

In established graphs, the agents work in an uncomplicated task environment, in which the environment is fully observable, deterministic, episodic and discrete [50]. The web requires that agents are capable of autonomous actions, situated in different environments where some are intelligent and, thus, adaptable to the environment. The environment has some characteristics that the agents can observe and act upon. For these agents, the web environment is a mix of characteristics, not a single set of characteristics.

For example, the web in a localized region, can be a fully observable environment since the agents can obtain complete, accurate and up-to-date information about the environment state [55]. The agents do not need to maintain any internal state to track the world [50] and can easily achieve the task based on the information in the environment. Thus, the agents have full access to their part of the environment and do not change their internal state to keep track of the external world. Moreover, the agents do not react to unexpected events since they decide the next step from the current step, which makes the environment deterministic. In this sense, the agents do not need to handle uncertainty.

The entire web, on the other hand, is a partially observable environment where only parts of the information is known and therefore, can be limitation for information collection. Environments such as the intranet and extranet are considered to be partially observable because of its nature of continuous growth [55]. However, the agents do not have to maintain their internal state to find satisfactory information. The agents working with services can use both the intranet and the extranet finding information. In those cases, the environment must be considered to be partially observable, but still able to find the significant information needed for finding solutions.

The web can be a deterministic environment since any action has a single guaranteed effect [55]. This implies that the agent has only one action to accomplish and that the next state of the environment is completely determined by the current state [50]. In such environments, which are also fully observable, the agents seldom have to deal with uncertainty. In a partially observable environment, the environment can appear to be stochastic. The agent cannot predict the behaviour of the environment, as in real-world cases, since the state, which will result from performing an action, is unclear. Even though the environment is deterministic, there can be stochastic elements that randomly appear. A partially observable environment with stochastic elements is what will be expected for the agents at the intranet and the extranet. The agents have a task of searching for services and information but the web and the information will vary with time, i.e., when actions on these are performed.

In an episodic task environment, the task is divided into episodes [50] and carried out, singly. Commonly, the agents' choice of actions only depends on the episode, where each agent performs a single task according to the atomic episode. These episodes also limit the number of actions the agent can take. On the web, each episode consists of agents performing an action and the different episodes are independent. The agents perform one task at the time without choosing the action. The task itself steers the agents by dividing it into smaller, single tasks, where each task is applied to an agent. The task is the query for services, which can be rather complex with lot of information to match, and each part needs to be assigned to agents. The query is inserted in slots for products, amount and price depending on the users' request. The results from the agents' actions are collected and assembled into a comprehensive solution. Thus, the results from carrying out episodes are compounded into a complete solution by storing the result of performing the task and combining the result to constitute more detailed and desirable information.

On the web, the environment cannot be argued to be static by considering the common definition, where static environment remains unchanged while the agents consider their course of action between states. That does not exclude changes, since the agents can cause changes in the environment. By contrast, a dynamic environment is expected to constantly change and is beyond the agents' control [55], which might not be completely true either. The environment can remain unchanged or change while agents are deliberating its contents [50]. In these environments, the agents need to interact with the environment and continuously check the surroundings to act properly. This characterizes the intranet and the extranet since these change on a daily bases. However, the environment might be static over smaller time intervals. Thus, the agents work in a semi-dynamic environment.

A semi-dynamic environment requires the agent to be autonomous and flexible [35] to respond to dynamic characteristics. The structure of the graph may not change often, but the properties in the environment changes frequently. These properties are static and dynamic, where the static characteristics are constant constraints in the links or vertex, such as new links or pages, and the dynamic characteristics are temporary obstacles, such as broken links, or errors in the page. The agents need to react according to both the constraints and the obstacles. Hence, the agents will consider continuous changes, which will affect the agents and make them monitor changes during each task. The event sets the agents off but the agents act according to the environment without external governance or control. Moreover, the agents have to adapt to the environment and learn about the static characteristics since it affects the speed of the agents. Agents that acknowledge the static properties only have to relearn those occasionally.

For the web, the discrete and continuous distinctions can be applied to the state of the environment and to the actions of the agents but also to how time is handled [50]. In discrete environments, the agents have a fixed and finite number of actions [55] and in continuous environments, there might be uncountable many states, arising from the continuous time problem. For each agent, there is a fixed and finite number of actions but the web, in overall, is almost infinite with uncountable many states. Continuous time is a problem for the intranet and the extranet agents because of the number of states and actions. This requires special treatment of the agents using an execution suspension to control the agents' performance. A common suspension of execution occurs when the agents have found information and returned with result. This limits the possibility of finding several solutions. A better solution in our system is an execution suspension for a short time interval followed by a resumption of the search.

7.3.3 The Intelligent Agents

The agents are intelligent in the sense that they observe the environment and collect information while performing tasks, thereby learning the environment, until the agents have finished their task. The information is data about the service and the facts in the pages that corresponds to the request. The ground construction of the intelligent agents is as following:

$$\forall IAg(x)\ \exists x(Task(x)\ \Lambda\ Environment\ Information\ (x))$$

For all intelligent agents *IAg(x)* in the system, the intelligent agents have an specific task to accomplish *Task(x)*. Thus, each agent has a unique task assigned, which they accomplish together. Since the task is finding the required information from several different pages, each agent is moving along the assigned edge and passes on the request to the agents. The task contains two different parts, the task assigned to the system and the agents performing the task. These are to be separated in our system.

Environment Information(x) is information about the semi-dynamic environment, i.e., both static and dynamic information. The collected environment information is data sorted as set of static characteristics and set of dynamic characteristics. Storing the data in two different sets support distinguish the characteristics in the environment. Hence, each agent includes more detailed information has the structure of four-tuple <t, a, s, d> where *t* denotes the assigned task, *a* represent the agent, and *s* denotes static characteristics and *d* symbolize dynamic characteristics. The main argument for the diving the sets is that static information, web site, does not need to be checked as often as dynamic information, page contents.

The task, t, is a compound statement including the request and the agents' position on the web. This includes the nodes on the web between which the agent is moving and identification of the link, or edge, and commonality. The agents are structured as follows:

agent(Web Node dispatcher, Commonality, Link Id, Web Node receiver)

One of the web nodes is the dispatcher and the other node is the receiver. More precisely, the *web node dispatcher* dispatches one agent, which at its arrival will pass the message to receiving node, i.e., *web node receiver*, which is the connection to all adjacent agents. While performing an event, the agent is moving between the nodes and collecting information, such as, bit rate, delay, jitter, risk of packet dropping and bit error rate; information that is stored and used for evaluation of optimal search.

The agent moves along the link, i.e., *link id*. Each link id is unique and denotes the connection between the nodes. The node is either a connection node, like a server, or a web page that might include information about the request of service. If it is a connection node, an agent is dispatched and proceeds to the web node receiver. Otherwise, a web page is found and scanned for information. If the matching service is found, the information is matched to the request from the user's input text string, and the information is sent back to the user. If not, the search continues.

The *commonality* is a set of desired properties used to connect several different edges together to constitute a commonality between several different links. This similarity can be used to solve a problem or find same services. The commonality can also support finding the agents with the same properties, such as static and/ or dynamic characteristics in the environment. It can be used to find several available services at the same web page. For example, the commonality can be using only the high bit rate transportation supplied by all the servers in the path or searching for several products supplied by the same company.

After the agents found the service, all the nodes, i.e., the nodes that have dispatched events, are sent as paths back to the user. Returning the result is a facility supplied by software at the web and can be effectively utilised by the meta-agents in our multi-agent system. The result of different paths can be calculated and compared to find the best solution but also analysed for errors and validity. For future browsing for same service, the paths can be used to direct the browsing to certain pages, that earlier supplied the particular service.

7.3.3.1 An Example of the Web with Agents

To illustrate the multi-agents on a web, we provide an example of a network symbolising the web with nodes and links, see Figure 7.2. On the web, the nodes are the web pages and the links are all the links there are between the pages. In this example, the web pages are denoted as letters, e.g., B, C and the link in between is called B-C, which is the intelligent agent. The agent's actual identification for the links is the IP-address between servers, routers and all the components involved in the message passing at the web at which the links are the actual wires connecting the servers. The users assign tasks, when asking for specified services with certain information, which launches the agents, where the first agent is starting from the users' web page. By this, the starting web page becomes the event generator and all the agents connected to this web page become the event handlers. Each agent's web receiver (like C in B-C) becomes the event destination.

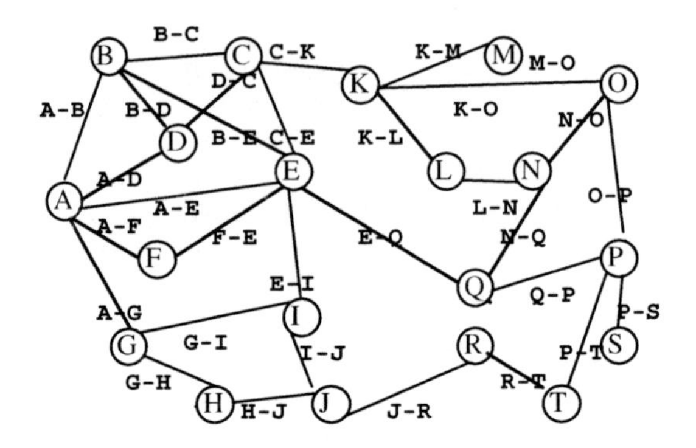

Fig. 7.2. Network with several intelligent agents. In: Håkansson and Hartung, 2008 [25].

Each agent holds the information about the nodes it works between, for example, from the node A, the connected agents are: the agent A-B, agent A-D, agent A-E, agent A-F and agent A-G. These agents are used to collect information and calculate the time for executing the agents to the event destination. More specifically, the agent A-B acts between the dispatching node A and the receiving node B and collects environment information used for calculating the time for message passing. As mentioned before, there are several paths through the web and the agents need to find all these paths where the agents, together, constitute paths of connecting and successfully performed agents.

For example, consider that the user is the node A and that he or she likes to find a service that can be supplied by node P, the A become the start node and the P to be the end node. The agents start moving from the start node, using all possible links from that node. It will be found that one of the paths through the network includes the nodes A, E, Q and P. Between these nodes there are several paths collecting information about the environment where one is includes the agent A-E, agent E-Q and agent Q-P acting between the A and E, E and Q, Q and P, respectively, see Figure 7.3.

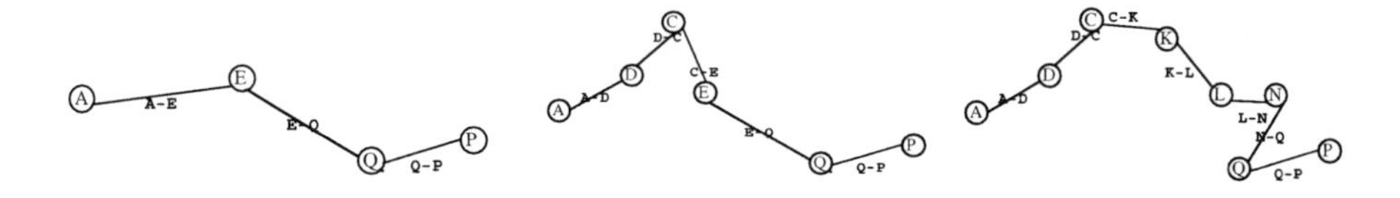

Fig. 7.3. Examples of paths for the same service at the web

As pointed out, there are a lot of alternative paths through the web and depending of the success of the agents, some of these paths are useful but also slower or faster depending on the environment. A path with a few numbers of nodes between the sender and the receiver is not guaranteed to be faster. Instead, an alternative path with more agents might be the faster. Still, the fastest paths may not be the most optimal solution, which necessitates a rigorous comparison of the different paths. Hence, all characteristics of the agents must be considered before calculating optimal path/paths.

The information stored in the agents can be more or less relevant to the result of optimal solution in the graph optimisation problems. The shortest computation time to reach a goal is, of course, the optimal solution. Nonetheless, it can be interesting to find out the reason for selecting the particular path. Therefore, the agents need to keep information about the static conditions, such as commonly used bit rate, that probably affect computing time for the paths. There can also be more radical problems, which are dynamic, like bit rate errors, queues and package dropping. The amount of information to be kept can be decided for each computation through the web. However, a lot of information can affect the time for computation, which is the reason for introducing agents at a meta-level.

7.3.4 The Meta-agents

The meta-agents form a horizontal layer on top of the ground-level agents to monitor the actions of the ground-level agents and collect these agents together with environment characteristics. The ground-level agents have a purely local view whereas meta-agents can capture a global view of the part of the network in which they are working. The different layers of ground-level agents and meta-level agents require some organisation of the agents, which support the linking between agents, directly or indirectly. The different layers correspond to hierarchies of agents. These hierarchies include event handling and communication used for agent cooperation [53] between the layers.

The meta-agents execute in the context of all the agents. At the top-level, the meta-agent holds the start and the goal of the computation. The meta-agents are assembled after the intelligent agents have successfully performed their tasks. The meta-agents constitute a chain of successful intelligent agents computed by accomplishing a task where the agents are connected back-to-back. The chain becomes a stack of events that are launched for a computation and used for searching the web pages. Thus, the meta-agents are used for computation, where each ground-level agent is an event searching for the pages. The events are activated in the same order, as they are stored in the meta-agent to find the web page again but also faster. For searching, the meta-agents also hold the query and the result from combining data from the page.

The structure makes it possible to use the meta-agents for monitoring and controlling the intelligent agents work but also guiding them through the network. The meta-agents keep track of time and information about the conditions the agents have collected from the environment. The information is passed as messages between the agents, i.e., at the ground-level and at the meta-level, in the network. Hence, the meta-agents work as a coordinating mechanism for the object-level agents, like in meta-level reasoning [8]. The meta-agents support communication to and from the ground-level agents and interact with other agents via a communication language for coordination [56; 18]. The communication occurs when an event is sent to the meta-agent.

In the system the meta-agents have the structure:

$$\forall MetaAg(x)\ \exists x(Initialpos(x) \wedge User\ query(x) \wedge Path(x), Endpos(x) \wedge Result(x))$$

For all meta-agents, each meta-agent has an initial position, *initialpos(x)*, a unique path and an end position, *endpos(x)*. The user query is a string of relevant words used for matching and the *path(x)* is the total route between the initial position and the end position used when finding the matching web page. The term *result(x)* contains the page that matches the query, together with a calculation of the amount and price. The path is structured as:

$<agent_1,...,agent_n>$.

The path comprises all the ground-level agents in the form of a two-tuple <ag id, c> where *ag id* is agent identification. The agent identification is used as launching mechanism for the intelligent four-tuple agents (as described above) and the tuple *c* is cost for computation of the current agent, used as a measurement for calculating the optimised solutions for the event-driven algorithm. The result is based on:

$<word_1,...\ word_n, total\ costs>$

The words are either product name, amount of the product and price. If possible, the meta-agent will calculate the combination of the amount and price to be able to present the total costs for the product.

To illustrate the construction of the meta-agents, we present an example of the procedure, corresponding to the left figure in Figure 7.3 above. The meta-level agent comprises the ground-level agents involved in finding the optimal solution while moving from A to P, see Figure 7.4. The meta-agent is located at the top in the figure and the ground-level agents are located beneath. The meta-agent incorporates the information about the sender, the path of agents, receiver, and the time for executing all the ground-level agents. The meta-agent also has the information about the web nodes (vertices) and conditions that the ground-level agents have collected during their execution, which is taken into account in the optimal solution later on. The ground-level agent to the left, agent A-E, has information about the initial web node dispatcher, the commonality of bit rate being 260, link id which is also the agent id and the receiver node. After executing, the ground-level agents also have collected information about the static and dynamic characteristics, which is part of the costs. The cost is sent as a

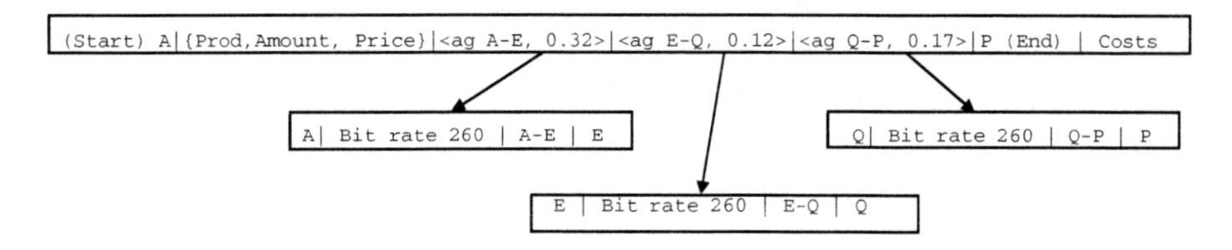

Fig. 7.4. Meta-level agents incorporating path of agents for a task

message of time to the meta-level agent, for example A-E 0.32. The ground-level agent to the right, i.e., agent Q-P, has information about the last receiving node, which is also the end node.

To the left in Figure 7.4, the first agent, i.e., agent A-E, is activated by the user. This agent reaches the E node, which causes the agent E-C, agent E-F, agent E-Q and agent E-I to start running. Only, the agent E-Q is captured in the current meta-agent while, the other agents will be captured in other meta-agents, i.e., agent E-C, agent E-F, and agent E-I. Additionally, for each agent, the cost for executing the links is hold by the meta-agent. Moreover, the total cost for the path is registered. Hence, the execution time for agents is: A-E agent is 0.32 msec, E-Q agent is 0.12 msec and Q-P agent 0.17 msec and for the meta-agent 0.51 msec.

The meta-agent has the logic to determine when a solution has been achieved and to present the solution as the result of the computation. The logic is a set of conditions and constraints that select the best paths to support scheduling the computation and determine the optimal solution at the given time. A simple approach to scheduling is based on the shortest path generated at any time. However, the shortest or the fastest may not be the best path for a given purpose. The best path is, of course, in the eye of the beholder and the condition for the path to be used, must be decided by the user. Calculating and comparing the several different paths through the network can provide several paths meeting the conditions given by the user.

The selection of the optimal solution is made by the meta-agents. The meta-agent enters each node of corresponding agents in the network and performs the calculation for the agents. Each meta-agent has access to all information contained in the ground-level agents, including all collected derived conclusions. The meta-agent simply adds the agents' performance in time, taken into account the static information and the dynamic information. The cost for the agents is the ground-level agent's computed costs determined from the static and dynamic characteristics in the environment. These costs can be used to compare the performance of the ground-level agents. The meta-agents compare the performance and select the fastest solutions. The calculation is made in two steps: First, each meta-agent performs the calculation for the each ground-level agent by using all information contained in the ground-level agents, including all derived conclusions. The meta-agents take into account the static information and the dynamic information, which is translated into time costs, and simply adds the costs in a total time cost for the ground-level agent. Secondly, the meta-agent takes all ground-level agents and adds the costs for these, which will become the total cost for the meta-agent. All the meta-agents' total costs are compared to obtain the fastest.

As it is possible to search for multiple solutions in the network, the computation can be made in parallel. Several ground-agents can start simultaneously to search the network for solution. For each computation in parallel, several meta-agents are created, i.e., unless the first agent reaches the end node and is also the fastest.

An example of a practical problem is to set up a set of paths for multiple deliveries to nodes in the network. This is a useful problem for airlines, trucking and other carriers. In this case, several meta-agents can be started simultaneously to search the network for solution. For each computation in parallel, several meta-agents may be created since the first agent to reach the end node might not be the fastest. Then, the meta-agents must be tested against each other.

The parallel computation can provide for the ground-level agents and meta-agents to use less computational time to find the solution. The event driven algorithm is designed for handling event driven approaches to implement parallel computation in the network, using agents.

7.3.5 The Search Technique

The search technique for the event-driven algorithm has been tried in a multi-agent system solving problems in connected and directed graphs. When using the system for a commence request of a service, the event-driven algorithm does not have any paths guiding its search through the network. Therefore, the events totally drive the agents around in the network.

Each agent has a routine, which makes it possible to receive the data about the environment and unexpected activities. To start working, the agents need event-handlers, which are checking for external events [15]. These event-handlers respond to an incoming event and dispatch the agents. When and where the event occurs is solely determined by the task to be accomplished. The user (the sender) provides the launching event and the destination (ultimate receiver). From the launching position, an event dispatch all the agents connected to that position to start computing. The connected agents are the ones that have links from the current launching position to receivers, which are the neighbours or vertices to the sender. The receiving vertex becomes the sender vertex when the event dispatches connecting agents. The agents carry on with the tasks until they reached the destination position (the target position), accomplishing the goal and, thereby, completing the task.

The event-driven algorithm follows the different agents but, since they work in parallel, we use the meta-level facility that keeps track of and stores the data from the different agents. The main tasks for these meta-agents are to collect successful task-achieved agents, but also to produce and maintain a path of events that constitute the event-driven algorithm. The algorithm uses the stored events and guides the agents' actions to work in an optimal way by selecting a shorter and faster path.

The agents were given the task of finding an optimal way between two different positions. It was found that the search technique for the intelligent agents resembles a mix of depth-first and breadth-first. Each individual meta-agent follows path in a more depth-first strategy. However at a branch point, meta-agents are created, providing an overall breadth-first strategy.

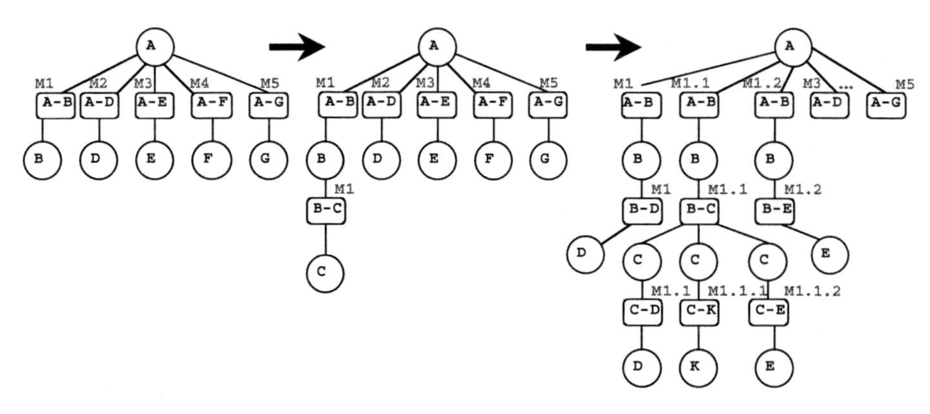

Fig. 7.5. An illustration of the algorithm of the system

The system starts to explore the intelligent ground-level agents associated to the start state, A, with breadth-first, see Figure 7.5 to the left (A-B, A-D, A-E, A-F and A-G). Five meta-agents are created (M1, M2, M3, M4 and M5) in where the ground-level agents become the first agents. From the meta-agents, the network continues to be explored with a depth-first tactic.

The system continues with the first explored ground-level agent (agent A-B, left branch in the figure in the middle) and next linked ground-level agent (agent B-D) is dispatched, expanding the meta-agent (M1) with the current ground-level agent. If the agent is dispatching several other agents, the first agent (agent A-B) is duplicated for each meta-agent (agent A-B in M1 and M1.1) see left branch in the figure to the right. Thus, it will be one meta-agent spawned for Ag A-B and Ag B-D (M1), one meta-agent spawned for Ag A-B and Ag B-C (M1.1) and one for Ag A-B and Ag B-E (M1.2). For each duplication, the number is expanded with adding an extra number, (1.1, 1.1.1, 1.1.1.1…), and increase that number, (1.1, 1.2, 1.1.1, 1.1.2 …) guaranteeing that the number will always be unique and there will never be a shortage of number but it will give an ordered growth of the numbering scheme. A benefit with the duplication of meta-agents is that it supports building up a unique set of agents while executing the system and, thereby, being able to apply parallel computing.

The algorithm works as follows:

```
While goal not found
  Visit first vertex or an enqueued vertex
  Use the event algorithm to discover all neighbours
  For all the neighbours
     Dispatch the agent and move to adjacent neighbour
     If successfully reached an vertex and a meta-agent
     does not exist
       Create a meta-agent
       Include current visited vertex
```

```
        If goal not found
           Use an event algorithm to discover vertices ad-
           jacent to the neighbour
           For all vertices and goal not found
              Enqueue look-ahead link
              If first vertex visited or only one link
                 Expand the meta-agent with current found
                 link
              Else
                 Duplicate meta-agent and expand the doubled
                 meta-agent with current found link
              Mark link as visited
        End for
     Mark the neighbour as visited
  End for
End of while
```

The algorithm contains three loops, one for continuing to look for a vertex to exploit, another for searching the neighbours[1] and the third for finding all the links connected to each neighbour, like a "look ahead" function. Along with finding the vertices, the meta-agents are created, expanded with a neighbour or duplicated for each neighbour. The number of meta-agents depends on the number of connected vertices.

The algorithm starts visiting the vertex, which the sender initiates the first time, and, secondly and thereafter, one of the enqueued vertices initiates this vertex. From the vertex, the algorithm searches for the neighbours connected to the vertex, where the neighbours are the ones that have links to the visited vertex. After the event algorithm discovered the neighbours, it returns a list of neighbours.

This list contains all the links are connected to the node (the visited vertex), which is traversed for finding different paths in the network. For each neighbour in the list, the attached agent is dispatched and moved to the neighbour. Thus, first time the agent moves from the first visited vertex, along the link, to the neighbour, or new vertex. Secondly, the agent moves from one of the enqueued vertices, see figure 7.6.

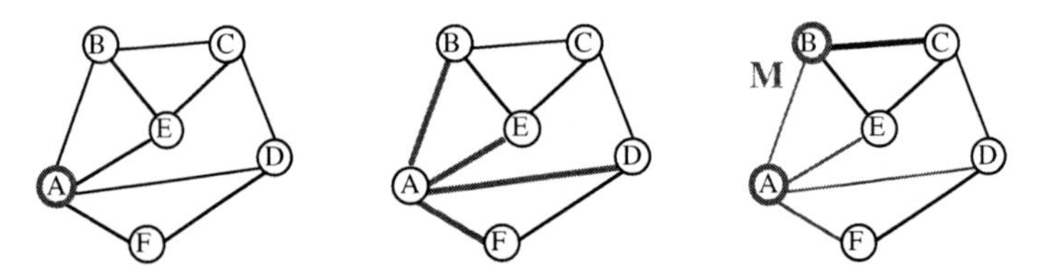

Fig. 7.6. Illustrating the first vertex and the list of neighbours

[1] The neighbours are just ordinary vertexes but for the readability of the algorithm, we denominated the vertexes differently.

When the agent successfully reached the neighbour, a meta-agent is created or an existing meta-agent is expanded with the current visited vertex and the agent. Moreover, if the goal is not found, the neighbour is explored to find vertices, i.e., use an event algorithm to discover vertices adjacent to the neighbour. Again a list of links is created and used to find all the connecting vertices. If a vertex is found, the link is enqueued and, for the first found link, the meta-agent (holding the current neighbour) is expanded with the agent. However, if several links are found, the meta-agent is expanded for the first dispatched agent but for the rest, the meta-agent is duplicated and then expanded with the ground-level agent. Irrespective of expanding an existing meta-agent or duplicate and expand a copied meta-agent, the link is enqueued, which means the agent is idle and marked for further exploration, see figure 7.6.

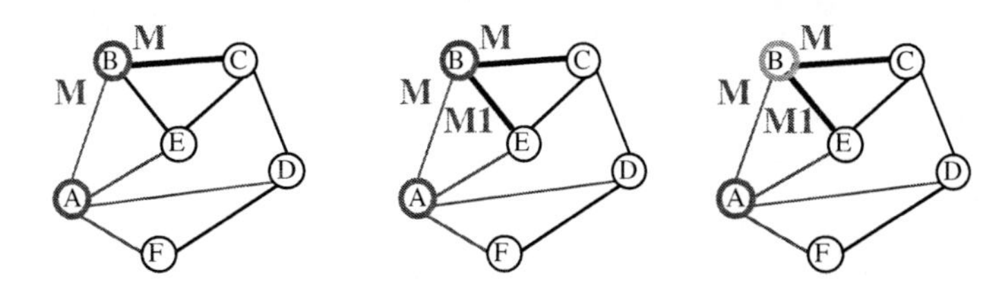

Fig. 7.7. Illustrating the first vertex and the list of neighbours

The enqueued vertex is easily found in the meta-agent since it holds the last position of the meta-agent. As a result of finding vertices, the meta-agent sends the last vertex back to the algorithm, which continues with next unexplored link, and thereby vertex. As long as the goal is not found, the algorithm continues with one of the enqueued vertices and starts all over again with visit the enqueued vertex. Conclusively, the algorithm examines two steps at the time.

The meta-agents are an interesting since they interact with multiple agents but also maintain more state information as the computation progresses. These meta-agents may need to migrate from processor to processor. Therefore, the meta-agents are assigned to a collection of processors, and can move from processor to process by their own choice.

7.3.6 Event-Driven Algorithm for Meta-agents

The algorithm acts as a search method to extract information from several different services and, once built up, the algorithm can guide future search and optimize the search for particular services. Optimizing the search includes both the path itself and the services being provided. So far, we have only looked at the creating paths through the web and next step is to select the successful target, i.e., the useful web pages. Not all the pages corresponding to the requested service provide relevant information back to the user. Therefore, the system needs to sort out unnecessary pages.

The event-driven algorithm works differently for the meta-agents than described above for the ground level agents. When the path of ground-level agents is built up, the path becomes an event queue handled by the event generator, or sender. Usually, an event generator is a component that generates a stream of events that the dispatcher processes [15]. However, our meta-agents administrate the event queue (the path of ground-level agents) and dispatch the ground-level agents according to the path. The order of the dispatching the agents follows the path of agents in the meta-agent. Thus, the path becomes the input stream from which the meta-agent uses information about the ground-level agents, dispatches agents and stores the result of the agents' performances, such as calculation time, in the meta-agent.

Once the matching web page is found, i.e., the page corresponding to the user's requested data, the meta-agent combines the data provided by the services and checks the constraints to guarantee fulfilling the user satisfaction. All the data to be combined must be in the user-defined task. An example of data can be a company that has a particular product in stock and that has a price for a requested quantity of the product that is within the acceptable costs range. Also the time for delivery of the product is important. When a company uses English in their query, the company can hit pages that are located on the other side of the world. If the agents find this supplier, the problem can be the delivery time. As a result, the information and the web page are returned to the user. However, the meta-agent must meet the challenge of finding information from several services in an acceptable time-range. The work includes extracting all the services providing the same type of service and optimally combining the different services to implement a complex requested service.

Besides finding one path through the network, we need to resend the request for several similar web pages. When resending the request, the search must be somewhat intelligent. For example, already visited web pages should not be revisited, avoided by using URL normalization. However, the servers can be revisited, but not the links. The reason is that the path A->D, can be more expensive than A->B->D. However, revisiting links is not preferred, especially, when going backwards in the network. Therefore, the vertex as a receiver must be marked but not when the vertex works as the sender. More specifically, incoming IP-addresses are marked but not outgoing IP-addresses.

Moreover, the content of the pages must be comparable. There are some techniques to be used for comparing the pages. For example, if the pages are tagged with semantic tags, those can be used to find similar pages. Other techniques are natural language processing techniques or keyword matching. However, ultimately, the user must decide the similarity between the agents. Now and then the user gets pages that are not satisfying the request and the user must have the ultimate control of the pages.

When finding the similar web pages, the paths in the meta-agent can be reused to distinguish the pages that are uninteresting and to continue the search for new pages. The search continues from the nodes that the meta-agents stopped at, i.e., when the first web page that matched the query was found. The event-driven algorithm starts from the enqueued agents and looks for new solutions. Thus, when the first web page is found, the event-driven algorithm must halt the search and not change anything until the user decides that all the needed web pages are found or the algorithm cannot find any more solutions matching to the request. If no services match the request, the event-driven algorithm must have a time-out handler or a break facility, which can be activated by the user.

One of the problems with reusing the results of already performed searches (paths in the meta-agents) is that the web changes fast. Over a short time period, the collected information can be out of date and combinations can vary too much to be useful. It can also have other problems such as the server is out of order or removed and/or the links are broken or removed. Then the event-driven algorithm must be used to backtrack through the path of agents. The backtracking must continue until the first solid agent is found. A problem is that even the closest agents to the web page sender may have changed.

When a new service is searched for, the search starts all over again, and starting building up new paths through the network. The same procedure starts all over where the event-driven algorithm finds the ground-level agents for the new task.

7.4 Conclusions and Further Work

The event-driven algorithm is an approach to searching for information in networks and graphs in multi-agent systems. The purpose for the event-driven algorithm is to find search paths within an acceptable time bound. Moreover, the event-driven algorithm has the benefit of driving its own way through the graph using agents as events. For this objective, we use ground-level agents and meta-agents, where the ground-level agents perform a task while the meta-agents collect information about the successful ground-level agents and provide paths between nodes. The ground-level agent systems have an event handler, which are trigged by the meta-agents. The meta-agents monitor the agents using an event-driven acting as a search method and extract the searching for information in networks.

The pseudo code for the algorithm is presented in the paper, following the computer system built for networks. The current example using the event-driven algorithm, is working with a network of agents. The algorithm is used in a small example of agents finding nodes in a network. The system builds up paths of the agents and stores the paths in the meta-agents. The meta-agents are used to calculate the total cost of each agent performing its task. However, these meta-agents are not used to compare the result against the others except for the costs. Next step in this research is to provide a large set of test cases and run these cases to other search algorithm. The test cases must cover more examples in graphs and network than the one we developed as a search technique in graphs.

There are several interesting approaches for building event-driven systems with meta-agents, especially, for the graph problems based on searching techniques. When patterns for basic search techniques are developed, then the problem itself can be overlaid over the search algorithm. The use of typical and well-known search approaches will make it easier for developers to apply them. The search approaches can use events to communicate between the agents and meta-agents. The meta-agents are charged with the oversight and direction of the search. The meta-agents drive the searching algorithm by assigning agents activities, for example, they can so the ranking of choices in a best first kind of algorithm. Also they will evaluate possible solutions.

The first logical approach to consider is, of course, a pure breadth-first search. After all, given a large collection of agents, it is easy to pursue the search problem as a parallel effort of multiple agents. This certainly seems at first inspection to be better than a depth-first base. While this can be effective is some cases, the speed-up is limited to the number of agents available.

A possible extension of the brute force breadth-first search is a best-first strategy. An approach that can be applied would be to let the meta-agents communicate to co-ordinate the search of the agents. For example, by applying a best-first strategy to the multiple agents. Here, we might have the available agents expand the best paths, up to the number of agents available. However, the communication cost will quickly impose a limiting effect on the amount of parallelism that can be achieved. This is a typical problem of multiprocessing systems.

A better approach would be to let the agents apply a best-first strategy locally, without using communication other than to report possible solutions as they are found. The result here is a combination of breadth-first limited by the number of agents. Then a depth-first approach should be used when the available agents are all busy. A logical approach in our meta-agent system would be to allow a group of agents to work under local coordination of single meta-agents. The meta-agent will monitor their progress and use events to redirect their activities.

The above system can extend to beam or branch and bound approaches, if some of the paths are eliminated from consideration using a bound on the number of paths retained and expanded.

Another possible approach is to apply heuristic or estimation functions to the search. These are easy extensions, provided the computation can be performed locally or semi-locally, that is in a meta-agent monitoring a small group of agents. This can result in either a heuristic or A* search strategy. The localization to a meta-agent may result is less that fully optimal search. For example, if the best paths pursed in some of the meta-agent groups are all very poor, it is a waste of effort. On the other hand, the parallelism can still leave the process more effective than a sequential algorithm.

The game searches, minmax and alpha-beta can be implemented in some interesting ways. One approach that seems interesting is a dataflow model. In this model, groups of meta-agents are assigned to levels in the search tree. So a set of levels is being developed in parallel. As paths are developed in level i, they are passed to agents in level $i+1$. The meta-agents in each level control the min or max functioning and also provide oversight of the expansion. Especially in alpha-beta pruning, the meta-agents carry a major portion of the algorithm. When level i is complete, the agents in that level are reassigned to the next level not yet expanded.

We can use multiple agents to try different heuristics in parallel. One of the issues with heuristics is their problem dependence. What works well in one case may fail in another. For this possibility, we can collect a large number of possible heuristics and then start multiple searches in parallel. Each search can try a best-first or a beam search approach using a heuristic. Then, the results are analyzed to find the best solution generated by any of the heuristics.

Bibliographic Remarks

Detailed discussion of graph algorithms is found in numerous texts. Cormen, *et al* [5] gives an introduction to algorithms by providing a detailed description of graph algorithm and efficiency. Also Gibbons [19] gives a good reference to some of the graph algorithms. Prim [48] presents a sequential minimum spanning tree algorithm and Dijkstra [13] a single-source shortest paths algorithm, which both can be applied to parallel settings. The parallel formulations of the algorithms are illustrated in Bentley [4], and Paige and Kruskal [44] and Kumar and Singh [33], respectively.

Extensive literature on search algorithms is available. The search algorithms used for solving discrete optimization problems are discussed by Kanal and Kumar [29], Skiena [52], Pearl [45] and Winston [54]. Many of these algorithms have been applied for parallel formulations [20]. Algorithms for parallel depth-first search have been formulated by Mornien and Vornberger [41], Peng *et al* [46] and Kumar and Rao [32] and for alpha-beta search by Baudet [3], Akl *et al* [2], and Finkel and Fishburn [16]. Algorithms for parallel best-first search with central strategy have been formulated by Crowder *et al* [12] and Mohan [40] and with distributed strategy by Kumar *et al* [34]. Heuristics have implications for exploiting parallelism [20]. For example, the TSP and Quadratic Assignment Problem (QAP) have been approached with parallel heuristics by Miller and Pekny [39] and Brungger *et al* [6].

Resource List

JADE - http://jade.tilab.com/
Tutorial: Events and event handlers - http://www.elated.com/articles/events-and-event-handlers/

References

1. Apelkrans, M., Håkansson, A.: Information coordination using Meta-Agents in Information Logistics Processes. In: 12th International Conference on Knowledge-Based and Intelligent Information & Engineering Systems, KES 2008, September 3-5, Zagreb, Croatia (2008)
2. Akl, S.G., Bernard, D.T., Jordan, R.J.: Design and implementation of a parallel tree search algorithm. IEEE transactions on Pattern Analysis and Machine Intelligence PAMI-4, 192–203 (1982)
3. Baudet, G.M.: The Design and Analysis of Algorithms for Asynchronous Multiprocessors. Ph D Thesis, Carnegie-Mellon University, Pittsburgh, PA (1978)
4. Bentley, J.L.: A parallel algorithm for constructing minimum spanning trees. Journal of the ACM 27(1), 51–59 (1980)
5. Berman, K., Paul, J.: Algorithms: Sequential, Parallel, and Distributed. Course Technology, 1st edn. (2004)
6. Brungger, A., Marzetta, J., Clausen, J., Perregaard, M.: Solving large-scale QAP problems in parallel with the search library zram. Journal of Parallel and Distributed Computing 50, 157–169 (1998)
7. Chakrabarti, S., van den Berg, M., Dom, B.: Focused crawling: A new approach to topic-specific web resource discovery. Computer Networks 31(11-16), 1623–1640 (1999)
8. Chelberg, D., Welch, L., Lakshmikumar, A., Gillen, M., Zhou, Q.: Meta-Reasoning For a Distributed Agent Architecture. In: System Theory, Proceedings of the 33rd Southeastern Symposium, pp. 377–381 (2001)
9. Cheong, F.-C.: Internet Agents: Spiders, Wanderers, Brokers, and Bots. New Riders Pub., Indianapolis (1996)
10. Cormen, T.H., Leiserson, C.E., Riverst, R.L., Stein, C.: Introduction to Algorithms, 2nd edn. MIT Press, Cambridge (2006)

11. Costantini, S.: Meta-reasoning: a survey. In: Kakas, A., Sadri, F. (eds.) Computational Logic: From Logic Programming into the Future: Special volume in honour of Bob Kowalski. Springer, Berlin (in print, 2002), `http://zen.ece.ohiou.edu/~robocup/papers/HTML/SSST/SSST.html`
12. Crowder, H., Johnson, E.L., Padberg, M.: Solving large-scale zero-one linear programming problem. Operations Research 2, 803–834 (1983)
13. Dijkstra, E.W.: A note on two problems in connexion with graphs. Numerische Mathematik 1, 269–271 (1959)
14. Dunkels, A., Schmidt, O., Voigt, T., Ali, M.: Protothreads: simplifying event-driven programming of memory-constrained embedded systems. In: Proceedings of the Fourth ACM Conference on Embedded Networked Sensor Systems, Boulder, Colorado, USA (2006)
15. Ferg, S.: Event-Driven Programming: Introduction, Tutorial, History (2006-02-08), `http://TutorialEventDrivenProgramming.sourceforge.net`
16. Finkel, R.A., Fishburn, J.P.: Parallelism in alpha-beta search. Artificial Intelligence 19, 89–106 (1982)
17. Gendron, B., Crainic, T.G.: Parallel Branch-And-Bound Algorithms: Survey and Synthesis Operations Research 42(6), 1042–1066 (1994)
18. Genesereth, M.R., Ketchpel, S.P.: Software Agents. Communication of the ACM 37(7), 18–21 (1994)
19. Gibbons, A.: Algorithmic graph theory. Cambridge University Press, Cambridge (1985)
20. Grama, A., Gupta, A., Karypis, G., Vipin, K.: Introduction to Parallel Computation, 2nd edn., Ch. 11. Pearon Addison Wesley, Reading (2003)
21. Hermans, B.: Intelligent Software Agents on the Internet, Ch. 1-3. First Monday 2(3) (1997)
22. Horowitz, E., Sahni, S., Rajasekaran, S.: Computer Algorithms C++: C++ and Pseudo code Versions, 2nd Rev. edn., December 15, p. 769. W. H. Freeman (1996)
23. Håkansson, A., Hartung, R.: Using Meta-Agents for Multi-Agents in Networks. In: Proceedings of the 2007 International Conference on Artificial Intelligence, ICAI 2007, vol. II, pp. 561–567. CSREA Press (2007)
24. Håkansson, A., Hartung, R.: Calculating optimal decision using Meta-level agents for Multi-Agents in Networks. In: Proceedings of Knowledge-Based and Intelligent Information & Engineering Systems, 11th International Conference. LNCS, pp. 180–188. Springer, Heidelberg (2007)
25. Håkansson, A., Hartung, R.: Autonomously creating a hierarchy of intelligent agents using clustering in a multi-agent system. In: The 2008 International Conference on Artificial Intelligence, ICAI 2008, Las Vegas, Nevada, USA, July 14–17 (submitted, 2008)
26. Håkansson, A., Hartung, R.L.: An approach to event-driven algorithm for intelligent agents in multi-agent systems. In: Nguyen, N.T., Jo, G.S., Howlett, R.J., Jain, L.C. (eds.) KES-AMSTA 2008. LNCS, vol. 4953, pp. 411–420. Springer, Heidelberg (2008)
27. Ishida, T.: Real-Time Search for Autonomous Agents and Multi-Agent Systems. Journal of Autonomous Agents and Multi-Agent Systems 1(2), 139–167 (1998)
28. Jade, Java Agent Development Framework (2008-03-15), `http://jade.tilab.com/`
29. Kanal, L.N., Kumar, V.: Search in Artificial Intelligence. Springer, New York (1988)
30. Kleinberg, J.: Authoritative sources in a hyperlinked environment. In: Proc. Ninth Ann. ACM-SIAM Symp. Discrete Algorithms, pp. 668–677. ACM Press, New York (1998)
31. Kruskal, J.B.: On the Shortest Spanning Subtree of a Graph and the Traveling Salesman Problem. Proceedings of the American Mathematical Society 7(1), 48–50 (1956)

32. Kumar, V., Rao, V.N.: Parallel depth-first search, part II: Analysis. International Journal of Parallel Programming 16(6), 501–519 (1987)
33. Kumar, V., Singh, V.: Scalability of parallel algorithms for the all-pairs shortest path problem. Journal of Parallel and Distributed Computing 13(2), 124–128 (1991)
34. Kumar, V., Ramesh, K., Rao, V.N.: Parallel best-first search of state-space graphs: A summary of results. In: Processings of the 1988 National Conference on Artificial Intelligence, pp. 122–126 (1988)
35. Luger, G.F.: Artificial Intelligence – Structures and strategies for Complex Problem Solving, 4th edn. Pearson Education, London (2002)
36. Maes, P., Guttman, R.H., Moukas, A.G.: Agents that buy and sell. Communications of the ACM archive 42(3), 81–92 (1999)
37. Menczer, F.: ARACHNID: Adaptive Retrieval Agents Choosing Heuristic Neighbourhoods for Information Discovery. In: Fisher, D. (ed.) Proceedings of the 14th International Conference on Machine Learning, ICML 1997 (1997)
38. Menczer, F., Pant, G., Srinivasan, P.: Topical Web Crawlers: Evaluating Adaptive Algorithms. ACM transactions on Internet Technology (TOIT) 4(4), 378–419 (2004)
39. Miller, D.L., Pekny, J.F.: The role of performance metrics for parallel mathematical programming algorithms. ORSA Journal on Computing 5(1) (1993)
40. Mohan, J.: Experience with two parallel programs solving the travelling salesman problem. In: Proceeding of the 1983 International Conference on Parallel Processing, pp. 191–193 (1983)
41. Mornien, B., Vornberger, O.: Parallel processing of combinatorial search trees. In: Proceedings of International Workshop on Parallel Algorithms and Architectures (1987)
42. Nisan, N.: Algorithms for Selfish Agents – Mechanism Design for Distributed Computation. In: Proceedings of Annual Symposium on Theoretical Aspects of Computer Science Trier, pp. 1–15 (1999)
43. Page, L.; Brin, S.; Motwani, R.; Winograd, T.: The PageRank Citation Ranking: Bringing Order to the Web (1999) (2008-03-09), `http://dbpubs.stanford.edu/pub/1999-66`
44. Paige, R.C., Kruskal, C.P.: Parallel algorithms for shortest path problems. In: Proceedings of 1989 International Conference of Parallel Processing, pp. 14–19 (1989)
45. Pearl, J.: Heuristics-Intelligent Search Strategies for Computer Problem Solving. Addison-Wesley, Reading (1984)
46. Peng, C.-H., Wang, B.-F., Wang, J.-S.: Recognizing unordered depth-first search of an undirected graph in parallel. IEEE Transactions on Parallel and Distributed Systems 11(6), 559–570 (2000)
47. Pfeifer, A., Ururahy, C., Rodriguez, N., Ierusalimschy, R.: Event-Driven Programming for Distributed Multimedia Applications. In: Distributed Computing Systems Workshops Proceedings. 22nd International Conference on Distributed Computing Systems, pp. 583–584 (2002)
48. Prim, R.C.: Shortest connection networks and some generalizations. Bell System Technical Journal 36, 1389–1401 (1957)
49. Rippel, E., Bar-Gill, A., Shimkin, N.: Fast Graph-Search Algorithms for General-Aviation Flight Trajectory Generation. Journal of guidance control and dynamics, American inst of aeronautics and astronautics 28(4), 801–811 (2005)
50. Russell, S., Norvig, P.: Artificial Intelligence: A Modern Approach. Prentice-Hall, Inc., Englewood Cliffs (1995)
51. Sedgewick, R.: Algorithms in Java, Part 5: Graph Algorithms, 3rd edn. Addison-Wesley Longman Publishing Co, Inc. (2003)

52. Skiena, S.: The Algorithm Design Manual, 1st edn. Springer, Heidelberg (1998)
53. Sauer, J., Appelrath, H.-J.: Scheduling the Supply Chain by Teams of Agents. In: The 36th Annual Hawaii International Conference on System Sciences, 81 (2003)
54. Winston, P.H.: Artificial Intelligence, 3rd edn. Addison-Wesley, Reading (1992)
55. Wooldridge, M.: An Introduction to MultiAgent Systems. John Wiley & Sons Ltd, Chichester (2002)
56. Wooldridge, M., Jennings, N.R.: Intelligent agents: Theory and practice. Knowledge Engineering Review 10(2) (1995)

8

A Generic Mobile Agent Framework towards Ambient Intelligence

Yung-Chuan Lee, Elham S. Khorasani, Shahram Rahimi, and Sujatha Nulu

Southern Illinois University
Carbondale, IL 62901 USA
{ylee,elhams,rahimi,snulu}@cs.siu.edu

Abstract. Recent advances of computing and networking technology have shifted computing convention from stationary to mobile. A forecast conducted by Gartner indicates that global mobile handset sales will increase exponentially in 2008 by 10% from 1.3 billion units in 2007 [13]. Furthermore, the popularity of wireless networking topology including both Wi-Fi access point and cellular mobile telecommunications enables users with constant access to online connection and further intensifies the demand of mobile devices. This provides the fundamental elements for creating ubiquitous environments of computing, networking, and interfacing that is both aware of and reactive to the presence of people. Such an environment is defined as Ambient Intelligence (AmI). Existing approaches that attempt to understand AmI environment mainly focus on how to seamlessly integrate hardware, i.e. mobile device and sensors, into human society and intelligently provide personalized knowledge and services. This has, however, left many essential issues unanswered, especially in regards to the integrity and performance of such dynamically distributed environments. In this chapter, we formulate a generic framework in which an AmI environment is generalized to consist only of users with devices, hosts where services are provided, and directory servers that act as information desks to users and hosts. In the proposed framework, the mobile agent notion is utilized to provide autonomous reasoning, learning, mobility, and collaboration features to construct AmI systems. Performance issues of load balancing and communication overhead in such a framework are then examined and analyzed against existing AmI techniques. Because of the dynamicity posed by AmI environments and the complexity of migrations and communications among agents, hosts and directory servers, it is necessary to provide a mechanism to warrant the accuracy and enhance the reliability of such an environment. Software verification employs formal methods of mathematically provable formulations to perform program analysis and model checking. Thus, general formulation using π-calculus is included to model the proposed approach to provide verification mechanism. In conclusion, the implementation of the proposed framework simulation using NetLogo is currently ongoing to visually demonstrate its feasibility. Some implementation details are included to stimulate comprehension of this framework.

8.1 Introduction

Ambient Intelligence (AmI) is defined as a ubiquitous environment of computing, networking, and interfacing that is aware of and reactive to the presence of people. It provides personalized knowledge and services to each individual by intelligently interacting with the environment and the individuals [6][7]. With the advancement of

L.C. Jain, N.T. Nguyen (Eds.): Knowl. Proc. & Dec. Mak. in Agent-Based Sys., SCI 170, pp. 175–192.
springerlink.com

mobile methodology, the technology disappears into our surroundings and only the user interface remains observable. In an ambient intelligence environment, people are supported in carrying out their activities in an easy and natural way, using intelligence that is embedded in the environment. By utilizing sensors and other small devices, AmI enabled systems can retrieve the profile of a user and provide relevant information and services to her autonomously, and even further, intelligently learn from the interactions with the user to refine her profile. Moreover, users in AmI environments may also manually request for information and services they desired.

Recent technology advances have made this vision one step closer to reality. Moore's Law states that the number of transistors placed on an integrated circuit can be inexpensively doubled every two years. Although researchers believe that Moore's Law may not continue to be followed as the physical limitation of a transistor approaches, the Law can be extended beyond the number of transistors to the design of transistors or placement of transistors [5]. Thus, integrated circuits have improved exponentially in recent years with smaller size, more functionality, and less energy consumption. This extends mobile devices such as cell phones, personal media players (PMP), mobile Internet devices (MID), ultra portable PCs (UMPC) and personal digital assistants (PDA) from single applications to multimodal tools including media playback capability, wireless networking, office/Internet productivities, gaming, or location awareness with GPS capability.

A forecast conducted by Gartner suggests that mobile devices can potentially reach a 10% increase in 2008 from 1.3 billion units in 2007 [13]. Furthermore, the popularity of wireless networking topology, including both Wi-Fi access point and cellular mobile telecommunications, provides users with constant access to online connection and further intensifies the demand of mobile devices. Finally, LiMo Foundation, founded from the beginning of 2007 and supported by all major mobile technology leaders, has committed to establishing a hardware-independent and open Linux-based platform that enables developers to contribute their innovations [10]. This provides the fundamental elements for creating AmI environments of computing, networking, and interfacing that is aware of and reactive to the presence of people.

The AmI environment presents a mixture of autonomy, learning, parallel and distributed computing domains [15]. Both traditional client-server approaches and mobile agent technology are capable of being applied toward the implementation of AmI environments; however, the traditional client-server approach suffers from message-passing methodology in which a network connection is mostly required through out the requested session. Multi-agent systems (MAS), on the other hand, utilizes a deploy-collect methodology in which the connection can be terminated after the deployment of mobile agents and shares many of its characteristics with AmI environments [2]. MAS facilitates the design and development of AmI environments by providing features such as autonomous reasoning, learning, mobility, and collaboration among others. Agents can migrate into different hosts to perform computation, while communicating with other agents, and then carry the results back to its origin. Many researchers also share the same belief that MAS is the Holy Grail to completely revealing the promising future of AmI environments [4][14][15][16][19].

While several MAS-based AmI architectures have been proposed, none have provided any mechanisms to appropriately evaluate the integrity and accuracy of its model. Because of the dynamicity posed by AmI environment and the complexity of the criteria for migrations and communications among agents and hosts, integrity verification becomes essential to maintaining the AmI system's reliability and scalability.

Software verification employs formal methods such as π-calculus [11][12] that are mathematically-provable formulations to perform program analysis and model checking. It provides a mechanism to warrant the correctness of a program and enhances the reliability at each stage of the software life cycle [8][18]. With the help of formal methods, a dynamic and complex system can be mathematically formulated and can then be analyzed to verify its correctness if necessary. A concise formulation with π-calculus is included in this chapter to establish the preliminary work on incorporating a verification procedure as part of the proposed framework.

In summary, this chapter presents a framework that utilizes mobile agents for ambient intelligence in a distributed ubiquitous environment to provide users with personalized knowledge and intelligent interactions, as well as to sustain expeditious performance under dynamic resource demands. Furthermore, general formulations using π-calculus is included to model the proposed system and to construct the foundation for future implementation of system verification. Some important features such as integrity verification on a given system, communication verification among agents and performance verification of a giving scenario will be included in the initial implementation. Some of the implementation details with NetLogo will be included at the end of this chapter for reference.

8.1.1 State-of-the-Art Approaches

The AmI domain has been gaining more and more attentions recently because of the advancement of its technology; therefore, some research studies related to the utilization of mobile agents in ambient intelligent environment will be discussed here. In creating location-aware services to provide personalized information to users, Satoh employed RFID technology and software agents in his approach [16]. Each user is assumed to carry a RFID tag and can be uniquely identified by RFID readers in each equipped location server. According to the user profile, the location server then assigns the user to a host within its coverage area. Then a mobile agent is spawned to assist the user and moves from one host to another to "follow" the user.

A similar approach utilizing RFID technology was developed by the AMILAB research group to apply software agents in an attempt to create an AmI manufacturing environment [14]. In this approach, AMICO architecture was formulated to interact intelligently with users to provide user-specified context information and machine functionalities. Users are identified through fingerprint sensors and location-awareness is supported through RFID tags and sensor network. Instrumental and data devices, voice recognition and wearable computing provide context-awareness to interact with users. Various specific types of agent are introduced in AMICO to interact with users and evolve through those interactions.

Yong and colleagues proposed a context-aware AmI application system based on a multi-agent architecture in which a smart agent introduced to perceive users' needs and coordinate aware-agent and executive-agent groups [19]. The aware-agent group contains different agents to recognize and locate outside objects or information that are then imported to the smart agent to analysis. The executive-agent group provides services and equipment controls according to requests from the smart agent. Thus, the smart agent is the center of the approach that acts as a human brain to process input information and send appropriate commands to execute.

In addition, several approaches have focused on facilitating people in daily tasks. Kidd and colleagues presented "the aware home project" to study how ubiquitous computing could assist people in daily life [9]. Because of the different characteristics of office and home environments, where activities in office environments are more goal-oriented while in home environments are more flexible, the project concentrated on developing a methodology that discovers the useful applications from the latest advances for each application.

Hagras and colleagues implemented an ambient-intelligence environment, iDorm, using embedded sensors, actuators, and software agents [4]. Based on user preferences, the embedded agents of iDorm proactively and seamlessly adjust the environment. Users can interact with the agents in the embedded controller, robots or mobile devices to control the environment. The system evolves from those interactions to provide a more precise user-friendly living environment.

The current state-of-the-art approaches are domain-specific and cannot be flexibly applied to other domains. Moreover, none of the mentioned approaches have addressed the performance of the system as well as load balancing among agencies in an AmI environment. Because of the autonomy of the agents and the dynamicity of the environment, any approach should take the performance criteria into consideration to provide users with responsive services. Furthermore, the complexity of such architectures highlights the importance of rigorous verification of the integrity of the system. Hence, we propose a verifiable generic agent-based AmI system with a particular focus on optimized communication costs and load balancing among agents and agencies.

8.1.2 π-Calculus Overview

Since the proposed system is modeled in π-Calculus, this section provides a brief overview to the main entities of π-Calculus. The interested reader is referred to [11][12] for more information on pi-calculus syntax and reduction rules.

In π-Calculus, two main entities are specified: "names" and "processes" (or "agents"). "Name" is defined as a channel or a value that can be transferred by a channel. We follow the naming rule and syntax in [11] in which *u, v, w, x, y, z* range over names and *A, B, C, ...* range over process (agent) identifiers.

The syntax of process is defined as follows:

$$P ::= 0 \mid \Sigma_{i \in I} \lambda_i.P_i \mid \overline{y}x.P \mid \overline{y}(x).P \mid \overline{x}(K) \mid x(U) \mid y(x).P \mid \tau.P \mid P_1 \mid P_2 \mid P_1 + P_2 \mid (x)P \mid [x = y]P \mid A(y_1,\ldots, y_n) \mid !P . \tag{8.1}$$

- 0: Agent P does not do anything.
- $\Sigma_{i \in I} \lambda_i.P_i$: Agent P will behave as either one of $\lambda_i.P_i$ where $i \in I$, but not more than one, and then behaves like P_i. If $I = \Phi$, P actually behaves like 0. Here λ_I denotes any actions that could take place in P (such as τ, $\overline{y}(x)$, and so on).
- $\overline{y}x.P$: Agent P sends free name x out along channel y and then behaves like P.
- $\overline{y}(x).P$: Agent P sends bound name x out along channel y and then behaves like P.

- $y(x).P$: Agent P receives name x out along channel y and then behaves like P.
- $\tau.P$: Agent P performs the silent action τ and then behaves like P (τ is a silent prefix).
- $P_1 \mid P_2$: Agent P has P_1 and P_2 executing in parallel independently or cooperatively. For instance, if $P_1 = \tau_1.P_1$ and $P_2 = \tau_2.P_2$, then P_1 and P_2 may behave independently. Otherwise, if $P_1 = \overline{y}(x).P_1$ and $P_2 = y(x).P_2$, then P_1 will send x tp P_2 through channel y.
- $P_1 + P_2$: Agent P will exercise either P_1 or P_2 but not both.
- $(x)P$: Agent P does not change except for that x in P becomes private and restricted to P. That means any outside communication through channel x will be prohibited.
- $[x = y]P$: If $x = y$, this agent behaves like P, otherwise like 0
- $A(y_1,..., y_n)$: This is an agent identifier in which $y_1,..., y_n$ are free names occurring in agent P.
- $!P$: This is a replication and can be thought of as an infinite composition $P|P|P|...$, i.e. $P = P|!P$.

In addition to the basic π-Calculus, higher-order π-Calculus has the ability to send and receive processes (agents). So in higher-order π-Calculus, $\overline{x}(K)$ means "send a name or process K through channel x" and $x(U)$ means "receive name or process U through channel x". Based on the above syntax, different components of the proposed framework are defined.

8.2 Proposed Framework

Our framework for ambient environments allows users to obtain their personalized information such as their profile, preferences, interests and habits, while engaging in minimum interactions with the environment. As the user moves, this information will be available to him/her at the new location. The framework presented here aims to be flexible, high performing and easy to implement.

Fig. 8.1 illustrates the overview of the framework. The system is formed from multiple geographically distributed environments. Each environment may provide different services to users such as banking, shopping, etc. Each environment can be generalized with three main components: (1) mobile devices, (2) hosts or agencies and (3) directory service centers (*DSC*). Different environments are connected through their directory service centers. In the following sections, we describe each of the elements in Fig. 8.1 in more detail. As mentioned before, we model our ambient framework using higher order π-calculus in order to provide a mathematical infrastructure for evaluation and verification of the system. More specifically, an environment is a cluster consisting of one *DSC*, multiple mobile devices and multiple hosts, and can be formulated as follows:

$$ENVIRONMENT = DSC \mid MD_1 \mid ... \mid MD_n \mid HOST_1 \mid ... \mid HOST_k . \quad (8.2)$$

Where *DSC* is the Directory Service Center, MD_1 ... MD_n represents mobile devices and $HOST_1$ ... $HOST_k$ indicates host/agencies in the environment.

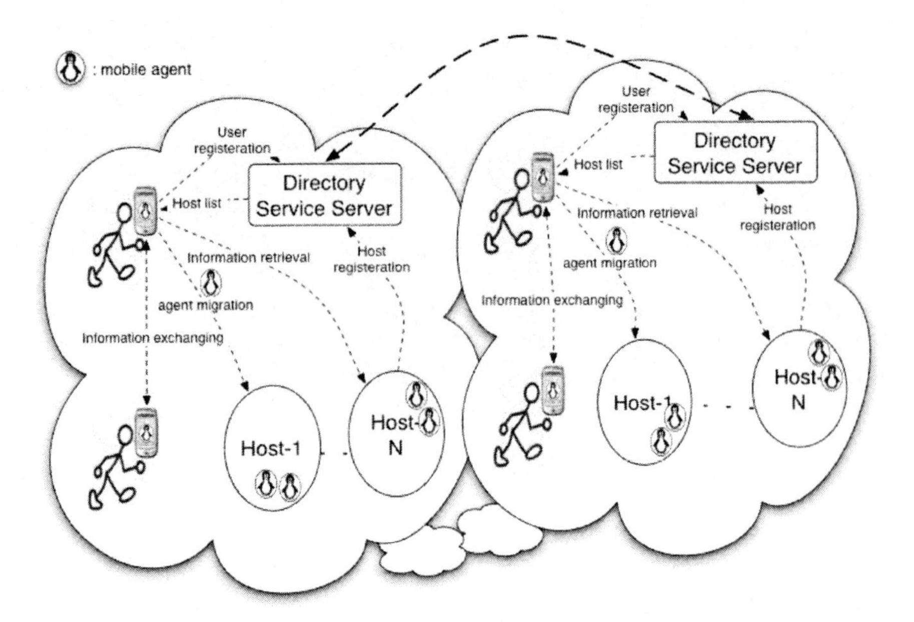

Fig. 8.1. This is an overview of the proposed framework. Some possible communication actions among mobile devices, hosts, and directory service servers are shown in dashed lines.

8.2.1 Mobile Devices

For simplicity, we assumed that each user carries at least one computing device such as tablet PCs, PDAs, cell phones, notebooks or even wearable computing devices. The user carries these portable devices to store his/her personalized information as well as to communicate with the environment over the wireless media. We refer to these portable devices as mobile devices. These devices are dedicated to a single person; therefore, the user is responsible for the security of his/her personalized information stored in his/her mobile device. Due to the limited capabilities of the portable devices in terms of CPU power, amount of memory and input/output facilities, mobile agent technology is employed. The mobile agent can merge into a suitable host (also called agency) in an environment to perform various services and computation for its owner. Additionally, it can use the environmental information which is provided by the agency.

Fig. 8.2 depicts the internal structure of an adequate mobile device. It consists of a registration module, administrative module, learning module, load balancing module and mobile device database. A mobile device (MD_i) consists of a registration module (MD_REG_i), administration module ($ADMIN_i$), load-balancing module (LB_i), learning module ($LEARN_i$) and agent database (MD_DB_i) and can be formulated as follows:

$$MD_i = MD\_REG_i \mid ADMIN_i \mid LB_i \mid LEARN_i \mid MD\_DB_i \,. \tag{8.3}$$

The registration module (MD_REG_i) is responsible for registering the mobile device to the new environment as the user goes from one environment to another.

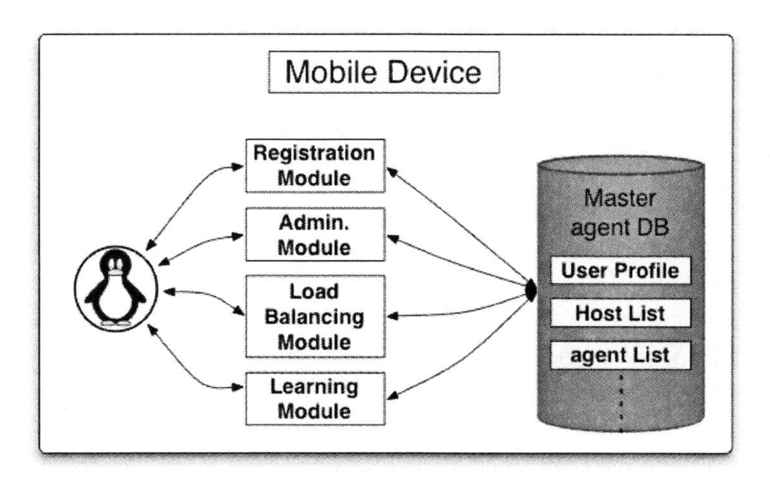

Fig. 8.2. This is the fundamental structure of a mobile device in the proposed framework. A master agent resides in the device and executes different modules to assist its user.

Whenever a user comes to a new environment, her mobile device sends its ID and the physical network adapter address to the *DSC* of the current environment. The *DSC* then decides whether to accept or reject the mobile device registration request due to its security considerations. The module can then be represented in π-calculus as follows:

$$MD\_REG_i = \nu(x, id)(\ \overline{y}(x)\ \ \overline{x}(id)).0\ . \tag{8.4}$$

Where y is a communication channel between MD_REG_i and *DSC.*

The mobile device acts as a master agent. It can create numerous mobile agents and dispatches them to different hosts. The administrative module ($ADMIN_i$) is responsible for creating and keeping track of the mobile agents that belong to the mobile device and reside in different hosts or agencies. This module can be modeled with π-calculus as follows, in which each process is further explained in details:

$$ADMIN_i = MERGE_i \mid TERMINATE_i \mid INQUIRY_i \mid RETRIEVE_i \mid UPDATE_i\ . \tag{8.5}$$

- *MERGE*: This process creates a mobile agent and sends its requirements to *DSC* through channel a. These requirements specify the agency capabilities that are needed by the mobile agent to perform its services. If the request is approved, *DSC* will send back a list of candidate hosts ($h_1, h_2, \ldots, h_n$) that fulfill the requirements. This list contains the agency's network address and other necessary information. The *MERGE* process then sends the list of candidate hosts to the load-balancing module (LB_i) through channel y and receives the address of the most appropriate host (h_{opt}) to which the mobile agent immigrates.

$$MERGE_i = \nu(AGENT, req, c)(\tau_{create}.\ \ \overline{a}(c).\ \ \overline{c}(req).c(h_1, h_2,\ldots, h_n).\ \ \overline{y}(c).\ \ \overline{c}(h_1, h_2,\ldots, h_n).\ c(h_{opt}).\ \ \overline{h_{opt}}(AGENT)).\ MERGE_i\ . \tag{8.6}$$

Where τ_{create} is the internal action for creating *AGENT*; *req* is the host capabilities needed by *AGENT* that is sent to *DSC* through the private channel c. $h_1, h_2,\ldots, h_n$ is the list of the hosts received from *DSC* and h_{opt} is the preferred host computed by load-balancing module for the migrating agent.

- $RETRIEVE_i$: This process retrieves the mobile agent from the environment to the mobile device. This includes retrieving the current address of the mobile agent from the database, sending a *ret* signal to it, and receiving the agent along a private channel.

$$RETRIEVE_i = \nu(c)(\ \overline{h_i}(c).\ \ \overline{c}ret.\ c(AGENT)).\ RETRIEVE_i\ . \tag{8.7}$$

Where h_i *is the current address of the agent.*

- $TERMINATE_i$: This process sends a termination signal, *term*, to the mobile agent to cease it. After sending the termination signal, the record of this mobile agent should be deleted from the mobile device database.
- $INQUIRY_i$: This process enquiries the mobile agent about the completeness of its current task by sending an *inq* signal.

$$\begin{aligned} TERMINATE_i &= \ \overline{b_{ij}}term.\ TERMINATE_i\ . \\ INQUIRY_i &= \ \overline{b_{ij}}ing.\ m_i(res).\ INQUIRY_i\ . \\ AGENT_{ij} &= b_{ij}(m).\ ([m = term]\ \tau_{term}.0\ |[m = ing]\ \ \overline{b_{ij}}(res).\ AGENT_{ij})\ . \end{aligned} \tag{8.8}$$

Where b_{ij} is the channel of communication between the mobile device (MD_i) and the $AGENT_{ij}$, τ_{term} is the internal action for terminating the agent, and *res* is the result of inquiry.

- $UPDATE_i$: This process updates agents' lists in the database of the mobile device. For each mobile agent that belongs to the mobile device, there exists a record in the database that contains the mobile agent's ID and its current address. If the mobile agent migrates from one host to another, it will send its new address to the mobile device and the *UPDATE* process updates the record of this mobile agent in the database.

The load-balancing module (LB_i) is responsible for spreading the communication and computation loads among the hosts in the environment to get a close-optimal utilization, minimum computation and communication delay. Whenever the admin module creates a mobile agent, it forwards the list of the candidate hosts ($h_1, h_2,\ldots, h_n$) for that mobile agent to the load-balancing module through the channel (y). The load-balancing module then sends a message (m) to each host ($HOST_i$) through the channels (h_i) in the list to obtain the hosts' resource utilization and to determine the response time ($resp_i$) for each. It estimates the cost of migrating the agent to each of these hosts and provides the most underutilized one to the administrator module.

$$\begin{aligned} LB_i = (y(c).\ c(h_1,\ h_2,\ldots,\ h_n).\ (\ \overline{h_1}m.h_1(resp_1)\ |\ \ \overline{h_2}m.h_2(resp_2)\ |\ \ldots\ | \\ \overline{h_m}m.h_m(resp_m)).\ \tau_{opt}.\ \ \overline{c}(h_{opt})).\ LB_i\ . \end{aligned} \tag{8.9}$$

Where h_{opt} is the selected host for the agent's migration and τ_{opt} is the internal action for computing the underutilized host based on the response time and process load.

The user profile as well as hosts and agents' information are stored in the database of the mobile device. Users are responsible for updating their sensitive information explicitly, such as username/password, credit card information, etc., and are responsible for the security of this information. The other information of the user profile, such as preferences and habits, are automatically updated by the learning module based on the user's interactions with the environment. The learning module concentrates on the behavior of the user and employs computing with words and gesture recognition methodologies to achieve its objectives. Because the complexity of learning cannot be fully comprehended within a few paragraphs and the main purpose of this paper is to sketch the blueprint of the framework, details on the learning module will be delineated in an upcoming paper.

8.2.2 Host/Agency

The agencies provide the facilities for the mobile agents to be executed and to perform various services for their owners. Java Application Development Framework (JADE) [1] is the platform of our choice to provide the runtime environment for mobile agents and the agencies to execute. Fig. 8.3 illustrates the host's structure. It consists of a user-history database ($HOST\_DB_i$), registration module ($HOST\_REG_i$), information module ($INFORMATION_i$) and JADE system ($AGENT\_EXEC_i$) and is modeled as follows:

$$HOST_i = HOST\_REG_i \mid AGENT\_EXEC_i \mid INFORMATION_i \mid HOST\_DB_i \,. \tag{8.10}$$

The registration module ($HOST\_REG_i$) registers the host with the directory service center. It sends a registry request to the DSC that contains the agency's network address and its device profile.

$$HOST\_REG_i = \nu(d)\ \overline{h_i}(d).\ \ \overline{d}(prf)\,. \tag{8.11}$$

Where h_i is the channel between $HOST_i$ and DSC, and prf is the device profile.

The $AGENT\_EXEC_i$ process receives and executes the agents (τ_{exec}). It can also return the agent back to its mobile device upon receiving the ret signal.

$$AGENT\_EXEC_i = !(\ h_i(AGENT).\ \tau_{exec} \mid h_i(c).\ c(m).\ [m = ret]\ \overline{c}AGENT\,. \tag{8.12}$$

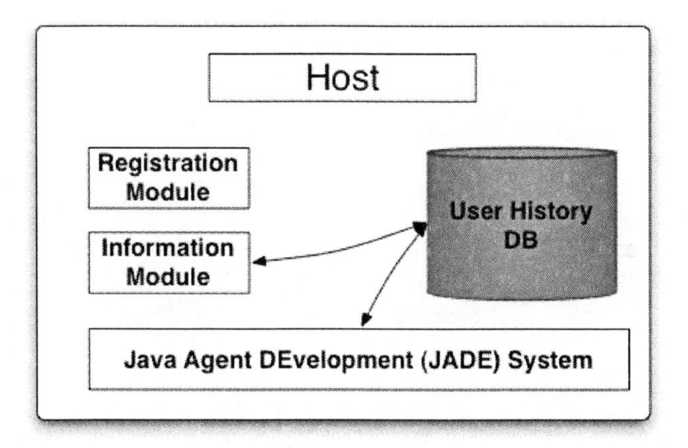

Fig. 8.3. This is the fundamental structure of a host in the proposed framework. The agent functionalities, such as migration, execution, or communication, are supported by JADE.

When a mobile device communicates with an agency, either by dispatching an agent to the agency or by remote method invocation (RMI), the agency records the user history in the user history DB. The information update module stores and updates the user's history according to the user activities. This history could include the user interaction with the agency, or in the case of agent migration, it could be the result of the execution of the mobile agent. After migration, the agency can request the mobile device for the summary of the agent execution results and update the user's history based on this information. The user can decide whether to provide this information to the agencies or not, by configuring its mobile device.

By employing the above approach, the environment could learn from the user activities, and in consequence, it can access its past information stored in the agencies the next time the user comes to the environment. The user should first authenticate itself to the host to be able to access its history.

8.2.3 Directory Service Center

The Directory Service Center (*DSC*) is responsible for managing mobile devices, host/agencies and their intercommunication as well as communicating with other *DSC*s in other environments. Fig. 8.4 depicts the *DSC* structure. It consists of a registration module (*DSC_REG*), communication module (*COMM*) and registration Database (*DSC_DB*), and is formulated as follows:

$$DSC = COMM \mid DSC\_REG \mid DSC\_DB. \tag{8.13}$$

The communication module (*COMM*) manages the communication among the mobile agents as well as the communication between the mobile device and the hosts. If the mobile device asks for a service that is not provided by a host in the same environment, then *DSC* broadcasts a message to *DSC*s in other environments to obtain the network address of a host that provides such a service. It then forwards the network address of that agency to the mobile device so that the mobile device can communicate directly with the agency.

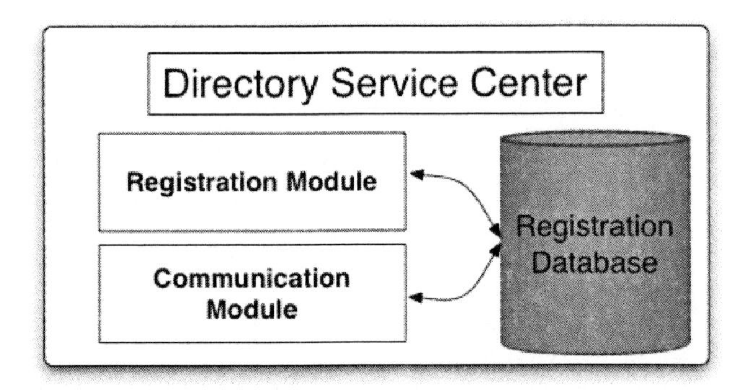

Fig. 8.4. This is the fundamental structure of a directory service center (*DSC*) in the proposed framework. A *DSC* can be best summarized as a Yellow Book, which provides the necessary information for contacting a service or person.

8.2.4 Communication among Mobile Agents

The mobile agents' communication is more complicated because they can proactively migrate from one host to another. Ensuring that a user can interact with the agent whenever necessary becomes crucial. One simple and straightforward solution is to generate a unique ID for each mobile agent per mobile device. Thus, each mobile agent's ID is a combination of its mobile device's ID with a unique number. A simple example is given in this section to demonstrate this concept.

The mobile agents' network addresses may change upon each migration. To address this issue, our proposed approach employs hierarchical methodology for mobile agent communication in our framework. The proposed framework benefits from a two-level hierarchy for tracking agent locations.

Fig. 8.5 demonstrates this hierarchy. Each mobile agent belongs to a mobile device. Upon migration, the agent notifies its mobile device regarding its new location. The directory service center stores the network address of all mobile devices that are registered. Each mobile agent is associated with a unique ID. This ID consists of two parts: (1) its master-agent-ID that shows to which mobile device this agent belongs, and (2) the agent-ID. As described earlier, the mobile device acts as a master agent and keeps track of the location of its agents, and the directory service center records the network addresses of all mobile devices in the environment.

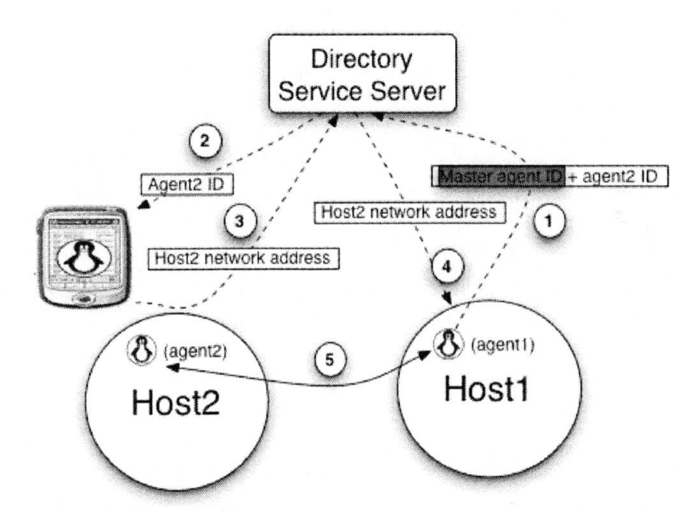

Fig. 8.5. This is an example to demonstrate communications among mobile agents. The numbered dashed lines show the order of communication among the *DSC*, the *Host* and the mobile device. The line 5 is the communication channel between *agent1* and *agent2*.

According to Fig. 8.5, where the numbers show the sequence of communications, when *agent1* needs to communicate with *agent2*, it sends *agent2*'s ID to *DSC* and asks for the location of *agnet2*. The directory server extracts the master-agent-ID part, sends a request to the mobile device associated with the master-agent-ID and asks for the location of *agent2*. As mentioned, the mobile device has a record of the actual location of the host in which agent2 resides. This location is then sent back to the

DSC by the master agent. Consequently, *DSC* forwards this address to *agent1*. Now *agent1* can communicate directly with *agent2*.

This communication scheme can be represented as follows:

$$COMM = DSC\_COMM \mid MD\_COMM_i \mid TAGENT\_COMM \mid SAGENT\_COMM \,. \tag{8.14}$$

$$DSC\_COMM = v(ma_1, ma_2, \ldots, ma_m)!(c(ma,sub).([ma = ma_1]\ \overline{d_1}sub.\ d_1(A) \mid [ma = ma_2]\ \overline{d_2}sub.\ d_2(A) \mid \ldots \mid [ma = ma_m]\ \overline{d_m}sub.\ d_m(A)).\ \overline{c}A).\ DSC\_COMM \,.$$

$$MD\_COMM_i = !(d_i(sub).\ \overline{d_i}(A)) \,.$$

$$SAGENT\_COMM = v(m)\ \overline{c}(ma,sub).\ c(A).\ !\ \overline{A}(m) \,.$$

$$TAGENT\_COMM = !A(m) \,.$$

Where *DSC_COMM, MD_COMM*$_i$*, TAGENT_COMM,* and *SAGENT_COMM* are the communication processes of *DSC*, master agent, target mobile agent, and source mobile agent, respectively. *ma* is the master-agent-ID, *sub* represents the mobile-agent-ID, and *A* is the address of the target agent.

8.3 Current Implementation

To examine our hypothesis, we are actively implementing the simulation system with NetLogo. NetLogo is a cross-platform and programmable multi-agent modeling environment for natural and social phenomena simulation [17]. Each Agent is created through turtle keyword in NetLogo and can follow instructions independently and simultaneously. In this section, some implementation details are explained with corresponding code sections.

The following defines *DSCs, hosts* and *users*, breeds that in turn can be used to initiate a set of instances in each breed, and each instance has its own ID variable. Instances of users, breed has two more variables, *NStatus* and *visited*, to keep track of current network status and list of visited *DSC*s. Furthermore, in the setup procedure (*to setup*), each breed is first assigned with different shapes to represent itself in the NetLogo world and is then created with the *create* keyword where number of instances for each breed is defined in the GUI interface that can be referred to in Fig. 8.6. Other than setting up a unique ID when initiating each instance, each host needs to be carefully positioned inside the "wireless signal" coverage (*DSC_radio_coverage*) in order to announce services to the *DSC*. The coverage area is further colored grey to better distinguish from the NetLogo world background.

After the setup, NetLogo will execute the go procedure (*to go*) repeatedly in an infinite loop fashion when the *go* button is clicked; however, the execution can be paused by clicking the go button again. Thus, the go procedure is more like the main function under C/C++. A section of codes that enabled the user agents to explore the environment is provided below:

```
Breed [users user]
Breed [hosts host]
Breed [DSCs DSC]
```

```
turtle-own [ID]
users-own [NStatus visited]
   .
   .
to setup
     .
     .
  set-default-shape DSCs "face happy"
  set-default-shape users "person"
  set-default-shape hosts "house"

  create DSCs number_of_DSCs
    [setxy random-xcor random-ycor
     ask DSCs
       [set ID who + (random 9007199254740992)]
     ask patches in-radius DSC_radio_coverage
       [set pcolor radioCoverage]
    ]
  create hosts number_of_hosts
    [let tempDSC one-of DSCs
     set xcor ([xcor] of tempDSC) + (random
DSC_radio_coverage)
     set ycor ([ycor] of tempDSC) + (random
DSC_radio_coverage)
     ask hosts
       [set ID who + (random 9007199254740992)]
     ask patches in-radius host_radio_coverage
       [set pcolor radioCoverage]
    ]
  create users number_of_users
    [setxy random-xcor random-ycor
     ask users
       [set ID who + (random 9007199254740992)
           .
           .
       ]
    ]
     .
     .
end

to go
  ask users
    [let rand random 4
     if rand = 0 [fd 2.1]
     if rand = 1 [bk 2.1]
     if rand = 2 [rt 90]
     if rand = 3 [lt 90]
     ifelse [pcolor] of patch-here = radioCoverage
       [set [NStatus] of self 1
        set [visited] of self 1
```

```
            ]
            [set [NStatus] of self 0]
               .
               .
               .
        ]
        .
        .
        .
  end
```

Due to the simplicity of NetLogo, we could not find any library or functionality to simulate the communications among the agents (i.e. *users*, *hosts*, and *DSCs*) without

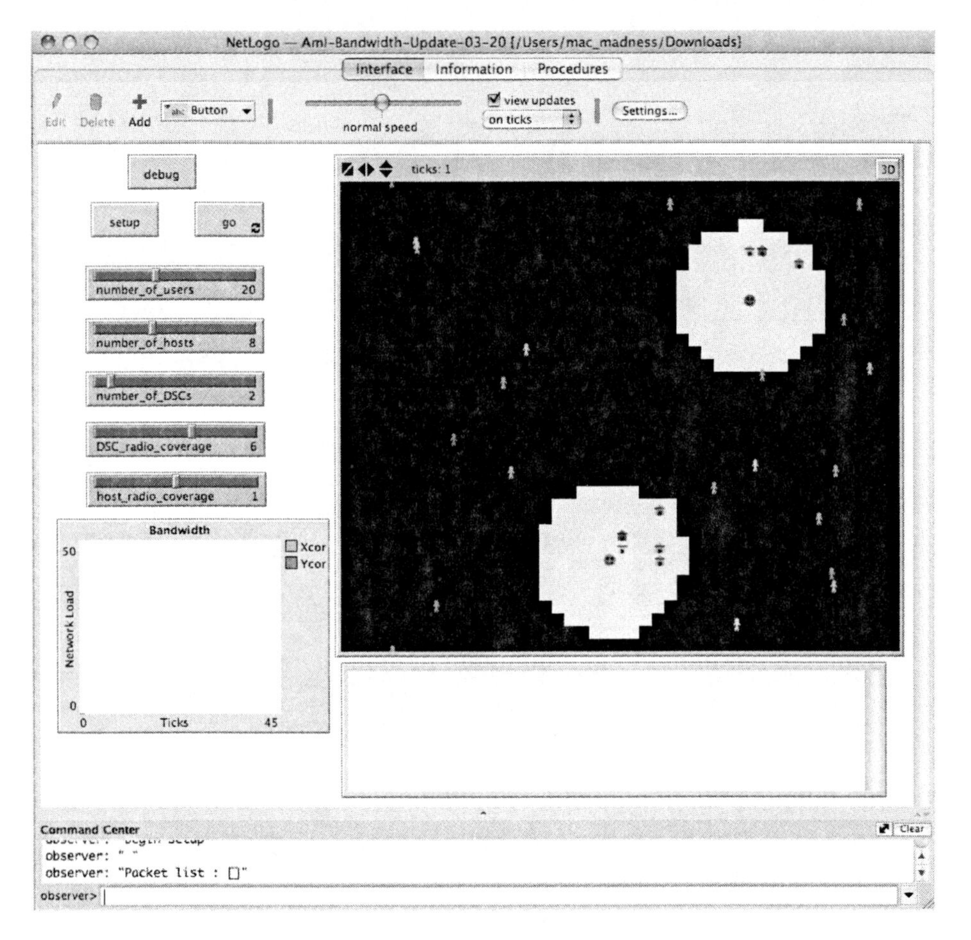

Fig. 8.6. This is a screenshot of current implementation in NetLogo. Setup button can initiate the Some possible communication actions among mobile devices, hosts, and directory service servers are shown in dashed lines.

setting up HubNet. HubNet is a powerful module for constructing a multi-devices environment to allow people to participant as agents through computing devices; however, it is too overwhelming to utilize HubNet for the initial implementation. Therefore, a simulated networking mechanism is currently under development to facilitate the observation among agents to study the performance.

8.4 Conclusion and Future Direction

By utilizing the proposed framework, we believe that a more intelligent and responsive interface between users and computing infrastructure can be implemented in a seamless environment. The framework employs mobile agents to eliminate hard-coded and fixed features of computing. This achieves an intelligence capability in which an intelligent infrastructure is more dependable, manageable, adaptable and affordable [3]. Mobile agents are deployed by demand and "live" in computing infrastructures to interact with users intelligently and intuitively, to provide personalized information, and to assist users on both daily and specific tasks. The proposed framework is designed to have performance advantages over the current state-of-the-art approaches when the number of the users increases.

In the near future, a simulated system with mentioned characteristics will be implemented for the framework to demonstrate the practicality and advantages of the proposed approach. π-Calculus will be employed to evaluate the system's integrity, validation, and performance.

The authentication and other security issues need to be studied further to validate the reliability and stability of our approach. Currently, we assume that the authentication method uses public key cryptography and the public key of each user and hosts are repopulated in the directory service centers when a regular user or host administrator physically registers to join the system. Furthermore, a pluggable module protocol will be study in the next revision of proposed framework to not only ease the addition of new hardware to existing AmI implementation, but to the extent of accomplishing multimodal interface by loading appropriate modules into a single device to dynamically support individuals with varying tasks. Finally, to find the optimal network utilization in our framework, communication protocols among agents and migrations of agents among agencies will be further examined.

References

1. Bellifemine, F., Caire, G., Trucco, T., Rimassa, G.: Jade Programmer's Guide, `http://jade.tilab.com/doc/programmersguide.pdf`
2. Braun, P., Rossak, W.R.: Mobile Agents: Basic Concepts, Mobility Models, and the Tracy Toolkit. Morgan Kaufmann, San Fransisco (2004)
3. Ferguson, R., Charrington, S.: Building an Intelligent IT Infrastructure. Intelligent Enterprise 7(18), 18 (2004)
4. Hagras, H., Callaghan, V., Colley, M., Clarke, G., Pounds-Cornish, A., Duman, H.: Creating an Ambient-Intelligence Environment Using Embedded Agents. IEEE Intelligent Systems 19, 12–20 (2004)

5. Hiremane, R.: From Moore's Law to Intel Innovation-Prediction to Reality. Technology@Intel Magazine (April 2005)
6. ISTAG: Scenarios for Ambient Intelligence in 2010, `http://www.cordis.lu/istag.htm`
7. ISTAG: Ambient Intelligence: from vision to reality, `http://www.cordis.lu/istag.htm`
8. Jackson, M.: What Can We Expect from Program Verification? Computer 39(10), 65–71 (2006)
9. Kidd, C., Abowd, G., Atkeson, C., Essa, I., MacIntyre, B., Mynatt, E., Starner, T.: The Aware Home: A Living Laboratory for Ubiquitous Computing Research. In: The Proceedings of the Second International Workshop on Cooperative Buildings, pp. 191–198 (October 1999)
10. LiMo Foundation: Introduction, Overview & Market Positioning. Mobile World Congress (February 2008)
11. Milner, R., Parrow, J., Walker, D.: A Calculus of Mobile Processes - Part I. LFCS Report-89-85. University of Edinburgh (June 1989)
12. Milner, R., Parrow, J., Walker, D.: A Calculus of Mobile Processes - Part II. LFCS Report-89-86. University of Edinburgh (June 1989)
13. Parker, A.: Mobile phone players ring the changes. Financial Times (February 28, 2008)
14. Perez, M.A., Susperregi, L., Maurtua, I., Ibarguren, A., Tekniker, F., Sierra, B.: Software Agents for Ambient Intelligence based Manufacturing. In: IEEE Workshop on Distributed Intelligent Systems: Collective Intelligence and Its Application, pp. 139–144 (2006)
15. Remagnino, P., Foresti, G.L.: Ambient Intelligence: A New Multidisciplinary Paradigm. IEEE Transactions on Systems, Man and Cybermetrics 35, 1–6 (2005)
16. Satoh, I.: Software Agents for Ambient Intelligence. In: IEEE International Conference on Systems man and Cybernetics, pp. 1147–1152 (2004)
17. Wilensky, U.: NetLogo. Center for Connected Learning and Computer-Based Modeling, Northwestern University, Evanston, IL (1999), `http://ccl.northwestern.edu/netlogo/`
18. Woodcock, J.: First Steps in the Verified Software Grand Challenge. Computer 39(10), 57–64 (2006)
19. Zhang, Y., Hou, Y., Huang, Z., Li, H., Chen, R.: A context- aware AmI system based on MAS model. In: IEEE Proceedings of the 2006 International Conference on Intelligent Information Hiding and Multimedia Signal Processing, pp. 703–706 (December 2006)

Resources

For more information regarding ambient intelligence, please refer to the following publications:

1. Abowd, G. D., Mynatt, E. D.: Charting past, present, and future research in ubiquitous computing. ACM Transactions Computer-Human Interaction, Vol. 7. 1 (March 2000), 29-58
2. Alexopoulos, D., Soldatos, J., Kormentzas, G., Skianis, C.: UbiXML: programmable management of ubiquitous computing resources. International Journal of Network Management, Vol. 17. 6 (November 2007) 415-435
3. Cabri, G., Ferrari, L., Leonardi, L., Zambonelli, F.: The LAICA project: supporting ambient intelligence via agents and ad-hoc middleware. Enabling Technologies: Infrastructure for Collaborative Enterprise. (June 2005) 39-44

4. Charif-Djebbar, Y., Sabouret, N.: An agent interaction protocol for ambient intelligence. IEEE Intelligent Environments, Vol. 1. (July 2006) 275-284
5. Dey, A. K., Hamid, R., Beckmann, C., Li, I., Hsu, D.: a CAPpella: programming by demonstration of context-aware applications. In Proceedings of the SIGCHI Conference on Human Factors in Computing Systems. (April 2004) 33-40
6. Dooley, J.; Callaghan, V., Hagras, H., Bull, P., Rohlfing, D.: Ambient intelligence - Knowledge representation, processing and distribution in intelligent inhabited environments. IEEE Intelligent Environments, Vol. 2. (July 2006) 51-59
7. Doorn, M., Loenen, E., Vries, A. P.: Deconstructing ambient intelligence into ambient narratives: the intelligent shop window. In Proceedings of the 1st international Conference on Ambient Media and Systems. (February 2008) 1-8
8. Eccles, D. W., Groth, P. T.: Wolves, football, and ambient computing: facilitating collaboration in problem solving systems through the study of human and animal groups. In Proceedings of the Third Nordic Conference on Human-Computer interaction, Vol. 82. (October 2004) 269-275
9. Grimm, R., Anderson, T., Bershad, B., Wetherall, D.: A system architecture for pervasive computing. In Proceedings of the 9th Workshop on ACM SIGOPS European Workshop: Beyond the Pc: New Challenges For the Operating System. (September 2000) 177-182
10. Grimm, R., Davis, J., Lemar, E., Macbeth, A., Swanson, S., Anderson, T., Bershad, B., Borriello, G., Gribble, S., Wetherall, D.: System support for pervasive applications. ACM Transactions Computer Systems, Vol. 22. 4 (November 2004) 421-486
11. Higel, S., O'Donnell, T., Wade, V.: Towards a natural interface to adaptive service composition. In Proceedings of the 1st international Symposium on information and Communication Technologies, Vol. 49. (September 2003) 169-174
12. Iliasov, A., Romanovsky, A., Budi Arief, Laibinis, L., Troubitsyna, E.: On Rigorous Design and Implementation of Fault Tolerant Ambient Systems. Object and Component-Oriented Real-Time Distributed Computing. (May 2007) 141-145
13. Ipiña, D., Mendonça, P., Hopper, A.: TRIP: A Low-Cost Vision-Based Location System for Ubiquitous Computing. Personal Ubiquitous Computing, Vol. 6. 3 (January 2002) 206-219
14. Loke, S. W., Krishnaswamy, S., Naing, T. T.: Service domains for ambient services: concept and experimentation. Mobile Networks and Applications, Vol. 10. 4 (August 2005) 395-404
15. Mamei, M., Zambonelli, F.: Pervasive pheromone-based interaction with RFID tags. ACM Transactions on Autonomous and Adaptive Systems, Vol. 2. 2 (June 2007) article 4
16. Mark W.: The Computer for the Twenty-First Century. Scientific American. (September 1991) 94-104
17. Mark W.: Some Computer Science Problems in Ubiquitous Computing. Communications of the ACM. (July 1993)
18. Mark W.: Hot Topics: Ubiquitous Computing. IEEE Computer. (October 1993)
19. Merdes, M., Malaka, R., Suliman, D., Paech, B., Brenner, D., Atkinson, C.: Ubiquitous RATs: how resource-aware run-time tests can improve ubiquitous software systems. In Proceedings of the 6th international Workshop on Software Engineering and Middleware. (November 2006) 55-62
20. Moraitis, P., Spanoudakis, N.: Argumentation-Based Agent Interaction in an Ambient-Intelligence Context. IEEE Intelligent Systems, Vol. 22. 6 (November 2007) 84-93
21. Nehmer, J., Becker, M., Karshmer, A., Lamm, R.: Living assistance systems: an ambient intelligence approach. In Proceedings of the 28th international Conference on Software Engineering. (SMay 2006) 43-50

22. Nielson, H. R., Nielson, F.: Shape analysis for mobile ambients. In Proceedings of the 27th ACM SIGPLAN-SIGACT Symposium on Principles of Programming Languages. (January 2000) 142-154
23. Nygard, K. E., Xu, D., Pikalek, J., Lundell, M.: Multi-agent designs for ambient systems. In Proceedings of the 1st international Conference on Ambient Media and Systems. (February 2008) 1-6
24. Obermair, C., Ploderer, B., Reitberger, W., Tscheligi, M.: Cues in the environment: a design principle for ambient intelligence. In CHI '06 Extended Abstracts on Human Factors in Computing Systems. (April 2006) 1157-1162
25. Perez, M. A., Susperregi, L., Maurtua, I., Ibarguren, A., Sierra, B.: Software Agents for Ambient Intelligence based Manufacturing. Distributed Intelligent Systems: Collective Intelligence and Its Applications. (June 2006) 139-144
26. Ramos, C., Augusto, J. C., Shapiro, D.: Ambient Intelligence—the Next Step for Artificial Intelligence. IEEE Intelligent Systems, Vol. 23. 2 (March 2008) 15-18
27. Ren, K., Lou, W.: Privacy-enhanced, attack-resilient access control in pervasive computing environments with optional context authentication capability. Mobile Networks and Applications, Vol. 12. 1 (January 2007) 79-92
28. Satoh, I.: A location model for ambient intelligence. In Proceedings of the 2005 Joint Conference on Smart Objects and Ambient intelligence: innovative Context-Aware Services: Usages and Technologies, Vol. 121. (October 2005) 195-200
29. Scholz, J., Grigg, M., Prekop, P., Burnett, M.: Development of the software infrastructure for a ubiquitous computing environment: the DSTO iRoom. In Proceedings of the Australasian information Security Workshop Conference on ACSW Frontiers 2003, Vol. 21. (January 2003) 169-176
30. Sekiguchi, M., Naito, H., Ueda, A., Ozaki, T., Yamasawa, M.: "UBWALL", ubiquitous wall changes an ordinary wall into the smart ambience. In Proceedings of the 2005 Joint Conference on Smart Objects and Ambient intelligence: innovative Context-Aware Services: Usages and Technologies, Vol. 121. (October 2005) 47-50
31. Stokic, D., Neves-Silva, R., Marques, M., Reimer, P., Ibarbia, J. A.: Ambient Intelligence Based System for Life-cycle Management of Complex Manufacturing and Assembly Lines. Industrial Informatics, 2007 5th IEEE International Conference on, Vol. 2. (June 2007) 1197-1202
32. Sukthankar, R.: Towards Ambient Projection for Intelligent Environments. Computer Vision for Interactive and Intelligent Environment. (November 2005) 162-172
33. Weber, W.: Ambient intelligence: industrial research on a visionary concept. In Proceedings of the 2003 international Symposium on Low Power Electronics and Design. (August 2003) 247-251
34. Wedde, H. F., Lischka, M.: Role-based access control in ambient and remote space. In Proceedings of the Ninth ACM Symposium on Access Control Models and Technologies. (June 2004) 21-30

9
Developing Actionable Trading Strategies

Longbing Cao

Faculty of Information Technology
University of Technology, Sydney, Australia
lbcao@it.uts.edu.au

Abstract. Actionable trading strategies for trading agents determine the potential of the simulated models in real-life markets. The development of actionable strategies is a non-trivial task, which needs to consider real-life constraints and organizational factors in the market. In this paper, we first analyze such constraints on developing actionable trading strategies. Further we propose an actionable trading strategy development framework. These points are deployed into developing a series of actionable trading strategies through optimizing, enhancing, discovering and integrating actionable trading strategies. We demonstrate working case studies in market data. These approaches and their performance are evaluated from both technical and business perspectives. Actionable trading strategies have potential to supporting smart trading decision for brokerage firms and financial companies.

9.1 Introduction

In financial markets, traders always pursue profitable trading strategies (also called trading rules) to make good return on investment. For instance, an experienced trader may use an appropriate pairs trading strategy, namely taking a long position in the stock of one company and shorting the stock of another in the same sector. A long position reflects the view the stock price will rise; shorting reflects the opposite. This strategy may statistically increase profit while decrease risk compared to a naive strategy putting all money on one stock. In practice, an actionable strategy can not only maximize the profit or return, but also result in the proper management of risk and trading costs.

Artificial financial market [1,9,14,22] provides an economic, convenient and effective electronic marketplace (also called e-market) for the development and back-testing of actionable strategies taking by trading agents without losing a cent. A typical simulation drive is the Trading Agent Competition [11,22], for instance, research work on auction-oriented protocol and strategy design [15], bidding strategy [16], design tradeoffs [12], and multi-attribute dynamic pricing [13] of trading agents. However, existing trading agent research presents a prevailing atmosphere of academia. This is embodied in aspects such as artificial data, abstract trading strategy and market mechanism design, and simple evaluation metrics. In addition, little research has been done on strategy optimization in continuous e-markets, while which consist of our daily financial life [1,9].

L.C. Jain, N.T. Nguyen (Eds.): Knowl. Proc. & Dec. Mak. in Agent-Based Sys., SCI 170, pp. 193–215.
springerlink.com

The above atmosphere has led to a big gap between research and business expectation. As a result, the developed techniques are not necessarily of business interest or cannot support business decision-making. In fact, the development of actionable strategies is a non-trivial task due to domain knowledge, constraints and expectation in the market [5]. Very few studies on continuous e-markets have been conducted for actionable trading strategies in the above constrained practical scenarios [3,18,6]. Therefore, it is a very practical challenge and driving force to narrow down the gap towards workable trading strategies for action-taking to business advantage.

An *actionable* trading strategy can assist trading agents in determining *right actions at right time with right price and volume on right instruments to maximize the profit while minimize the risk* [7,8]. The development of actionable strategies targets an appropriate combination or optimization of relevant attributes such as target market, timing, actions, pricing, sizing and traded objects based on proper business and technical measurement. The above combination and optimization in actionable trading strategy development should consider certain market microstructure and dynamics, domain knowledge and justification, as well as investors' aims and expectation. These form the constrained environment in developing actionable strategies for trading agents.

In this paper, we discuss lessons learnt in actionable trading strategy development in continuous e-markets based on our years of research and practical development. The main contributions consist of (1) discussing real-life constraints that need to be cared in trading agent research, (2) proposing an actionable trading strategy framework, and (3) investigating a series of approaches to actionable strategy development.

We first identify some important institutional features and constraints on designing trading strategies for trading agents. For instance, varying combinations of organizational factors form different market microstructure. The basic ideas include that the development of actionable strategies in constrained scenarios is an iteratively in-depth strategy discovery and refinement process where the involvement of domain knowledge and the human-agent cooperation are essential. The involvement of domain experts and their knowledge can assist in understanding, analyzing and developing highly practical trading strategies. In-depth strategy discovery, refinement and parallel supports can effectively improve the actionable capabilities of an identified strategy.

Following the above ideas, we study a few effective techniques for developing actionable trading strategies in continuous e-market context. These include designing and discovering quality trading strategies, and enhancing the actionable performance of a trading strategy through analyzing its relationship with target stocks. In addition, parallel computing is also imposed on efficiently mining actionable strategies. All of these methods are simulated and back-tested in an agent service-based artificial financial market F-Trade [6] with online connection to multiple market data. The experiments show that the introduced techniques have the potential for improving the actionability of trading strategies when they are deployed into the real market.

9.2 What Is Actionable Trading Strategy

9.2.1 Trading Strategies

Intelligent agent technology is very useful and increasingly used for developing, backtesting and evaluating automated trading techniques and program trading strategies in e-market places [22] without market costs and risks before they are deployed into the business world [15]. In fact, with the involvement of business lines in agent-based computational finance and economics studies, trading agent has potential to be customized for financial market requirements. The idea is to extend and integrate the concepts of trading agents, agent-based financial and economic computation and data mining with trading strategy development in finance to design and discover appropriate trading strategies for trading agents. Classic agent intelligence such as autonomy, adaptation, collaboration and computation is also encouraged in aspects such as automated trading and trading agent collaboration for strategy integration. In this way, trading agents can dedicate to the development of financial trading strategies for market use.

A trading strategy indicates when a trading agent can take what trading actions under certain market situation. For instance, the following illustrates a general Moving Average (MA) based trading strategy.

EXAMPLE 9.1. (MA Trading Strategy). An MA trading strategy is based on the calculation of moving average of security prices over a specified period of time. Let n be the length (i.e. number of prices) of MA in a time period, $P_{I,i}$ (or for short P_i) be the price of an instrument I at the time of No. i ($i < n$) price occurrence. An MA at time t (which corresponds to No. n price) is calculated as $MA_t(n)$:

$$MA_t(n) = \frac{1}{n}\sum_{i=0}^{n-1} P_i. \tag{9.1}$$

A simple MA strategy is to compare the current price P_t of security I with its MA value: MA_t. Based on the conditions met, a MA strategy generates 'sell' (denoted by −1), 'buy' (denoted by 1) or 'hold' (denoted by 0) trading signal at time t. If P_t rises above $MA_t(n)$, a buy signal is triggered, the security is then bought and held until the price falls below MA, at which time a sell signal is generated and the security is sold. For any other cases, a hold signal is triggered.

The pseudo code of MA strategy for generating trading signal sequence S is represented as follows.

$$\begin{cases} S = 1: & \text{if } P_t > MA_t(n) \text{ and } P_i < MA_i(n) \\ & \forall\, i \in \{1, \cdots, n-1\} \\ S = -1: & \text{if } P_t < MA_t(n) \text{ and } P_i > MA_i(n) \\ & \forall\, i \in \{1, \cdots, n-1\} \\ S = 0: & \text{otherwise} \end{cases} \tag{9.2}$$

This strategy is usually not workable when it is employed into the real world. To satisfy the real-life needs, market organizational factors, domain knowledge, constraints, trader preference and business expectation are some key factors that must be involved in developing actionable trading strategies for workable trading agents.

9.2.2 Actionable Trading Strategies

As the above discussed, searching actionable trading strategies for trading agents is a process to identify trading patterns that can reflect the 'most appropriate' combination of purchase timing, position, pricing, sizing and objects to be traded under certain market situations and interest-driving forces [21]. To this end, trading agents may cooperate with each other to either search the 'optimal' solutions from a huge amount of searchable strategy space denoted by a trading pattern. In some other cases, they collaborate to synthesize multiple trading strategy fragments favored by individual agents into an integrative strategy satisfying general concerns of each agent as well as global expectation representing trader's interest.

In addition, data mining can play a critical role in actionable strategy searching and trading pattern identification. Data mining in finance has potential in identifying not only trading signals, but also patterns indicating either iterative or repeatable occurrences. Therefore, developing actionable trading strategies for trading agents [13, 17] is an interaction and collaboration process between agents and data mining [19]. The aim and objective of this process is to develop smart strategies for trading agents to take actions in the market that can satisfy trader's expectation under certain market environment.

DEFINITION 9.1 (Trading Strategy)**.** A trading strategy actually represents a set of individual instances , which is a tuple defined as follows.

$$\begin{aligned} \Omega &= \{s_1, s_2, \ldots, s_m\} \\ &= \{(t, b, p, v, i) | t \in T, b \in B, p \in P, v \in V, i \in I)\} \end{aligned} \tag{9.3}$$

s_1 to s_m are instantiated trading strategies. Each of them is represented by instantiated parameters of t, b, p, v and an instrument i to be traded. $T = \{t_1, t_2, \ldots, t_m\}$ is a set of appropriate time points when trading signals are triggered; B = {buy, sell, hold} is the set of possible behavior (i.e., trading actions) executed by trading agents. $P = \{p_1, p_2, \ldots, p_m\}$ and $V = \{v_1, v_2, \ldots, v_m\}$ are the sets of trading prices and volumes matching with corresponding trading times. $I = \{i_1, i_2, \ldots, i_m\}$ is a set of target instruments traded.

With the consideration of environment complexities and trader's favorite, the optimization of trading strategies is to search a combination set Ω' in the whole candidate set Ω, in order to achieve both user-preferred technical (*tech_int*()) and business-favored (*biz_int*()) interestingness metrics [15] in an 'optimal' or 'sub-optimal' manner. Here 'optimal' refers to the maximal/minimal (in some cases, smaller is better) values of technical and business interestingness metrics under certain market conditions and user preferences. In some situations, it is impossible or too costly to obtain 'optimal' results. For such cases, certain 'sub-optimal' results are also acceptable. In this case, the sub-set Ω' indicates 'appropriate' parameter combinations of trading strategies that can support trading agents to take actions to their owner's advantage. As a result, in some sense, trading strategy optimization is to extract actionable strategies with multiple attributes towards multi-objective optimization in constrained market environment.

DEFINITION 9.2 (Actionable Trading Strategy)**.** An actionable trading strategy set Ω' is to achieve the following objectives:

$$tech\_\text{int}() \rightarrow optimal\{tech\_\text{int}()\} \tag{9.4}$$

$$biz\_\text{int}() \rightarrow optimal\{biz\_\text{int}()\} \tag{9.5}$$

while satisfying the following conditions:

$$\begin{aligned} \Omega' &= \{w_1, w_2, \ldots, w_n\} \\ \Omega' &\subset \Omega \\ m &> n \end{aligned} \tag{9.6}$$

where w_i (i = 1, 2, . . . , n) is an instance of actionable trading strategies satisfying general *tech_int*() and *biz_int*() metrics.

The performance of actionable trading strategies should satisfy expected technical interestingness as well as business expectations under multi-attribute constraints. In the formulas (4,5), the predicate 'optimal' is to find certain parameter combination associated with either a 'maximal' or 'minimal' optimization objective. For instance, benefit is maximized while cost is minimized. Further, the performance needs to be evaluated in terms of the background market microstructure and dynamics. Only in this way the developed trading agents can assist traders in taking right actions at right times with right prices and volumes on right instruments. For instance, an actionable moving average based strategy, say $MA(x,y)$ is a function with appropriate x and y to reach the best of expected business performance in certain market data. As a result, it generates a subset $\Omega'(MA)$ of general moving average strategies $\Omega(MA)$.

Under different situations, technical interestingness and business expectation need to be instantiated into corresponding forms. For instance, in pair mining of trading strategies [25], coefficient and sharpe ratio are used for strategy selection.

In this paper, trading agent's performance is evaluated toward enhancing benefits while reducing cost and risk of host agents when they execute certain strategies in the market. This involves strategy actionability as discussed in the next section.

9.3 Constraints on Actionable Trading Strategy Development

Typically, actionable trading strategy development must be based on a good understanding of organizational factors associated with a market. Otherwise it is not possible to accurately evaluate actionability. In real-world actionable pattern mining, underlying environment is more or less constrained [11]. Constraints may be broadly embodied in terms of data, domain, interestingness and deployment aspects [15].

9.3.1 Domain Constraint

Market organization factors relevant to trading strategy development consist of the following fundamental entities: $M = \{I, A, O, T, R, E\}$. Table 1 briefly explains these entities and their impact on strategy actionability. In particular, the orderbook form O is further represented by attributes T, B, P and V in Ω, i.e., $O = \{(t, b, p, v) | t \in T, b \in B, p \in P, v \in V\}$. The elements in M form the constrained market environment of trading strategy development.

In practice, any particular actionable trading strategy development needs to be discovered in an instantiated market niche m ($m \in M$). This market niche specifies particular constraints, which are embodied through the elements in Ω and M, on strategy definition, description, representation, mining, evaluation and deployment. Such varying constraints greatly impact the development and performance of actionable trading strategy. Their consideration in trading strategy development can reduce the search space mining.

Constraints surrounding the development and performance of actionable trading strategy set Ω' in a particular market data set form a constraint set:

Table 9.1. Domain factors and its impact to actionability

Organizational factors	Impact to actionability
Traded *instruments* I, such as stock or derivatives, I={*stock, option, feature, …*}	Varying instruments determine different data, analytical methods and objectives
Market *participants* A, A={*broker, market maker, mutual funds,…*}	Traders have the final right to evaluate and deploy discovered trading evidence to their advantage
Orderbook forms O, O={*limit, market, quote, block, stop*}	Order type determines what data set (e.g., orderbook) to be mined, as well as particular business interestingness
Trading *session*, indicated by timeframe T showing whether a market includes call market or continuous session	Setting up the focusing session can prune order transactions
Market *rules* R, e.g., restrictions on order execution defined by exchange	They determine pattern validity of discovered trading patterns when deployed
Execution system E, e.g., a trading engine is order or quote-driven	It limits pattern type and deployment manner after migrated to a real trading system

$$\sum = \{ \delta_i^k \mid c_i \in C, k \in N\} \tag{9.7}$$

δ_i^k stands for the k-th constraint attribute of a constraint type c_i. $C=\{M, D\}$ is a set of constraint type covering all types of constraints in market microstructure M and data D in the mining niche. N is the number of constraint attributes for a specific type i.

Correspondingly, actionable trading pattern set Ω' is a conditional function of $\sum$, which is described as

$$\Omega' = \{(\omega, \delta) \mid \omega \in \Omega,\ \delta \in \{(\delta_i^k, a) \mid \delta_i^k \in \sum, a \in A\}\} \tag{9.8}$$

ω is an 'optimal' trading strategy instance, and δ indicates specific constraints on the developed strategy that is recommended to a trader a.

EXAMPLE 9.2. For the rule $MA(r)$ discussed in Example 9.1, let it be deployed to trade BHP (BHP Billiton Limited) in the order-driven ASX market by a broker. S/he instantiates it into a form of $MA(5)$, which is a five-transaction moving average, and set δ_0=AUD$25.890 in trading BHP on 24 January 2007. In this situation, s/he believes that the instantiated rule is one of the most dependable $MA(r)$. Here M is instantiated into $\{ \delta_M^k \}$ = {*stock, broker, market order, continuous session, order-driven*}.

To work out the actionable set Ω', efforts in many aspects are essential. To this end, concerns and expectations of business people play inevitable roles.

9.3.2 Data Constraint

The second type of constraints on trading strategy development is data constraint. Huge quantities of historical data can play especially important role in strategy modeling. We model strategies using data mining to discover interesting and actionable trading patterns. Data constraint set D consists of the following factors D={*quantity, attribute, location, format, frequency, privacy, compression*}. Exchange intraday data stream normally presents characteristics such as high quantities, high frequency and multiple attributes of data. Some exchanges specify user-defined data format or follow a standard format such as FIX protocol [12]. Data may be distributed in multiple clusters with compression for big exchanges. Data constraint seriously affects the development and performance of trading strategies. For instance, the efficiency of complex strategy simulation and modeling may involve parallel supports on multiple sources, parallel I/O, parallel algorithms and memory storage.

9.3.3 Deliverable Constraint

Often deliverable trading strategies are not actionable to the real market even though they are sensible to research. This simulation may be due to the interestingness gaps between academia and business. What makes this trading strategy more interesting than the other? This is determined by *deliverable constraint*, which is embodied through interestingness, namely a set Int ={$Tint$, $Bint$}. Trading strategy interestingness covers both technical interestingness $Tint$ and business interestingness $Bint$. In the real world, simply emphasizing technical interestingness such as statistical measures of validity is not adequate for designing strategies. Social and economic interestingness (we refer to *business interestingness*), for instance, *profit*, *return* and *return on investment*, should be considered in assessing whether a strategy is actionable or not. Integrative interestingness measures are expected which should integrate both business and technical interestingness. Satisfying the integrative interestingness can benefit the actionability of trading agents in the real world.

There may be some other type of constraints such as dimension/level constraints. These ubiquitous constraints form a multi-constraint scenario, namely {M, D, Int, ...}, for actionable strategy design. We think that the actionable trading strategy development and optimization in continuous e-markets should be studied in the above constrained scenario.

9.3.4 Human-Agent Cooperation

The constraint-based context and the actionable requirement of trading strategies determine that the development process is more likely to be human involved rather than automated. Human involvement is embodied through cooperation between humans (including investors and financial traders) and trading agents. This is achieved through the compensation between human intelligence (e.g., domain knowledge and experience) and agent intelligence. Therefore, trading strategy development likely presents as a human-agent-cooperated interactive discovery process.

The role of humans may be embodied in the full period of development from market microstructure design, problem definition, data preprocessing, feature selection, simulation modeling, strategy modeling and learning to the evaluation, refinement and interpretation of discovered strategies and resulting outcomes. For instance, the experience and meta-knowledge of domain experts can guide or assist with the selection of features and strategy modeling, adding trading factors into the modeling, designing interestingness measures by injecting traders' concerns, and quickly evaluating results. This can largely improve the effectiveness of designing trading strategies.

To support human involvement, human agent cooperation is essential. Interaction often takes explicit form, for instance, setting up interaction interfaces to tune the parameters of trading strategies. Interaction interfaces may take various forms, such as visual interfaces, virtual reality technique, multi-modal and mobile agents. On the other hand, interaction may also follow implicit mechanisms, such as accessing a knowledge base or communicating with a user assistant agent. In interactive trading strategy optimization, the performance of the discovered strategies highly relies on interaction quality in terms of the representability of domain knowledge, and the flexibility, user-friendliness and run-time capability of interfaces.

9.4 Methods for Developing Actionable Trading Strategies

Following the ideas introduced in Section 9.3, this section illustrates some of approaches for developing actionable trading strategies. They consist of optimizing trading strategies, extracting in-depth trading strategies, discovering trading strategies, developing trading strategy correlated with instruments, and integrating multiple strategies. Their promising business performance is demonstrated in tick-by-tick data.

9.4.1 Optimizing Trading Strategies

There are often huge quantities of variations and modifications of a generic trading strategy by parameterization. For instance, MA(2, 50, 0.01) and MA(10, 50, 0.01) refer to two different strategies. However, it is not clear to a trader which specific rule is actionable for his or her particular investment situation. In this case, trading strategy optimization may generate an optimal trading rule from the generic rule set.

Optimizing trading strategies is to find trading strategies with better target performance. This can be through developing varying optimization methods. Genetic Algorithm (GA) is a valid optimization technique, which can be used for searching combinations of trading strategy parameters satisfying user-specified performance [18]. However, a simple use of GA may not necessarily lead to trading strategies of business interest. To this end, domain knowledge must be considered in fitness function design, search space and speed design, etc. The fitness function we used for strategy optimization is Sharpe Ratio (*SR*).

$$SR = (R_p - R_f) / \sigma_p, \tag{9.9}$$

R_p is the expected portfolio return, R_f is the risk free rate, and σ_p is the portfolio standard deviation. When *SR* is higher, it indicates higher return but lower risk.

Fig. 9.1 illustrates some results of GA-based trading strategy optimization. The trading strategy is Filter Rule Base, namely FR(δ). It actually indicates a generic class of correlated trading strategies, by which you go long on the day that the price rises by δ% and hold until the price falls δ%, at which time you close out and go short, where $\delta \in [0,1]$ is the percentage price movement of highest high and lowest low.

```
TRADING STRAGE 1: A generic strategy FR(δ)
At time point t, get high(t) and low(t)
IF price(t-1) > high(t-1)
    high(t)= price(t-1)
ELSE
    high(t)= high(t-1)
IF price(t-1) < low(t-1)
    low(t)= price(t-1)
ELSE
    low(t)= low(t-1)
Generate trading signals
            IF price(t) < high(t)*(1- δ)
                Generate SELL signal
            IF price(t) > low(t)*(1+ δ)
                Generate BUY signal
```

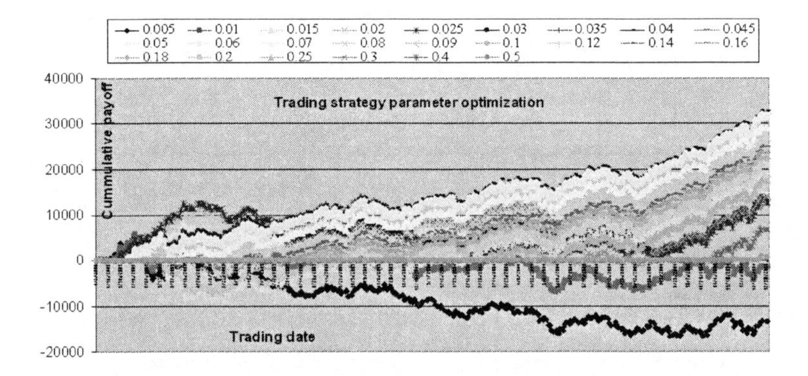

Fig. 9.1. Some results of GA-based trading strategy optimization

In this rule, there is only one parameter δ, which can be used for optimization because δ is hard to be managed well in real-life market. Fig. 9.1 shows the optimization results of the stock Australian Commonwealth Bank (CBA) in Australian Stock Exchange (ASX) in 2003~2004. It shows that from 14 July 2003, the cumulative payoff with $\delta = 0.04$ always beats other δ.

9.4.2 Extracting In-Depth Trading Strategies

In many real-life cases, a given trading strategy may not work well due to missing considerations of some organizational factors and constraints. To this end, we need to enhance a trading strategy by involving real-life constraints and factors. For instance, the above rule FR(δ) does not consider the noise impact of false trading signals and dynamic difference between high and low sides. These aspects can be reflected into the rule by introducing new parameters.

Extracting in-depth trading strategies is not a trivial task. It needs to consider domain knowledge and expert advice, massive back-testing in historical data, and mining hidden trading patterns in market data. Otherwise a developed strategy likely does not make sense to business. For instance, we create an Enhanced Filter Rule FR(t, δ_H, δ_L, h, d) as follows.

TRADING STRATEGY 2: An enhanced FR(t, δ_H, δ_L, h, d)
At time point t, get $high(t)$ and $low(t)$
IF $price(t-1) > high(t-1)$
    $high(t) = price(t-1)$
ELSE
    $high(t) = high(t-1)$
IF $price(t-1) < low(t-1)$
    $low(t) = price(t-1)$
ELSE
    $low(t) = low(t-1)$
Generate trading signals
        IF $price(t) < high(t)*(1- \delta_H)$
            Generate *SELL signal*
            IF $position(t-1)$ <> 0 & $hold(t-1) = h$
                $position(t) = 1$
        IF $price(t) > low(t)*(1+ \delta_L)$
            Generate *BUY signal*
            IF $position(t-1)$ <> 0 & $hold(t-1) = h$
                $position(t) = -1$

This enhanced version considers the following domain-specific aspects, which make it more adaptive to the real market dynamics compared with the generic rule MA(sr, lr, δ).

- More filters are imposed on the generic FR to filter out false trading signals which would result in losses, say fixed percentage band filter δ_H and δ_L for high and low price movement respectively, and time hold filter h;

4	A Trade Date	B Trade Price	C Maximum Price	D Minimum Price	E Sell Cond Met	F Buy Cond Met	G Trade Signal Indicator	H Trades	I Position Hold Days	J Positions	K Payoff ($)	L Payoff($) Cumulative
5	1/2/2003	3027	0	0	0	0	0	0	0	0	0	0
6	1/3/2003	3066	0	0	0	0	0	0	0	0	0	0
7	1/6/2003	3079	0	0	0	0	0	0	0	0	0	0
8	1/7/2003	3088	0	0	0	0	0	0	0	0	0	0
9	1/8/2003	3072	0	0	0	0	0	0	0	0	0	0
10	1/9/2003	3065	3088	3027	0	0	0	0	0	0	0	0
11	1/10/2003	3066	3088	3065	0	0	0	0	0	0	0	0
12	1/13/2003	3063	3088	3065	-1	0	-1	-1	0	0	0	0
13	1/14/2003	3077	3088	3063	0	0	-1	0	1	-1	-350	-350
14	1/15/2003	3075	3077	3063	0	0	-1	0	2	-1	50	-300
15	1/16/2003	3062	3077	3063	-1	0	-1	0	3	-1	325	25
16	1/17/2003	3055	3077	3062	-1	0	-1	0	4	-1	175	200
17	1/20/2003	3058	3077	3055	0	0	-1	0	5	-1	-75	125
18	1/21/2003	3054	3077	3055	-1	0	-1	0	6	-1	100	225
19	1/22/2003	3016	3075	3054	-1	0	-1	0	7	-1	950	1175
20	1/23/2003	3016	3062	3016	0	0	-1	0	8	-1	0	1175
21	1/24/2003	3021	3058	3016	0	0	-1	0	9	-1	-125	1050
22	1/27/2003	3001	3058	3016	-1	0	-1	0	10	-1	500	1550
23	1/28/2003	2952	3054	3001	-1	0	-1	0	11	-1	1225	2775
24	1/29/2003	2924	3021	2952	-1	0	-1	0	12	-1	700	3475
25	1/30/2003	2947	3021	2924	0	0	-1	0	13	-1	-575	2900
26	1/31/2003	2943	3021	2924	0	0	-1	0	14	-1	100	3000
27	2/3/2003	2933	3001	2924	0	0	-1	0	15	-1	250	3250
28	2/4/2003	2934	2952	2924	0	0	-1	0	0	0	0	3250
29	2/5/2003	2892	2947	2924	-1	0	-1	0	0	0	0	3250
30	2/6/2003	2881	2947	2892	-1	0	-1	0	0	0	0	3250
31	2/7/2003	2887	2943	2881	0	0	-1	0	0	0	0	3250
32	2/10/2003	2870	2934	2881	-1	0	-1	0	0	0	0	3250
33	2/11/2003	2871	2934	2870	0	0	-1	0	0	0	0	3250
34	2/12/2003	2874	2892	2870	0	0	-1	0	0	0	0	3250
35	2/13/2003	2818	2887	2870	-1	0	-1	0	0	0	0	3250
36	2/14/2003	2812	2887	2818	-1	0	-1	0	0	0	0	3250
37	2/17/2003	2839	2874	2812	0	0	-1	0	0	0	0	3250
38	2/18/2003	2845	2874	2812	0	0	-1	0	0	0	0	3250
39	2/19/2003	2854	2874	2812	0	0	-1	0	0	0	0	3250
40	2/20/2003	2803	2854	2812	-1	0	-1	0	0	0	0	3250
41	2/21/2003	2810	2854	2803	0	0	-1	0	0	0	0	3250
42	2/24/2003	2846	2854	2803	0	0	-1	0	0	0	0	3250
43	2/25/2003	2790	2854	2803	-1	0	-1	0	0	0	0	3250
44	2/26/2003	2825	2854	2790	0	0	-1	0	0	0	0	3250
45	2/27/2003	2783	2846	2790	-1	0	-1	0	0	0	0	3250
46	2/28/2003	2794	2846	2783	0	0	-1	0	0	0	0	3250
47	3/3/2003	2816	2846	2783	0	0	-1	0	0	0	0	3250
48	3/4/2003	2803	2825	2783	0	0	-1	0	0	0	0	3250
49	3/5/2003	2771	2825	2783	-1	0	-1	0	0	0	0	3250
50	3/6/2003	2769	2816	2771	-1	0	-1	0	0	0	0	3250
51	3/7/2003	2732	2816	2769	-1	0	-1	0	0	0	0	3250
52	3/10/2003	2733	2816	2732	0	0	-1	0	0	0	0	3250
53	3/11/2003	2713	2803	2732	-1	0	-1	0	0	0	0	3250
54	3/12/2003	2710	2771	2713	-1	0	-1	0	0	0	0	3250

Fig. 9.2. Some results of enhanced trading strategy FR

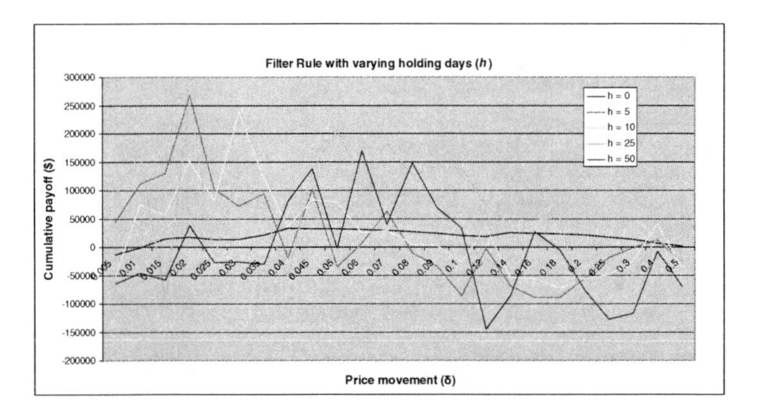

Fig. 9.3. Performance comparison between base and enhanced trading strategies

- The fixed band filter δ_H (or δ_L) requires the buy or sell signal to exceed *high* or *low* by a fixed multiplicative band δ_H (or δ_L);
- The time hold filter h requires the buy or sell signal to hold the long or short position for a pre-specified number of transactions or time h to effectively ignore all other signals generated during that time;

Fig. 9.2 shows the trading results of a trading agent taking the above strategy in ASX data 2003~2004.

Fig. 9.3 further shows the performance difference between a base rule and its enhanced version. It indicates that the involvement of domain knowledge and organizational constraints can to most extent enhance the business performance (cumulative payoff in our case) of trading strategies.

9.4.3 Discovering Trading Strategies

Another method of trading strategy development is through mining trading patterns in stock data. For instance, based on the domain assumption that some instruments are associated with each other, we can discover trading patterns effective on multiple correlated instruments. The following illustrates an effective pair trading strategy. It indicates that a trading agent can go long with one instrument while short another.

TRADING STRATEGY 3: A pair trading strategy PT{*S*, *T*}

C1. Calculating the *tech_int*(), e.g., coefficient ρ, of two stocks S and T considering market index;

C2. Determining stock pairs according to *tech_int*() and *biz_int()* defined through cooperation with traders, market aspects such as market sectors, volatility, liquidity and index are considered;

C3. Designing trading strategy to trade stock *S* and *T* alternatively by training it in in-sample data:
IF P_S - (*P_T >= d_0, THEN buy *T* and sell *S*
IF P_S - (*P_T <= - d_0, THEN sell *T* while buy *S*
Where P_S and P_T are prices of *S* and *T*, weight (and distance d_0 are business factors that are optimized and tuned by the system with user guide

C4. Generating trading signals by checking the above rules in out-of-sample data,

C5. Iteratively evaluating and refining the strategy by calculating *biz_int*(), e.g., *return* in our case, and considering impacts of volatility, liquidity and index
C6. Deploying the refined strategy to the market

Fig. 9.4 shows pair trading signals generated in trading Australia stock pairs CBA and GMF in ASX market. Fig. 9.5 further shows the impact of business factors – distance and weight on return and the number of triggered signals.

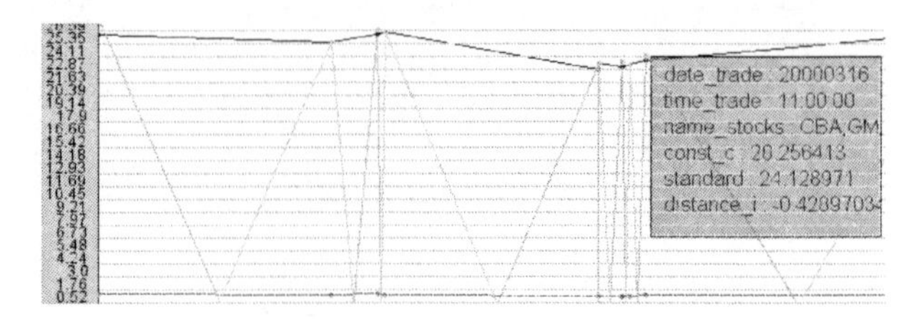

Fig. 9.4. Some results of discovered trading strategy
(CBA and GMF, ASX data from 1 Jan 2000 to 20 Jun 2000)

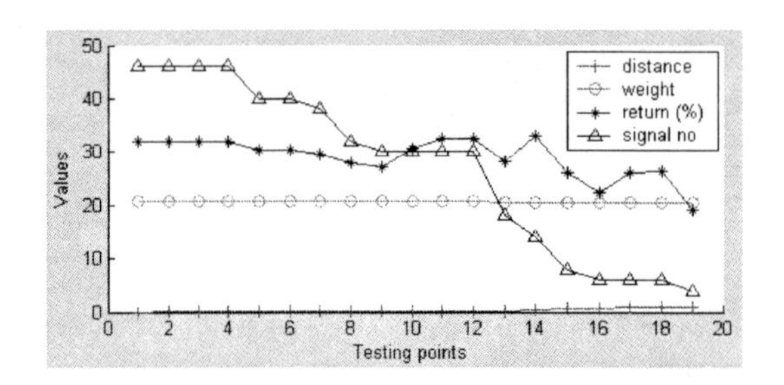

Fig. 9.5. Relation between d_0, , *return* and *signal number*
(ASX intraday data, from 1 Jan 2000 to 20 Jun 2000)

In discovering pair trading strategy for trading agents on ASX Top 32 stocks from January 1997 to June 2002 in F-Trade [3,4,6], the trading agent got the following findings.

- Pair relationship between stocks and the combination of the above four factors interesting to trading cannot just be determined by technical measures such as coefficient ρ. They are also highly affected by stock movement such as volatility and liquidity. High volatility improves return while high liquidity balances the market impact on return.
- All 13 correlated stocks mined in Top 32 ASX come from different sectors. This finding means that pairs are not necessary from the same sector as presumed by financial researchers.

The pairs trading strategies discovered have been deployed into the famous exchange surveillance system SMARTS [24].

9.4.4 Developing Trading Strategies Correlated with Instruments

In real-life market, some trading rules are tested to be more profitable to trade a class of stocks, while others are more suitable for other stocks. This triggers another method to improve trading strategies that is through analyzing the relationship between a trading strategy and its tradable instruments. We developed the following algorithms to discover the correlations between trading strategies and instruments, and then let a trading agent trade those identified associated instruments only.

TRADING STRATEGY 4: Improving trading strategies by analyzing correlation between strategies and stocks

C1. Mining actionable rules for an individual stock;
C2. Mining highly correlated rule-stock pairs by high dimension reduction;
C3. Evaluating and refining the rule-stock pairs by considering traders' concerns;
C4. Recommending actionable rule-stock pairs.

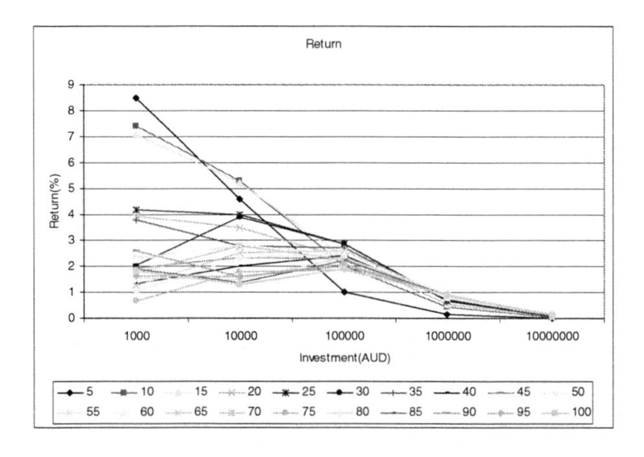

Fig. 9.6. Return on investment of trading strategy-stock pairs

In discovering trading strategy-stock pairs, traders were invited to give suggestions on designing features, interestingness measure and parameter optimization. They also helped us design mechanisms for evaluating and refining rule-stock pairs. Taking the ASX as an instance, six types of trading rules such as MA and Channel Breakout and 27 ASX stocks such as ANZ and TEL were chosen for the experiments in the intraday training data from 1 January 2001 to 31 January 2001 and the testing set from 1 February 2001 to 28 February 2001. Five different investment plans were conducted on the above rules and stocks. In organizing pairs, we ranked them based on return, and generate 5% pair, 10% pair, and so forth from the whole pair set. The 5% pair means that return for trading these pairs is the top 5% in the whole pair set. Fig. 9.6 illustrates returns for different investment plans on different pair groups (where the digital legend below the graph refers to pair percentage, e.g., 5 means top 5% pairs). These results show that trading top pairs can get high return but with high risk as well, while trading middle-level pairs can obtain lower return with lower risk. Thus, massively trading a basket of stocks with a lower ranking pair may be an investment strategy more interesting to investors who have big money. As a result, they can very likely obtain superior profit with low risk.

9.4.5 Integrating Multiple Strategies

In real-life trading, trading strategies can be categorized into many classes. To financial experts, different classes of trading strategies indicate varying fundamental principles of the market model and mechanisms. As a result, a trading agent may take serial positions generated by a specific trading strategy, which instantiates a class of trading strategies. It may also take concurrent positions created by multiple trading strategies. We will not discuss multi-strategies taking by a trading agent in this section. Rather, we are interested in identifying the most suitable trading strategy from all available strategies or by integrating multiple strategies. This leads to the following methods.

METHOD 1. A trading agent identifies and takes the best trading strategy s_o from all parameter combinations of a trading strategy s.

For Method 1, we can use optimization techniques like Genetic Algorithms [5] to search for the strategy with highest benefit b_s but lowest cost c_s. We won't address it in this paper.

METHOD 2. A trading agent identifies and takes the best trading strategy s_c from a trading strategy class c. A trading strategy class may consist of several types of trading strategies.

For instance, double Moving Average *MA* is a common trading strategy. It can further be instantiated into *MA-B*, *MA-C* and *MA-D*, where *B*, *C* and *D* represent different organizational factors and constrained filters considered in the designing the MA strategies. Following the Method 2, we need to identify the golden rule of the *MA* family {*MA*, *MA-B*, *MA-C*, *MA-D*}.

METHOD 3. A trading agent identifies and takes the best trading strategy s_{ci} (*ci* is the trading strategy class *i*, i = 1, 2, …) from each class *ci*. The agent follows such best strategies from a group of trading strategy classes.

For instance, there are the following trading strategy classes: FR, MA, CB, SR and OBV. For each of them, there exist a few types of instances just like MA discussed in the above. An agent will discover the golden strategies for them individually, and then follow them to take trading positions concurrently in the market. The strategy development process is as follows.

Given a trading strategy *a*, a trading strategy class *ci* (i=1, 2, ...), $a \in ci$, b_a and c_a are the benefit and the cost of a trading agent in executing the strategy *a*,
A. Data preparation:
Separating the source data into two data sets in terms of:
1). Splitting two years of data for training to identify best trading strategies;
2). Picking up the follow-up year of data to deploy the identified strategies;
3). Searching optimal strategies as discussed in part B;
4). Sliding the 2-year training and the 1-year deploying data windows one year forward to extract data sets as in 1) and 2), and repeating the operations of searching optimal strategies;

B. Searching optimal strategies:

1). Searching for the strategy instance a' of strategy a with $\max(b_{a'})$ of its positions;

2). Searching for the strategy a'' of strategy a in its class ci with $\max(b_{a''})$ and $\min(c_{a''})$ in class ci when its positions are executed;

3). Searching for all strategies a''_1, a''_2, ... (i=1, 2, ...) in all strategy classes satisfying conditions in step 2);

4). Generating the positions of a trading agent taking all strategies identified in step 3), respectively;

5). Checking the benefits and costs of a trading agent executing the above positions, and compare to find the strategy with highest benefits while as low as possible cost/benefit ratio;

6). Executing multiple strategies concurrently.

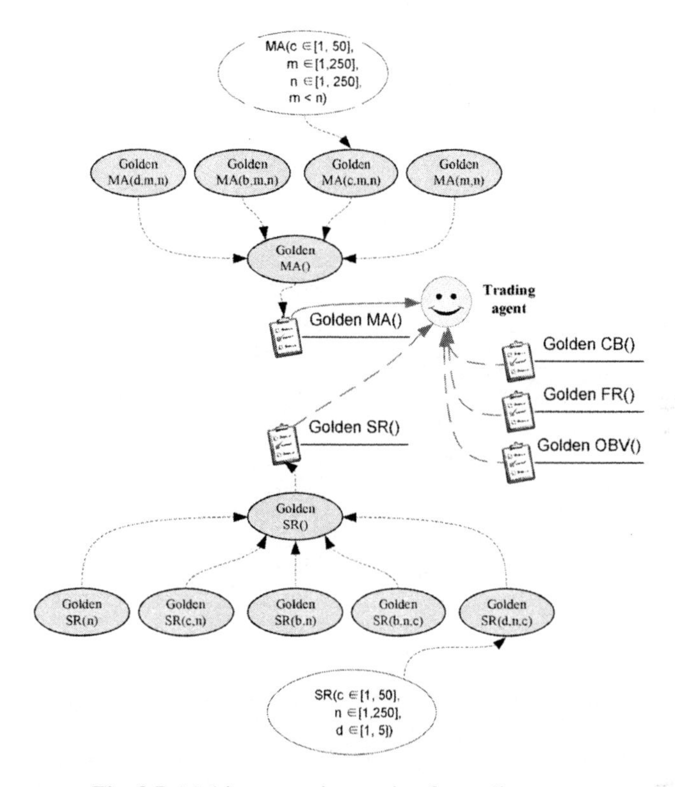

Fig. 9.7. Multi-strategy integration for trading agents

Fig. 9.7 further illustrates the process of a trading agent selecting golden trading strategies from individual strategy class.

Trading agent can integrate all golden trading strategies and execute them concurrently in the market. In this case, Table 9.2 shows the positions recommended by each golden strategy.

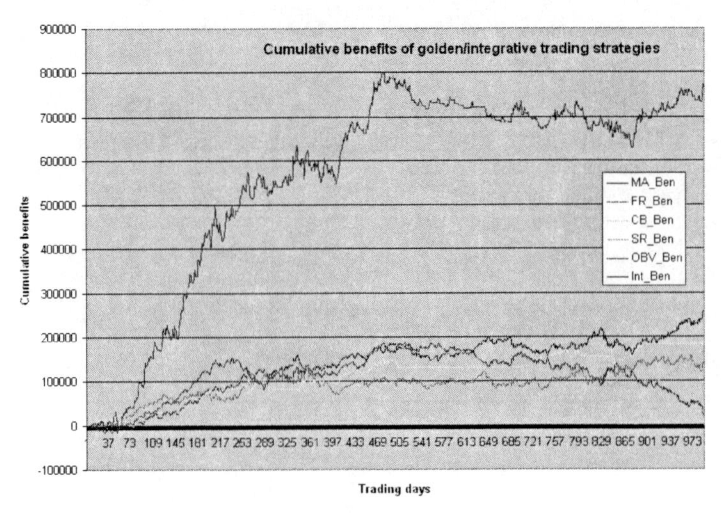

Fig. 9.8. Cumulative benefits of each trading strategies
(Year: 2003-2006, Market: Hongkong, Strategies: MA, FR, CB, SR, OBV, and Integrative)

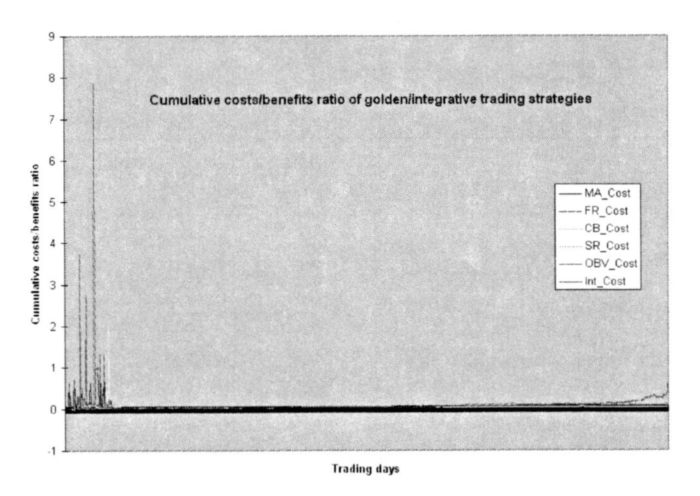

Fig. 9.9. Cumulative cost/benefit ratio of each golden trading strategies
(Year: 2003-2006, Market: Hongkong, Strategies: MA, FR, CB, SR, OBV, and Integrative)

Table 9.2. Trading agent positions recommended by five trading strategy classes (excerpt)
(Strategy class: MA, FR, CB, SR, OBV; Data: Hongkong; Year: 2006)

Date	Position_MA	Position_FR	Position_CB	Position_SR	Position_OBV
2006-11-16	1	1	0	1	1
2006-11-17	1	1	0	1	1
2006-11-20	1	1	0	1	1
2006-11-21	-1	-1	0	1	1
2006-11-22	-1	-1	0	1	1

Figs 9.8 and 9.9 show the cumulative benefits and cost/benefit ratios of a trading agent taking positions recommended by golden trading strategies as shown in Table 9.2 in the market, where *Int_Ben* and *Int_Cost* are the benefit and cost obtained when a trading agent executes all golden positions concurrently.

9.5 Multiagent-Based Actionable Trading Strategy Development

In this section, we illustrate some of the process and results in optimizing strategies through Evolutionary Trading Agents and integrating strategies via Collaborative Trading Agents.

Given a trading strategy s, a trading strategy class S_i (i=1, 2, . . .), $s \in S_i$, α_s and β_s are the benefit and cost of a trading agent in executing the strategy *s*. The development process of integrating strategies through trading agent collaboration is as follows.

Part A. Data Manager Agent prepares data:
0). UserAgent receives trader's input requests;
1). DataManager agent splits two years of data for training;
2). RepresentativeAgent invokes EvolutionaryAgents to identify locally golden trading strategies with highest $\gamma_{\alpha\beta}$ as discussed in part B;
3). DataManager agent splits another three years of data following the training windows for testing;
4). RepresentativeAgent invokes EvolutionaryAgents to test the identified golden strategies as discussed in part B;
5). DataManager agent slides the 2-year training and the 3-year deploying data windows one year forward to extract data sets as in A:1) and A:3);
6). RepresentativeAgent invokes EvolutionaryAgents to repeat the operations of searching golden strategies;

Part B. Evolutionary Trading Agents search golden strategies:
1). EvolutionaryAgent calls a StrategyAgent *s* in class S_i and searches strategy instance *s*' with max($\alpha_{s'}$) for *s*' positions;
2). EvolutionaryAgent calls a StrategyAgent s and searches strategy *s*'' with max($\gamma_{\alpha\beta}$,*s*) when *s*'' positions are executed;
3). RepresentativeAgent invokes EvolutionaryAgents to search all strategies s_i'' ((i = 1, 2, . . .)) in all strategy classes satisfying conditions in step B:2) respectively;

Part C. Collaborative Trading Agents aggregate golden strategies:
1). PositionAgents extract all positions from EvolutionaryAgents with all strategies identified in step B:3) for RepresentativeAgent ;
2). EvaluationAgents check the benefits, costs and benefit-cost ratio of each RepresentativeAgent executing the above positions;
3). DecisionAgents filter out strategies with low $\gamma_{\alpha\beta}$,*s* for each strategy class *i*;
4). CoordinatorAgents call all RepresentativeAgents to execute the above filtered strategies concurrently to generate the final outcomes.

Experiments of trading agent collaboration for multi-strategy integration in stock market data have been conducted as follows:

- Trading strategies: MA, FR, CB, SR, and OBV as shown in Table 9.2;
- Markets and stocks: selected stocks from ASX, Hongkong, London, New York, and Japan;
- Interday trade data consisting of times, prices, volumes from 1/1/1998 to 30/12/2006;
- Training data: 2-year sliding window, say 1/1/1998-30/12/1999;
- Testing data: 1-year sliding window, say 1/1/2000-30/12/2000.

Fig. 9.3 illustrates some optimization results of $FR(\delta)$ using evolutionary trading agents. For a single strategy, for instance, $FR(\delta)$, even though there is only one parameter δ in this rule, it is hard to find the most favorite δ in a real-life market. As shown in Fig. 9.3, evolutionary trading agent is helpful for searching such a golden δ.

Table 9.3 further shows the signals, positions, benefits and costs of trading agents following MA-BMN Strategy, which is an identified golden strategy by evolutionary trading agents in 2004 Hongkong Exchange data.

Table 9.2 shows the positions recommended by each golden strategy identified by Collaborative Trading Agents in 2006 Hongkong United Exchange data. Fig. 9.8 shows the cumulative benefits MA Ben, FR Ben, CB Ben, SR Ben, OBV Ben of trading agents taking positions recommended by golden trading strategies MA, FR, CB, SR, OBV as shown in Table 9.2, as well as the benefit (*Int_Ben*) of executing all golden positions concurrently recommended by Collaborative Trading Agent in 2003-2006 Hongkong United Exchange data. A large amount of tests in five markets of data have shown that trading agents following our recommended golden trading strategies can obtain higher benefit-cost ratios (except FR in the first few days). In

Table 9.3. Output excerpt of a trading strategy (Strategy: MA-BMN; Data: 2004)

Date	Price	Sell	Buy	Position	($) Benefit	($) Cost
2004-8-16	3466	-1	0	-1	9200	103
2004-8-17	3480	-1	0	-1	8850	106.5
2004-8-18	3472	-1	0	-1	9150	108.5
2004-8-19	3481	-1	0	-1	8825	110.75
2004-8-20	3494	0	0	-1	8500	114

Table 9.4. Lift comparison between random chosen strategies and golden strategies

Lift	MA-CMN	FR-XY	OBV-B	CB-NXC	SR-NC
Random	10%	0	20%	10%	10%
Optimized	70%	80%	80%	90%	100%

particular, collaborative trading agents concurrently executing positions recommended by individual golden strategies can greatly increase benefits while control very low costs compared with those taking positions recommended by either an individual strategy or randomly chosen strategies only (see Table 9.4, lift [25] measures how good a trading strategy is in all split data sets).

9.6 Evaluation of Actionable Trading Strategies

9.6.1 Technical Performance

Following the two-way significance framework [5], we evaluate the performance of identified trading strategies from both technical and business perspectives. Taking the trading strategy-stock pair identification as an instance, this section presents some of methods for evaluating identified trading evidence.

From technical side, here we evaluate the survival probability of a trained pattern in test set. Let D be the number of total trading strategies (e.g., rule-stock pairs, deleting any overlap patterns) found in training and test sets satisfying both technical and business interestingness. Let A and B be the number of total strategies in training and test data sets, respectively. AB be the number of pairs existing in both training and test sets, which satisfying both technical and business interestingness. We define the following statistical measures to assess the technical performance of extracted trading strategies.

Definition 9.3. (Probability of trading strategy) The following probability functions are defined for trading strategy in or across training and test data sets: P(A)=A/D, P(B)=B/D, P(AB)=AB/D, P(A|B)=P(AB)/P(B), P(B|A)=P(AB)/P(A).

Definition 9.4. (Survival metrics of trading strategy) The following metrics: Sur_Supp, Sur_Conf, All_Conf and Sur_Cos are defined for measuring the survival performance of trained trading strategy in test data.

$$Sur\_Supp = P(AB) \tag{9.10}$$

$$Sur\_Conf = \max(P(A \mid B), P(B \mid A)) \tag{9.11}$$

$$All\_Conf = P(AB) / \max(P(A), P(B)) \tag{9.12}$$

$$Sur\_Cos = P(AB) / \sqrt{P(A)P(B)} \tag{9.13}$$

Larger values of the above metrics denote more robust performance when deployed in real data.

Taking the identification of trading strategy-stock pairs for trading agents as an instance, we specify top x% pairs from training or test set in terms of satisfying user-specified business interestingness thresholds. We call them *crisp pairs*. Table 9.5 lists Sur_Supp and Sur_Conf of top 5% crisp strategy-stock pairs (ASX 2001 orderbook data from April to October 2001, transaction costs = 0%). In this case, there exists the following relation among relevant metrics.

$$Sur\_Conf = All\_Conf = Sur\_Cos \tag{9.14}$$

Table 9.5. Statistical performance of top 5% crisp rule-stock pairs (ASX 2001 orderbook)

	Apr	May	Jun	Jul	Aug	Sept
Sur_Supp	6.7%	14.3%	23%	14.3%	6.7%	14.3%
Sur_Conf	12.5%	25%	37.5%	25%	12.5%	25%

In this top 5% crisp pairs, we find 11 pairs also survived in test set in June 2001. Among these pairs, two of them are relatively frequent pairs: Rule 1 – Stock 14, Rule 2 – Stock 24 (in this particular pair set, their association supports are larger than 20%). Here the numbers of trading rules and stocks are decoded in terms of commercial reasons.

9.6.2 Difference from Considering Business Interestingness

Here we show the difference resulting from the consideration of business interestingness and the proposed fuzzy two-way significance method in selecting and measuring actionable knowledge. The experiment series for business performance evaluation are as follows. Three groups of trading rule-stock pairs are extracted based on different methods.

METHOD 4: trading rule-stock pairs are identified based on the best technical interest only.

METHOD 5: pairs are identified based on the fuzzy two-way significance method discussed in [5].

METHOD 6: pairs are extracted on the basis of equally weighted sum of technical and business interestingness.

It is worthy of noting that the comparison of each particular pair's performance may not be fair. To eliminate the noise triggered by pair variation and uncertainty, we trade the top 10% pairs from each of groups as a whole based on same investment in the same market. We then calculate the monthly return obtained in trading the whole bundle of top 10% pairs in each group. Fig. 9.10 shows the monthly return of these three groups from Mar to Oct in ASX 2001 orderbook data (total investment = AUD100k).

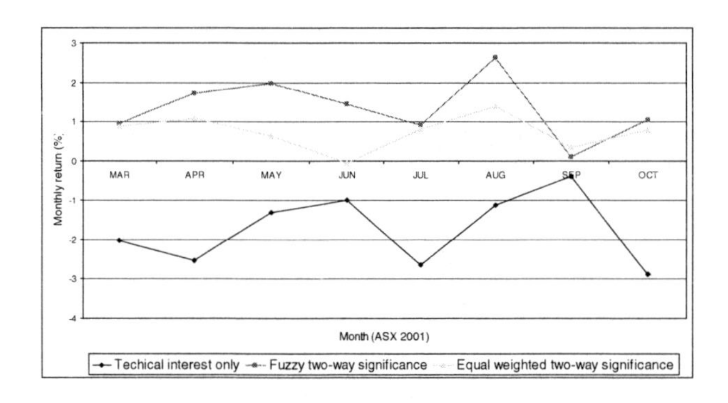

Fig. 9.10. Monthly return *TR* of top 10% rule-stock pairs
(Technical interest only vs. fuzzy two-way significance vs. equally weighted two-way significance)

It is shown that the business performance of those pairs only based on technical interestingness is much worse than that of pairs extracted by a two-way significance approach. The averaged monthly return of *technical interest only* is -1.737%, which is much lower than that of *fuzzy two-way* (1.343%) and *equally weighted two-way* (0.731%) methods. Pairs based on the fuzzy two-way significance approach get better business performance than the equally weighted two-way strategy.

9.6.3 Satisfying Business People's Expectations

In this part, we further demonstrate the effectiveness of our proposed domain-driven actionable knowledge extraction approach in satisfying business expectations. From the *objective* perspective, we calculate and evaluate economic performance such as *trade return* (*TR*), *index return* (*IR*), and *sharpe ratio* (*SR*). On the *subjective* side, empirical and psychoanalytic indicators are studied. We check the possibility of "beating transaction costs", "beating market return" of trading our extracted patterns in the market. These are used as real-world benchmarks of judging whether traders can take actions on identified trading patterns or not. *Beating transaction costs* means that specific economic measures, say *TR*, must positively surpass user-specified thresholds after deducting the impact of transaction costs on each trading transaction. *Beating market return* means that the actual monthly mean trade return *TR* generated by a trading rule must be better than the market index return *IR*.

Extensive experiments have been conducted to measure the business performance of trading rule-stock pairs identified in terms of trader's expectations and market dynamics. With trading rule-stock pairs identified by the fuzzy two-way approach, the monthly trade return *TR* gained by trading different levels of trading rule-stock pairs after deducting 0.25% transaction costs in Feb 2001 is positive. It demonstrates that trading on these mined trading rule-stock pairs likely beats transaction costs. Traders welcome this encouraging result because their experience tells them "transaction costs can 'kill' most of trading rules" in the market trading.

Fig. 9.11 further compares the market index return *IR* with the monthly trade return *TR* of top 5% pairs in ASX 2001 orderbook data after deducting 0.25% transaction costs. This result shows that under the bear market situation in 2001 ASX market,

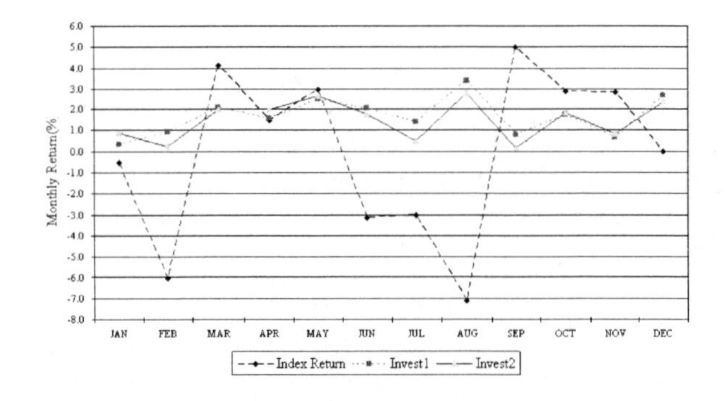

Fig. 9.11. The monthly returns *TR* vs. *IR*

trading these mined trading rule-stock pairs (with averaged *TR*= 1.563%) could beat not only transaction costs, but the market index return (averaged *IR*= -0.00773%). From another perspective, it demonstrates that the discovered rule-stock pairs using the proposed method are very promising for supporting traders' decision making.

9.7 Conclusions

Trading strategies can play important roles in supporting traders' decision in the market. However, the traditional studies on trading strategy development are not necessarily of business interest. To make trading strategies workable and support action-taken in the market, it is necessary to consider real-life constraints and organizational factors.

This paper has presented a systematic view of developing actionable trading strategies. It covers the definition of strategy actionability, a framework for actionable trading strategy development, and approaches for developing actionable strategies. We have demonstrated a number of approaches to developing actionable trading strategies. The identified trading strategies have been tested in continuous market data, and presented promising performance from not only technical but also business perspectives. For instance, our results can beat transaction costs and market index return, which are viewed as practical challenges.

Actionable trading strategies can enhance the workable capability of trading agents. Our future work is on collaboration and integration amongst trading agents to develop workable trading strategies.

Acknowledgement

This work is sponsored by Australian Research Council Discovery Grant (DP0773412, LP0775041, DP0667060), and UTS internal grants.

References

[1] Arthur, W.B., Durlauf, S.N., Lane, D.A.: The Economy as an Evolving Complex System II, vol. 27, p. 583. Addison-Wesley, Santa Fe Institute (1997)

[2] Cao, L.: Multi-strategy integration for actionable trading agents. In: ADMI 2007 workshop joint with IAT2007, pp. 487-490 (2007)

[3] Cao, L., Wang, J., Lin, L., Zhang, C.: Agent Services-Based Infrastructure for Online Assessment of Trading Strategies. In: Proc. of IAT 2004, pp. 345–349. IEEE press, Los Alamitos (2004)

[4] Cao, L., Ni, J., Wang, J., Zhang, C.: Agent services-driven plug-and-play in F-TRADE. In: Webb, G.I., Yu, X. (eds.) AI 2004. LNCS (LNAI), vol. 3339, pp. 917–922. Springer, Heidelberg (2004)

[5] Cao, L., Zhang, C.: Two-Way Significance of Knowledge Actionability. Int. J. of Business Intelligence and Data Mining 4 (2007)

[6] Cao, L.B., Zhang, C.: F-Trade: An Agent-Mining Symbiont for Financial Services. In: AAMAS 2007 (2007)

[7] Cao, L., Luo, C., Zhang, C.: Developing actionable trading strategies for trading agents. In: IAT 2007, pp. 72–75 (2007)
[8] Cao, L., He, T.: Developing actionable trading agents, Knowledge and Information Systems: An International Journal (2008)
[9] Chan, T.: Artificial Markets and Intelligent Agents. PhD thesis, Massachusetts Institute of Technology (2001)
[10] Cheng, S.-F., Leung, E., Lochner, K.M., O'Malley, K., Reeves, D.M., Schvartzman, L.J., Wellman, M.P., Walverine: A Walrasian Trading Agent. In: Proc. of AAMAS 2003, pp. 465–472. ACM, New York (2003)
[11] Trading Agent Competition, `http://www.sics.se/tac/`
[12] Ioannis, A., Vetsikas, Selman, B.: A principled study of the design tradeoffs for autonomous trading agents. In: Proc. of AAMAS 2003, pp. 473–480 (2003)
[13] Dasgupta, P., Hashimoto, Y.: Multi-attribute dynamic pricing for online markets using intelligent agents. In: Proc. of AAMAS 2004, pp. 277–284. ACM, New York (2004)
[14] Esteva, E., et al.: AMELI–An Agent-Based Middleware for Electronic Institutions. In: Proc. of AAMAS 2004, pp. 236–243 (2004)
[15] David, E., Azoulay-Schwartz, R., Kraus, S.: Protocols and strategies for automated multi-attribute auctions. In: Proc. of AAMAS 2002, pp. 77–85 (2002)
[16] David, E., Azoulay-Schwartz, R., Kraus, S.: Bidders' strategy for multi-attribute sequential English auction with deadline. In: Proc. of AAMAS 2003, pp. 457–464 (2003)
[17] FIX protocol, `http://www.fixprotocol.org/`
[18] Lin, L., Cao, L.B.: Mining in-depth patterns in stock market. Int. J. of Intelligent Systems Technologies and Applications (2007)
[19] Madhavan, A.: Market Microstructure: A Survey. Journal of Financial Markets, 205–258 (2000)
[20] Omiecinski, E.: Alternative Interest Measures for Mining Associations. IEEE Transactions on Knowledge and Data Engineering 15, 57–69 (2003)
[21] Wan, H.A., Hunter, A.: On Artificial Adaptive Agents Models of Stock Markets. Simulation 68(5), 279–289
[22] Wellman, M.P., et al.: Designing the Market Game for a Trading Agent Competition. IEEE Internet Computing 5(2), 43–51 (2001)
[23] Zhang, C., Zhang, Z., Cao, L.: Agents and data mining: Mutual enhancement by integration. In: Gorodetsky, V., Liu, J., Skormin, V.A. (eds.) AIS-ADM 2005. LNCS (LNAI), vol. 3505, pp. 50–61. Springer, Heidelberg (2005)
[24] SMARTS, `http://www.smarts.com.au`
[25] Cao, L., Luo, D., Zhang, C.: Fuzzy genetic algorithms for pairs mining. In: Yang, Q., Webb, G. (eds.) PRICAI 2006. LNCS (LNAI), vol. 4099, pp. 711–720. Springer, Heidelberg (2006)

10

Agent Uncertainty Model and Quantum Mechanics Representation: Non-locality Modeling

Germano Resconi[1] and Boris Kovalerchuk[2]

[1] Dept. of Mathematics and Physics, Catholic University, Brescia, Italy, I-25121
resconi@numerica.it
[2] Dept. of Computer Science, Central Washington University,
Ellensburg, WA 98926-7520, USA
borisk@cwu.edu

Abstract. This work presents the Agent–based Uncertainty Theory (AUT) and its connection with quantum mechanics where agents are interpreted in terms of the particles. This connection serves a dual goal to justify AUT operations as physically meaningful and to provide a new explanatory mechanism for contradictory issues in quantum mechanics. The AUT is described in agent terms and then is interpreted in quantum mechanics terms. The AUT uses complex aggregations of crisp (non-fuzzy) conflicting judgments of agents. It gives a uniform representation and an operational empirical interpretation for several uncertainty theories such as rough set theory, fuzzy sets theory, intuitionistic fuzzy sets, evidence theory, and probability theory. To build such uniformity the AUT exploits the fact that agents as independent entities can give conflicting evaluations of the same attribute. The AUT many-valued logic is a derived from classical logic with local and global evaluations of proposition p. The evaluation by an individual agent is called a local evaluation and evaluation by a set of agents is called a global evaluation of p. The local operations AND, OR, NOT are classical logic operations, but global operations differ from them. Here a set of agents generates the vectors of logic values. In the AUT local evaluations of proposition p by different agents can be in conflict in contrast with the classical logic. In quantum mechanics, non-locality is related to the superposition of different positions of the particles and to representation of two particles in different positions as a single non-local particle with correlation or entanglement. The logic operations between vectors are the base of AUT many-valued logic combined with the superposition of agents' evaluations.

10.1 Introduction

Complex multi-agent agent systems (MAS) have many different aspects from the internal structure of the agent to the environment and communications between agents. This paper concentrates on logical aspects of reasoning by individual agents and a society of agents under uncertainty. The common approach in multi-agent modeling in software engineering assumes that agents should be rational to fulfill their mission [3, 14]. Any agent that maximizes chances of success without any logical conflict is considered as a *rational* one. The rationality of a software agent rests in the maximization of its chances of success based on agent's knowledge of its

L.C. Jain, N.T. Nguyen (Eds.): Knowl. Proc. & Dec. Mak. in Agent-Based Sys., SCI 170, pp. 217–246.
springerlink.com

environment and interaction with the environment and other agents. In fact, an agent's success *criteria, reasoning* and *acting* abilities as well as knowledge of the *environment*, can be quite conflicting and uncertain from the logic viewpoint. In such conflicting situation, the agent can be in a logical state that can be called *irrational.* Accordingly, agent's behavior is not rational.

In software engineering, the concept of self-conflict is applicable to the program verification process. If a software agent (program S) produces different results when run several times on the same input data D then S is in a self-conflict state. The reason can be that the non-verified program does not initialize working memory correctly and then reads from memory data C_t that are left from the previous computations.

Another area that is deeply interested in partially rational agents called *boundedly rational* agents is motivated by fundamental goals in *psychology and economics* [5,10]. The works in this area explore the psychology of intuitive beliefs and choices by examining their bounded rationality as Kahneman outlined in his Nobel Prize Lecture in 2002. In economics, bounded rationality means the use of (1) *multiple utility functions* instead of one global scalar utility function, (2) *limited types* of utility functions due to cost of information collection, and (3) *heuristics*.

Our concept of agent in a conflicting logical state (irrational state) is motivated by fundamental goals in the *artificial intelligence* area of logic and reasoning under uncertainty [6,7] with agents [4].

Quantum mechanics is another area where the concept of agents with limited rationality can be useful as we show in this paper. We envision a formal logicization of reasoning of irrational agents that do not follow rational reasoning in the frames of the classical logic and the probability theory and even fuzzy logic. This area is quite different from two previous areas as well as the areas that study expert reasoning simply because irrational agents hardly fit a definition of experts. We define a specific concept of an *ME-rational* (the agent that is rational relative to mutual exclusion) with a clear criterion how to distinguish such agents. Similarly, we define a concept of an *ME-irrational* agent as a *self-conflicting agent* as well as *contradictory agents* that contradict each other but without self-conflict in judgment. The concepts of conflict and self-conflict in judgments are critical for our study of irrational agents, because uncertainty often means that several contradictory statements are made. The concept of self-conflict itself is studied in paraconsistent logics [19] that reason with contradictory statements. Our approach differs from paraconsistent logic in the introduction of agent as fundamental entity for the evaluation.

We build our approach on the results of direct measurements or answers of agents not on the aggregated utility functions to be able to model a structure of contradictions explicitly. Typically current axiomatic theories of uncertainty lack such capability. They assume that initial uncertainty evaluations are already given outside of these theories.

Conducting experiments with devices or asking agents are two common ways to assign initial (basic) uncertainties. Agents are asked to evaluate a proposition p using a classical truth-value logic function v(p) with values in {False,True}. The frequencies of answers provided by a set of agents G are computed. Alternatively each agent is asked to evaluate p using a generalized truth-value logic function v(p) with values in [0,1], where 0 and 1 are classical False and True values and values between 0 and 1 represent intermediate degrees of truth. Both frequency and the subjective evaluations can be used to get such v(p) as it is done in construction of the fuzzy logic membership function (MF)[7]. In the context of conflicting evaluations of preferences by agents, we discuss the empirical base that includes a mechanism for identifying a type of logic and reasoning computations in the agent logic.

This chapter introduces a hierarchy of logics of uncertainty that is core of the proposed *Agent-based Uncertainty Theory* (AUT). It is a further development of our previous works [8-13] that contain extensive references to related work. The fundamental analysis and review of relevant issues can be found in [1-4, 33-34].

The new contributions include constructing and discovering:

- the agent uncertainty theory (AUT) to model the uncertainty in the society of agents;
- connection among different types of uncertainties by using agents;
- connection between classic logic and fuzzy logic using agents that provides a clear distinction of their domains;
- a new model of many-valued logic and uncertainty by using concepts of conflicts among agents and irrationality of agents;
- a way to model quantum mechanics by using agents and many-valued logic.

The rest of the chapter is structured as follows. Section 10.2 presents the fundamental concepts of conflicts among agents of different orders with the physical interpretation in quantum mechanics. The first order of conflict is defined in section 10.3 Sections 10.4 and 10.5 contain definitions of the conflict of the second order or self-conflict with introduction of agents' correlation in quantum mechanics. In these sections, we define a new negation operator that allows the contradiction to be true and the tautology to be false. Sections 10.6 and 10.7 demonstrate how the self-conflict concept can be used to give a new representation for the evidence theory and the rough sets. This representation helps a user to understand better advantages and disadvantages of these theories and use them more efficiently. Section 10.9 summarises the uniformity of the AUT presentation of different uncertainty theories and section 10.10 describes application areas. Section 10.11 concludes the chapter.

10.2 Concepts and Definitions

The probability calculus does not incorporate explicitly the concepts of irrationality or agent's state of logic conflict. It misses structural information at the level of individual objects, but preserves global information at the level of a set of objects. Given a dice the probability theory studies frequencies of the different faces E={e} as independent (elementary) events. This set of elementary events E has *no structure*. It is only required that elements of E are *mutually exclusive* and *complete*, that is no other alternative is possible. The order of its elements is irrelevant to probabilities of each element of E. No irrationality or conflict is allowed in this definition relative to mutual exclusion. The classical probability calculus does not provide a mechanism for modelling uncertainty when agents communicate (collaborates or conflict). Recent work by Halpern [6] is an important attempt to fill this gap.

Now we will provide more formal definition of AUT concepts. It is done first for individual agents then for sets of agents. Consider a set of agents $G=\{g_1, g_2,\ldots,g_n\}$.

Each agent g_k assigns binary true/false value v∈{True, false} to a proposition p. To show that v was assigned by the agent g_k we will use notation $g_k(p) = v_k$.

Definition. An agent g is called a *reasoning agent* if g assigns a truth-value v(p) to any proposition p from a set of propositions S and any logical formula based on S.

Definition. A set of reasoning agents G is called *totally consistent* for proposition p if any agent g from G={g}, always provides the same truth value for p.

In other words, all agents in G are in concord or logical coherence for the same proposition. Thus, changing the agent does not change logic value v(p) for a totally consistent set of agents. Here v(p) is *global* for G and has *no local variability* (independent on the individual agent's evaluation). The classical logic is applicable for such set of consistent (rational) agents.

Definition. A set of reasoning agents G is called *inconsistent for* the proposition p if there are two subset of agents G_1, G_2 such that agents from them provides different truth values for p.

Definition. Let S be a set propositions, S={p_1 , p_2 ,…,p_n} then set ¬S = {¬p_1 , ¬p_2 ,…, ¬p_n} is called a *complimentary set* of S.

Definition. A set of reasoning agents G is called *S-only-consistent* if agents {g} are consistent only for propositions in S={p_1 , p_2 ,…,p_n} and are inconsistent in the complimentary set ¬S.

The evaluations of p is a vector-function **v**(p)=(v_1(p), v_2(p)…,v_n(p)) for a set of agents G that we will represent as follows:

$$\mathbf{v}(p)=\begin{pmatrix} g_1 & g_2 & \cdots & g_{n-1} & g_n \\ v_1 & v_2 & \cdots & v_{n-1} & v_n \end{pmatrix} \tag{10.1}$$

An example of the logic evaluation by five agent is shown below for *p="A>B"*

$$f(p)=\begin{pmatrix} g_1 & g_2 & g_3 & g_4 & g_5 \\ true & false & true & false & true \end{pmatrix}$$

Here A > B is true for agents g_1,g_3, and g_5 and it is false for the agents g_2, and g_4.

Kolmogorov's axioms of the probability theory are based on a totally consistent set of agents for a set of statements on mutual exclusion of elementary events. It follows from the following definitions for a set of events E={e_1.e_2,…,e_n}.

Definition. Set E={e_1.e_2,…,e_n} is called a set of *elementary events* (or mutually exclusive events) for predicate E if

$$\forall\, e_i\, e_j \in A \;\; E(e_i) \vee E(e_j) = \text{True} \;\text{ and } E(e_i) \wedge E(e_j) = \text{False}.$$

In other words, event e_i is an *elementary event* if for any j, $j \neq i$ events e_i and e_j cannot happen simultaneously, that is probability $P(e_i \wedge e_j) = 0$ and $P(e_i \vee e_j) = P(e_i) + P(e_j)$. Property $P(e_i \wedge e_j)=0$ is the *mutual exclusion axiom* (ME- axiom).

Let $S=\{p_1,p_2,.., p_n\}$ be a set of statements, where $p_i=p(e_i)$=True if and only if event e_i is an elementary event. In probability theory, $p(e_i)$ is not associated with any specific agent. It is assumed to be a global property (applicable to all agents). In other words, statements $p(e_i)$ are totally consistent for all agents.

Definition. A set of agent $S = \{g_1.g_2,\ldots,g_n\}$ is called a *ME-rational* set of agents if

$$\forall\, e_i\, e_j \in A,\ \forall\, g_i \in S,\ p(e_i) \vee p(e_j) = \text{True and } p(e_i) \wedge p(e_j) = \text{False}.$$

In other words, these agents are *totally consistent* or *rational on Mutual Exclusion.* Previously we assumed that each agent assigns value v(p) and we did not model this process explicitly. Now we introduce a set of criteria $\mathbf{C}=\{C_1,C_2,\ldots,C_m\}$ by which an agent can decide if a proposition p is true or false, i.e., now $v(p) = v(p,C_i)$, which means that p is evaluated by using the criterion C_i.

Definition. Given a set of criteria **C**, agent g is in a *self-conflicting state* if

$$\exists\, C_i\,,\, C_j\ (C_i\,,\, C_j \in \mathbf{C})\ \&\ v(p,C_i) \neq v(p,C_j)$$

In other words, an agent is in a self–conflicting state if two criteria exist such that p is true for one of them and false for another one.

With the explicit set of criteria, the logic evaluation function **v**(p) is not vector-function any more, but it is expanded to be a matrix function as shown below:

$$\mathbf{v}(p) = \begin{bmatrix} & g_1 & g_2 & \cdots & g_n \\ C_1 & v_{1,1} & v_{1,2} & \cdots & v_{1,n} \\ C_2 & v_{2,1} & v_{2,2} & \cdots & v_{2,n} \\ \cdots & \cdots & \cdots & \cdots & \cdots \\ C_m & v_{m,1} & v_{m,2} & \cdots & v_{m,n} \end{bmatrix} \tag{10.2}$$

For example, four agents using four criteria can produce **v**(p) as follows:

$$\mathbf{v}(p) = \begin{bmatrix} & g_1 & g_2 & g_3 & g_4 \\ C_1 & true & false & false & true \\ C_2 & false & false & true & true \\ C_3 & true & false & true & true \\ C_4 & false & false & true & true \end{bmatrix}$$

Here agent g_1 is in a self-conflicting state for the criteria C_1 and C_2, p is true for C_1 and false for C_2. Similarly, g_3 is in self-conflict for the same criteria. It is also in conflict with g_1. Agents g_2 and g_4 are not in self-conflict, but in conflict with each other.

If p is a preference statement, p = (A> B) and, ¬p is an opposite preference, p = (A> B) then a self-conflicting agent can provide two or more contradictory preferences, (p,¬p) for proposition p by stating that p is true and false at the same time.

This can be an indication that g has two or more different competing criteria, C_1, C_2,...C_m to compute the logic value for proposition p. These criteria may or may not be known in advance. The agent may say that car A is better than car B, p = (A>B)= True, and also that B is better than A, ¬p = (B>A) = True. This agent is clearly in self-conflict. It can be a result of using implicitly two different criteria C_1 and C_2, where C_1 minimizes price and C_2 maximizes quality.

Note that introduction of **C** explains self-conflict, but does not remove it. The agent still needs to resolve the ultimate preference contradiction having a goal to buy only one and better car. It also does not resolve conflict among different agents if they need to buy a car jointly.

If agent g can modify criteria in **C** making them consistent for p then g resolves self-conflict. The agent can be in a logic conflict state because of inability to understand the complex *context* and to evaluate *criteria*. For example, in the stock market environment, some traders quite often do not understand a complex context and rational criteria of trading. These traders can be in logic conflict exhibiting chaotic, random, and impulsive behavior. They can sell and buy stocks, exhibiting logic conflicting states "sell" = p and "buy" = ¬p in the same market situation that appears as irrational behavior, which means that the statement $p \wedge \neg p$ can be true.

A logical structure of self-conflicting states and agents is much more complex than it is without self-conflict. In the case of m binary criteria C_1 , C_2,...C_m that used to evaluate the logic value for the same attribute, there are 2^m states and only two of them (all true or all false values) do not exhibit conflict between criteria.

10.3 Framework of First Order of Conflict Logic State

10.3.1 Definitions

Definition. A set of agents G is in a first order of conflicting logic state (*first order of conflict*, for short) if

$$\exists\, g_i\,,\, g_j\;\; (g_i\,,\, g_j \in G)\;\&\; v(p,g_i) \neq v(p,g_j).$$

In other words, there are agents g_i and g_j in G for which in

$$\mathbf{v}(p) = \begin{pmatrix} g_1 & g_2 & \dots & g_{n-1} & g_n \\ v_1 & v_2 & \dots & v_{n-1} & v_n \end{pmatrix}$$

exist different values v_i and v_j.

Consider n agents g_i that buy cars. There is preference relation or proposition p = (A> B) between cars to be assigned by each of these agents (potential buyers) based on some preference criterion C, p = True if *A*>*B* else p = False. If criterion C is explicitly stated then we can write $A >_C B$. Each agent g_i answers a questionnaire with two options offered for C:

(1) "*A*>*B* is true" and (2) "*A*>*B* is false".

Say 70% of agents marked the proposition p = "*A*>*B* is true", giving *m*(*A* >*B*)= 0.7, which measures the level of conflict among n agents. The situation for which a group of non self-conflicting agents assumes that A>B is true and a complementary group of agents assumes that the same A>B is false, is a situation of conflict/contradiction among agents. We denote this situation as a *first order of conflict* for logic statements.

Here each individual agent is completely *rational* and has *no self-conflict*. The conflict exists only *between different agents* when they evaluate the same proposition with only one criterion C. We will write g(p) or g(A>B) to identify that preference statement p=(A>B) is assigned by the agent g. We also will write $g(p,C_k)$ in the case when it is needed to identify a criterion C_k used to compute g(p).

Let G(*A*>*B*) be a subset of agents in G such that *A*>*B* = True and *G(A<B)* be a subset of agents G such that (*A*<*B*) = True:

$$G(A>B) = \{g \in G \mid A>B \text{ is true}\}, \quad G(A<B) = \{g \in G \mid A<B \text{ is true}\}.$$

A set of agents G is in the *first order of conflict* if

$$G(A>B) \cap G(A<B) = \varnothing \;, \; G(A>B) \neq \varnothing \,, G(A>B) \cup G(A<B) = G.$$

The following definition presents this idea in general terms.

Definition. A set of agents G is in the *First Order Conflict (FOC)* for proposition p if

$$G(p) \cap G(\neg p) = \varnothing \text{ , and } G(p) \neq \varnothing \text{ , } G(p) \cup G(\neg p) = G.$$

Example. Let $G = \{g_1,g_2,g_3,g_4\}$ and for two agents $G(p) = \{g_1,g_4\}$ from *G* proposition p is true and for two other agents $G(\neg p) = \{g_2,g_3\}$ proposition $\neg$p is true, and p is false where p = (A>B). It can be written as g_1(p)=True, g_2(p)=False, g_3(p)=false, g_4(p)=True. The set of agents G is in a conflict logic state. We record the vector of logic values of p provided by all four agents as

$$\mathbf{v}(p) = \begin{bmatrix} g_1 & g_2 & g_3 & g_4 \\ v(p) = true & v(p) = false & v(p) = false & v(p) = true \end{bmatrix}$$

Fig. 10.1 shows a set of 20 agents in the logic conflicting state, where 7 white agents are in the state True and 13 black agents are in the state False for the same proposition p = " A > B".

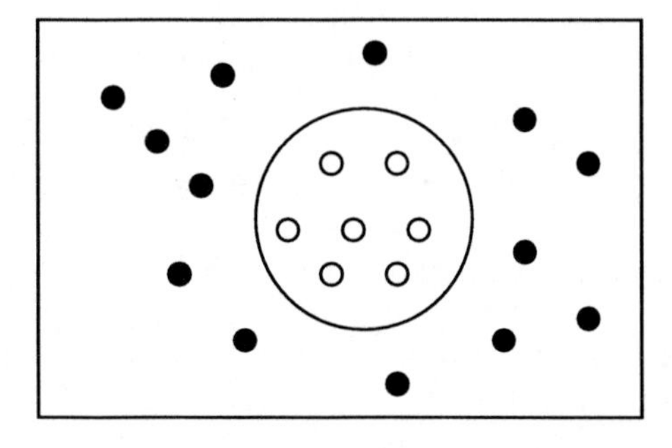

Fig. 10.1. A set of 20 agents in the first order of logic conflict

Below we show that at the first order of conflicts, AND and OR operations should not be the traditional classical logic operations, but should be *vector operations* in the space of the agents' evaluations (*agents space*). The vector operations reflect a structure of logic conflict among coherent individual agent evaluations.

10.3.2 Fusion Process

If a single decision must be made at the first order of conflict, then we must introduce a *fusion process* of the logic values of the proposition p given by all agents. A basic way to do this is to compute the weighted frequency of logic value given by all agents:

$$\mu(p) = w_1 v_1(p) + ... + w_n v_n(p) = \begin{bmatrix} v_1(p) \\ v_2(p) \\ \dots \\ v_n(p) \end{bmatrix}^T \begin{bmatrix} w_1 \\ w_2 \\ \dots \\ w_n \end{bmatrix} \tag{10.3}$$

where $\begin{bmatrix} v_1(p) \\ v_2(p) \\ \dots \\ v_n(p) \end{bmatrix}$ and $\begin{bmatrix} w_1 \\ w_2 \\ \dots \\ w_n \end{bmatrix}$ are two vectors in the space of the agents' evaluations.

The first vector contains all logic states (True/False) for all agents, the second vector (with property $\sum_{k=1}^{n} w_k = 1$) contains non-negative weights (utilities) that are given to each agent in the fusion process. In a simple frequency case, each weight is equal to 1/n.

At first glance, μ(p) is the same as used in the probability and utility theories. However, classical axioms of the probability theory have no references to agents producing initial uncertainty values and do not violate the mutual exclusion. As a result, formulas for assigning initial uncertainty values such as μ(A>B) are not a part of the theory. Value μ(p) can differ significantly from the classical relative frequency for some irrational agents.

A probability value can be agent's judgment or a result of physical experiments. In [2] agents are connected to logic and probability in the following way. In a given state s, the formula $P_j(\varphi)$ denotes the probability of the logic proposition φ according to the probability distribution for agent g_j in the state s. In this model, an individual agent gives the probability. All agents are autonomous and no conflict among agents and self-conflict is modeled explicitly. This formalization does not provide tools to fuse the contradictory knowledge of different agents that we described.

How we can justify formula (3) beyond the reference to the utility and probability theories? It can be done in two general ways by using concepts of descriptive and prescriptive theories to clarify this issue. To be descriptive (3) should have weight $\{w_i\}$ that model a specific task at hand and properly instantiated using available data.

To be prescriptive (3) must be inferred in a more general way, say, from axioms that are appropriate for a particular task. Below we define *vector logic operations* for the first order of conflict logic states **v**(p).

Consider μ_{max} = max (μ(p)). This maximum is reached for p such that v_i(p)=1 for all agents g_i. If in addition to this $\sum_{k=1}^{n} w_k = 1$ and $w_k \geq 0$ then μ_{max}(p)=1.

Definition. The *Agent Set Contradiction (ASC) index* λ(p) for proposition p is defined as follows

$$\lambda(p) = \begin{cases} 1-\mu(p)/\mu_{\max}(p), \text{ if } \mu(p) \geq 0.5\mu_{\max}(p) \\ \mu(p)/\mu_{\max}(p), \quad \text{if } \mu(p) < 0.5\mu_{\max}(p) \end{cases}$$

For instance, if μ(p) =0.7 then λ(p) =1-μ_{max}(p)= 0. 3 if μ_{max}(p)=1 and if μ(p) =0.3 then λ(p) = μ(p)=0. 3. Thus in both cases, λ(p) is the same showing the equal difference from the certain (True, False) values.

Definition. Agent g is *irrational for proposition* p if *Ir*(g,p)= 1, where *Ir*(g,p)= 1 ⇔ g(p)∧g(¬p)=True or g(p)∨g(¬p)=False.

Definition. The *Individual Agent Irrationality (IAI) index* γ(g,P) for a *set of h propositions* P is defined as follows:

$$\gamma(g, P) = \frac{1}{h} \sum_{i=1}^{h} Ir(g, p_i)$$

where *Ir*(g,p)= 1 ⇔ g(p)&g(¬p)=True or g(p)∨g(¬p)=False.

For instance, if γ(g,P)=0.5 then for 50% of propositions agent g exhibits irrational behavior. In this case it is better to avoid using the standard probabilistic technique, but if γ(g,P)=0.0001 then it can be very reasonable to use the probabilistic technique for propositions P for this individual agent.

Definition. Agent g is *irrational for proposition* p *and criterion* C_k if *Ir*(g,p,C_k) = 1, where *Ir*(g,p,C_k) = 1 ⇔ g(p,C_k) ∧ g(¬p,C_k)=True or g(p,C_k)∨g(¬p,C_k)=False.

Definition. The *Individual Agent Irrationality (IAI) index* γ(g,p,**C**) for *proposition* p and m criteria **C** is defined as follows:

$$\gamma(g, p, \mathbf{C}) = \frac{1}{m} \sum_{k=1}^{m} Ir(g, p, C_k)$$

where *Ir*(g,p,C_k)= 1 ⇔ g(p,C_k) ∧g(¬p,C_k)=True or g(p,C_k)∨g(¬p,C_k)=False.

For instance, if γ(g,P)=0.5 then a half of the agents are irrational at some extend, that is for 50% of agents at least one proposition exists for which g exhibits irrational behavior. If γ(p,**C**)=0.5 then

For instance, if $\gamma(g,p,\mathbf{C})=0.5$ then for 50% of criteria agent g exhibits irrational behavior for proposition p. This means that explicit identification of the criterion C_k does not resolves the contradiction for 50% of criteria. This is the indication that 50% of criteria do not capture the source of the irrationality specifically enough and further specification of criteria can be needed before a technique that is applicable to rational agents will be used. If $\gamma(g,p,\mathbf{C})=0.001$ then agent g has logic conflict with p only for 1% of criteria. Often this 1% can be ignored and the probabilistic technique cab be used.

Definition

$$\mathbf{v}(p \wedge q) = v_1(p) \wedge v_1(q), \ldots, v_n(p) \wedge v_n(q),$$

$$\mathbf{v}(p \vee q) = v_1(p) \vee v_1(q), \ldots, v_n(p) \vee v_n(q)$$

$$\mathbf{v}(\neg p) = \neg v_1(p), \ldots, \neg v_n(p),$$

where the symbols $\wedge$, $\vee$, $\neg$ in the right side of the equations are the classical AND, OR, and NOT operations.

These operations can be written also with explicit indication of agents involved in the first row:

$$\mathbf{v}(p \wedge q) = \begin{pmatrix} g_1 & g_2 & \cdots & g_{n-1} & g_n \\ v_1(p) \wedge v_1(q) & v_2(p) \wedge v_2(q) & \cdots & v_{n-1}(p) \wedge v_{n-1}(q) & v_n(p) \wedge v_n(q) \end{pmatrix}$$

$$\mathbf{v}(p \vee q) = \begin{pmatrix} g_1 & g_2 & \cdots & g_{n-1} & g_n \\ v_1(p) \vee v_1(q) & v_2(p) \vee v_2(q) & \cdots & v_{n-1}(p) \vee v_{n-1}(q) & v_n(p) \vee v_n(q) \end{pmatrix}$$

$$\mathbf{v}(\neg p) = \begin{pmatrix} g_1 & g_2 & \cdots & g_{n-1} & g_n \\ v_1(\neg p) & v_2(\neg p) & \cdots & v_{n-1}(\neg p) & v_n(\neg p) \end{pmatrix}$$

Below we present the important properties of sets of conflicting agents at the first order of conflicts. Let $|G(x)|$ be the numbers of agents for which x is true.

Statement 1. If G is a set of agents at the first order of conflicts and

$G(q)| \leq |G(p)|$ then

$$|G(p \wedge q)| = \min(|G(p)|, |G(q)|) - |G(\neg p \wedge q)|, \quad (10.4)$$

$$|G(p \vee q)| = \max(|G(p)|, |G(q)|) + |G(\neg p \wedge q)|. \quad (10.5)$$

If G is a set of agents at the first order of conflicts and

$|G(p)| \leq |G(q)|$ then

$$|G(p \wedge q)| = \min(|G(p)|, |G(q)|) - |G(\neg q \wedge p)|,$$

$$|G(p \vee q)| = \max(|G(p)|, |G(q)|) + |G(\neg q \wedge p)|.$$

and also $|G(p \vee q)| = |G(p) \cup G(q)|$, $G(p \wedge q)| = |G(p) \cap G(q)|$

Below we illustrate correctness of Statement 1 with two examples for three agents. In the example (a) three agents g_1, g_2, and g_3 provided answers $v_1(p)=1$, $v_2(p)=v_3(p)=0$ and $v_1(q)=v_2(q)=1$ and $v_3(q)=0$, respectively.

In the example (b) the answers are different: $v_1(p)=0$, $v_2(p)=v_3(p)=1$, and $v_1(q)=v_2(q)=0$, $v_3(q)=1$.

In example (a) only one agent (g_1) answered that p is true, thus $|G(p)|=1$. For q two agents (g_1,g_2) answered that q is true, thus $|G(q)|=2$ and $|G(p)| < |G(q)|$. Therefore, formula (4) is not applicable to this example. Its assumption $|G(q)| < |G(p)|$ is false.

In example (b) two agents (g_2, g_3) answered that p is true, thus $|G(p)|=2$. For q one agent (g_3) answered that q is true, thus $|G(q)|=1$. Hence, $|G(q)| < |G(p)|$ and (4) can be applied for this example. Its assumption is true. Here $\min(|G(p)|,|G(q)|) = 1$ and $|G(\neg p\wedge q)| = 0$ because $g_1(\neg p\wedge q) = 0$, $g_2(\neg p\wedge q) = 0$, $g_3(\neg p\wedge q) = 0$.
Thus, $\min(|G(p)| , |G(q)|) - |G(\neg p\wedge q)| = 1$. We have $|G(p\wedge q)| = 1$ on the left side of (4) because $g_1(p \wedge q)=0$, $g_2(p \wedge q)=0$, $g_3(p \wedge q)=1$. Thus, statement (1) is correct for example (b).

Corollary 10.1. If G is a set of agents at the first order of conflicts such that $G(q) \subset G(p)$ or $G(p) \subset G(q)$ then

$$G(\neg p \wedge q)| = \varnothing \text{ or } G(\neg q \wedge p)| = \varnothing$$
$$|G(p \wedge q)| = \min(|G(p)| , |G(q)|)$$
$$|G(p \vee q)| = \max(|G(p)| ,|G(q)|)$$

This follows from the statement 1. The corollary presents a well-known condition when the use of min, max operations has the clear justification.

Let $G^c(p)$ is a *complement* of G(p) in G: $G^c(p)= G \setminus G(p)$, $G= G(p) \cup G^c(p)$.

Statement 2. $G = G(p) \cup G^c(p)= G(p) \cup G (\neg p)$.

Corollary 10.2. $G (\neg p) = G^c(p)$. It follows directly from Statement 2.

Statement 3. If G is a set of agents at the first order of conflicts then

$$G (p \vee \neg p) = G(p) \cup G(\neg p) = G(p) \cup G^c(p) = G$$
$$G (p \wedge \neg p) = G(p) \cap G(\neg p) = G(p) \cap G^c(p) = \varnothing$$

It follows from the definition of the first order of conflict and statement 2. In other words, $G (p \wedge \neg p) = \varnothing$ corresponds to the contradiction $p \wedge \neg p$, that is always false and $G (p \vee \neg p)= G$ corresponds to the tautology $p \vee \neg p$, that is always true in the first order conflict.

Let $G_1 \oplus G_2$ be a *symmetric difference* of sets of agents G_1 and G_2,

$$G_1\oplus G_2 = (G_1 \cap G_2^c) \cup (G_1^c \cap G_2)$$

and let $p\oplus q$ be the *exclusive or* of propositions p and q,

$$p\oplus q = (p \wedge \neg q) \vee (\neg p \wedge q).$$

Consider, a set of agents G(p⊕q). It consists of agents for which values of p and q. Below we use the number of agents in set G(p⊕q) to define a differ from each other, that is

$$G(p \oplus q) = G((p \wedge \neg q) \vee (\neg p \wedge q)).$$

measure of difference between statements p and q and a measure of difference between sets of agents G(p) a G(q).

Definition. *Measure of difference* D(p,q) between statements p and q and a measure of difference D(G(p),G(q)) between sets of agents G(p) a G(q) are defined as follows:

$$D(p,q) = D(G(p),G(q)) = |G(p) \oplus G(q)|$$

Statement 4. D(p,q) = D(G(p),G(q)) is a distance, i.e., it satisfies distance axioms

$$D(p,q) \geq 0$$

$$D(p,q) = D(q,p)$$

$$D(p,q)+D(q,h) \geq D(p,h).$$

This follows from the properties of the symmetric difference ⊕ [Flament 1963].

Fig. 10.2 illustrates a set of agents G(p) for which p is true and a set of agents G(q) for which q is true. In Fig. 10.2(a) the number of agents for which truth values of p and q are different, $(\neg p \wedge q) \vee (p \wedge \neg q)$, is equal to 2. These agents are represented by white squares. Therefore the distance between G(p) and G(q) is 2. Fig. 10.2(b) shows other G(p) and G(q) sets with the number of the agents for which $\neg p \wedge q) \vee (p \wedge \neg q$ is true equal to 6 (agents shown as white squares and squares with the grid). Thus, the distance between the two sets is 6.

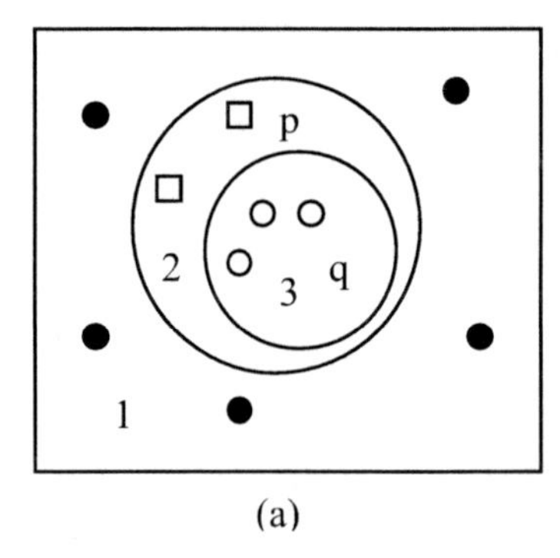

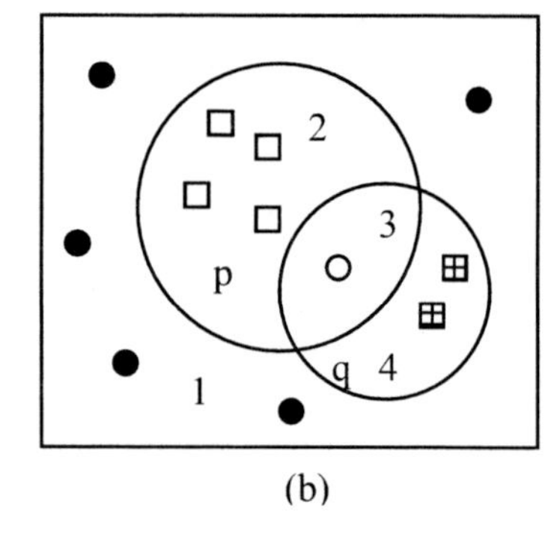

Fig. 10.2. A set of total 10 agents with two different splits to G(p) and G(q) subsets (a) and (b). These sprits produce different distances between G(p) and G(q). The distance in the case (a) is equal to 2 , the distance in the case (b) is equal to 6. Set 1 (black circles) consists of agents for which both p and q are false, Set 2 (white squares) consists of agents for which p is true but q is false. Set 3 (white circles) consists of agents for which p and q are true, and Set 4 (squares with grids) consists of agents for which p is false and q is true.

In Fig. 10.2(a), set 2 consists of 2 agents $|G((p \wedge \neg q)| = 2$ and set 4 is empty, $|G((\neg p \wedge \neg q)| = 0$, thus D(Set2, Set4)=2. This emptiness means that a set of agents with true p includes a set of agents with true q. In Fig. 10.2(b), set 2 consists of 4 agents $|G((p \wedge \neg q)| = 4$ and set 4 consists of 2 agents, $|G((\neg p \wedge \neg q)| = 2$, thus D(Set2,Set4)=6.

If G(p) includes G(q) or G(q) includes G(p) (case (a) in Fig. 10.2) then

$$|G(p \wedge q)| = |G(p) \cap G(q)| = \min(|G(p)| , |G(q)|) = 3 = |G(q)|$$
$$|G(p \vee q)| = |G(p) \cup G(q)| = \max(|G(p)| ,|G(q)|) = 5 = |G(p)|$$

In this case nothing is added or subtracted in (4) and (5),

$$|G(p \wedge q)| = \min(|G(p)| , |G(q)|) - |G(\neg p \wedge q)|,$$
$$|G(p \vee q)| = \max(|G(p)| ,|G(q)|) + |G(\neg p \wedge q)|,$$

Therefore, $|G(p \wedge q)|$ reaches its maximum value and $|G(p \vee q)|$ reaches its minimum value. In the case (b) in Fig. 10.2 the situation is different with the distance equal to 6, $|G(\neg p \wedge q)| = 2$ and

$$|G(p \wedge q)| = |G(p) \cap G(q)| = \min(|G(p)| , |G(q)|) - |G(\neg p \wedge q) | = 1$$
$$|G(p \vee q)| = |G(p) \cup G(q)| = \max(|G(p)| ,|G(q)|) + |G(\neg p \wedge q) | = 7$$

If sets of agents do not overlap then the distance between them reaches its maximum on these sets. Similarly, for AND operation the value of $|G(p \wedge q)|$ reaches its minimum and for OR operation the value of $|G(p \vee q)|$ reaches its maximum,

$$|G(p \wedge q)| = |G(p) \cap G(q)| = \min(|G(p)| , |G(q)|) - |G(\neg p \wedge q) | = 0$$
$$|G(p \vee q)| = |G(p) \cup G(q)| = \max(|G(p)| ,|G(q)|) + |G(\neg p \wedge q) | = |G(p)|+|G(q)|$$

Therefore, with the same $|G(p)|$ and $|G(q)|$ we may have different distances and different results for AND and OR operations among the sets. These operations depend on the internal structures of the agents' states. They are not simple truth-functional operations of the classical propositional logic. We remark that in this way t-norm and t-conorm (that are truth-functional) can be redefined to become context dependent.

The negation operator "$\neg$" for agents satisfies the De Morgan rule:

$$G [\neg (p \wedge q)] = G (p \wedge q)^C = [G(p) \cap G(q)]^C = G[\neg p \vee \neg q)] =$$
$$G(\neg p) \cup G(\neg q) = G^C(p) \cup G^C(q) = \max(G(\neg p) ,G(\neg q)) + G(p \wedge \neg q)$$
$$G [\neg (p \vee q)] = G(p \vee q)^C = G [\neg p \wedge \neg q)] = G(\neg p) \cap G(\neg q) =$$
$$G(p)^C \cap G(q)^C = \min(G(\neg p) ,G(\neg q)) - G(\neg p \wedge q),$$

and it can be used for reasoning with agents' logic evaluations that involves negation.

For complex computations of logic values provided by the agents we can use a *graph of the distances* among the sentences p_1 , p_2 ,….,p_N. For example, for three sentences p_1 , p_2 , and p_3, we have a graph of the distances shown in Fig. 10.3.

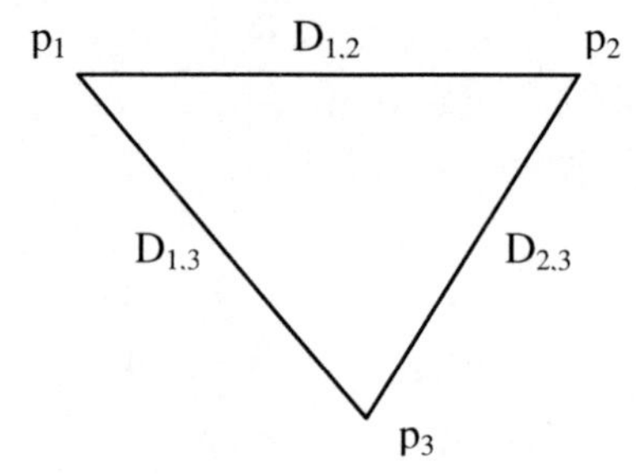

Fig. 10.3. Graph of the distances for three sentences p_1 , p_2 , and p_3

This graph has can be represented by a distance matrix

$$D = \begin{bmatrix} 0 & D_{1,2} & D_{1,3} \\ D_{1,2} & 0 & D_{2,3} \\ D_{1,3} & D_{2,3} & 0 \end{bmatrix}$$

which has a general form of a symmetric matrix,

$$D = \begin{bmatrix} 0 & D_{1,2} & \dots & D_{1,N} \\ D_{1,2} & 0 & \dots & D_{2,N} \\ \dots & \dots & \dots & .. \\ D_{1,N} & D_{2,N} & \dots & 0 \end{bmatrix}$$

Having the distances D_{ij} between propositions we can use them to compute complex expressions in agents' logic operation.

For instance, using $0 \le D(p, q) \le G(p) + G(q)$ and

$$G(p \wedge q) \equiv G(p \vee q) - G((\neg p \wedge q) \vee (p \wedge \neg q)) \equiv G(p \vee q) - D(p, q)$$

we can infer

$$D(p,q) = 0 \Rightarrow G(p \wedge q) = G(p \vee q) \tag{10.6}$$

and

$$D(p, q) = G(p) + G(q) \Rightarrow G(p \vee q) = G(p) + G(q) \text{ and } G(p \wedge q) = 0 \tag{10.7}$$

Similarly, we can infer:

$$G(\neg p \wedge q) \equiv D(p, q) - G(p \wedge \neg q) \tag{10.8}$$

for expression $\neg p \wedge q$,

$$G(\neg p \vee q) \equiv G(\neg p \wedge q)^C = N - (D(p, q) - G(p \wedge \neg q)) \tag{10.9}$$

for the implication rule $\neg p \vee q = p \rightarrow q$,

$$G(q) = G(p_1 \vee p_2) - D(p_1, p_2) \quad (10.10)$$

$$G(F) = G(q \vee p_3) - D(q, p_3) \quad (10.11)$$

for $F = p_1 \wedge p_2 \wedge p_3$ and $q = (p_1 \wedge p_2)$.

In this section, we formalized the concept of first order conflicting agents with the conflict only between different agents while each agent has no self-conflict. It is shown that for these conflicts AND and OR operations should be defined as vector operations in the agent space. Such operations preserve the coherence of individual agent evaluations. Next, we had shown that evaluations for the first order conflicting agents can differ from traditional evaluations that do not consider conflict among agents.

10.3.3 Quantum Mechanics Superposition and First Order Conflicts

The Mutual Exclusive (ME) principle states that an object cannot be in two different positions at the same time. In quantum mechanics, we have *non-local phenomena* for which the same particle can be in many different positions at the same time. This is a clear violation of the ME principle. The non-locality is essential for quantum phenomena. Given proposition p = "The particle is in position x in the space", agent g_i can say that p is true but agent g_j can say that the same p is false.

Individual states of the particle include its position, momentum and others. The *complete state* of the particle is a *superposition* of the quantum states of the particle w_i at different positions x_i in the space, This superposition can be a global wave function:

$$\psi = w_1 |x_1\rangle + w_2 |x_2\rangle + \ldots.. + w_n |x_n\rangle,$$

where $|x_i\rangle$ denotes the state at the position at x_i [26]. If we limit our consideration by an individual quantum state of the particle and ignore other states then we are in Mutual Exclusion situation of the classical logic (the particle is at x_i or not). In the same way in AUT, when we observe only one agent at the time the situation collapses to the classical logic that may not adequately represent many-valued logic properties. Having multiple states associated with the particle, we use AUT multi-valued logic as a way to model the multiplicity of states. This situation is very complex because measuring position x_i of the particle changes the superposition value ψ. However, this situation can be modeled by the first order conflict because each individual agent has no self-conflict.

10.4 Second Order of Conflicts, Self-conflicts, and Irrationality

10.4.1 Definitions and Examples

At the first order of logic conflict, we have $G(\neg p) = G^C(p)$ and fused $\mu(\neg p)$ satisfies the following statement 5.

Statement 5

$$\mu(\neg p) = w_1 v_1(\neg p) + ... + w_n v_n(\neg p) =$$

$$\begin{bmatrix} v_1(\neg p) \\ v_2(\neg p) \\ \cdots \\ v_n(\neg p) \end{bmatrix}^T \begin{bmatrix} w_1 \\ w_2 \\ \cdots \\ w_n \end{bmatrix} = \begin{bmatrix} 1-v_1(p) \\ 1-v_2(p) \\ \cdots \\ 1-v_n(p) \end{bmatrix}^T \begin{bmatrix} w_1 \\ w_2 \\ \cdots \\ w_n \end{bmatrix} = 1 - \mu(p)$$

This property is consistent with the known Zadeh's property $\mu(\neg p) + \mu(p) = 1$. Atanassov assumes in his model [16] three entities: property p , its negation ¬ p, and an indeterminacy s (about p or its negation) with the following definition of their relations,

$$\mu(\neg p) + \mu(p) + \mu(s) = 1.$$

Atanassov's model is also a *fuzzy partition* as defined by Ruspini [24]. Atanassov's model differs from Lukasiewicz's three-value logic that has a fixed μ(s) = 0.5, which is interpreted as possibility [15]. The simple observation suggests a potential conflict between Zadeh's and Atanassov's models noted in [15]. To give a meaning to Atanassov's model we introduce the second order of conflict or the *logical self-coflict* (LSC) among agents into this model.

We remark that at the first order of conflict the distance between p and ¬p is

$$D(p,\neg p) = |G(p)| + |G(\neg p)| = n,$$

where n is the total number of agents. Conceptually it follows from the fact that in the first order of conflict every agent uses only one criterion C to obtain the logic state true or false for the proposition p. In contrast, when an agent uses two criteria C_1 , C_2 to obtain the logic value for p we can get more complex situations, e.g.:

$$\begin{bmatrix} & C_1 & C_2 \\ State_1 & true & true \\ State_2 & true & false \\ State_3 & false & true \\ State_4 & false & false \end{bmatrix}$$

The states 1 and 4 are internally coherent states, where the agent is not in a self-conflict. In state 1, both criteria give true value, and in state 4, both criteria give false value for proposition p. The states 2 and 3 are self-conflicting logic states. In state 2, Criterion C_1 gives p true value, but criterion C_2 for the same p gives false value. Therefore, the same proposition is true and false in the same time and this generates agent's self-conflict. For state 3, the situation is opposite, where the agent gives the logic value false for C_1 and true for C_2. Here for the same proposition p we have two different logic values given by the same agent. It is clear that this is again a self-conflict situation. Thus, for two criteria we have four states of agents. An example is shown in Fig. 10.4.

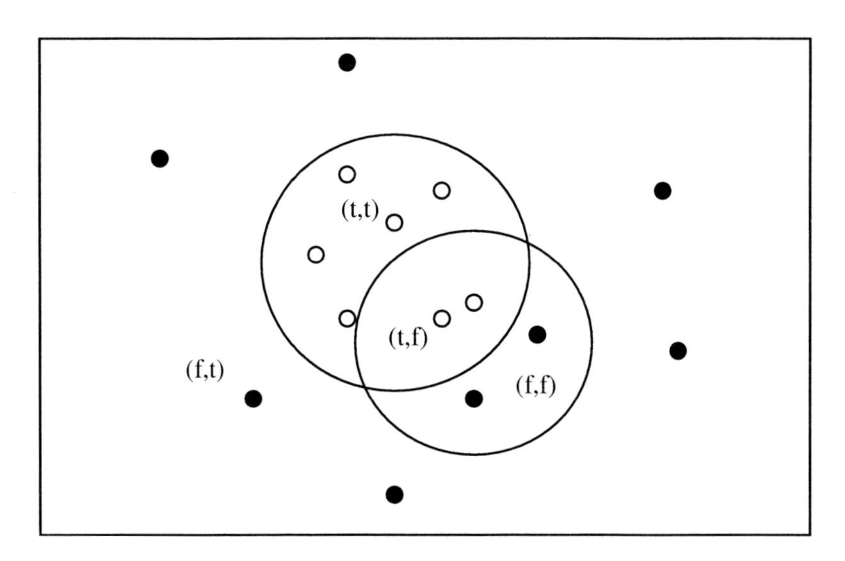

Fig. 10.4. Illustration of two criteria C_1 and C_2. White circles are agents for which p is true for criterion C_1 and black circles are agents for which p is false and $\neg$ p is true for C_1.

Let agents $g_1, g_2, \ldots, g_n$ be agents in G that use criteria C_1 and C_2 to assign logic values $v_{1,j}$ and $v_{2,j}$, respectively, which are true or false.

Definition. An agent g is a *self-conflicting agent* if $v(g,p,C_1) \neq v(g,p,C_2)$, where $v(g,p,C_1)$ and $v(g,p,C_2)$ are values assigned by agent g to proposition p using criterion C_1 or C_2, respectively.

Any self-conflicting agent can be denoted as an *irrational agent.* If p is preference relation and criteria C_1 and C_2 are known then the conflict is explicit and it has the explanation. It can be a conflict between cost (C_1) and reliability (C_2). If C_1 and C_2 are not known explicitly then the conflict is *implicit* and it has no explanations by criteria.

Definition. A set of agents G is in *the second order of conflict* if a self-conflicting agent g exists in G.

Formally, the complete evaluation by all agents for the two criteria is a matrix:

$$\mathbf{v}(p) = \begin{bmatrix} & g_1 & g_2 & \cdots & g_n \\ C_1 & v_{1,1} & v_{1,2} & & v_{1,n} \\ C_2 & v_{2,1} & v_{2,2} & & v_{2,n} \end{bmatrix}$$

The first row indicates agents, the second row contains the logic value for the first criterion, and the third row contains the logic value for the second criterion.

Below we show that at the second order of conflicts, the classical negation takes place if the agent uses one criterion at the time. Let agent g_k accepts proposition p as true using criterion C_1. In this case, two evaluations are possible when g_k uses C_2:

$$\mathbf{v}_1(p) = \begin{bmatrix} g_k \\ true \\ false \end{bmatrix}, \mathbf{v}_2(p) = \begin{bmatrix} g_k \\ true \\ true \end{bmatrix}$$

If agent g_k evaluates p using C_1 as false then we have two other cases:

$$\mathbf{v}_3(p) = \begin{bmatrix} g_k \\ false \\ false \end{bmatrix}, \mathbf{v}_4(p) = \begin{bmatrix} g_k \\ false \\ true \end{bmatrix}$$

If g_k ignores values provided by C_2 when negating results provided by using C_1 then it is classical logic situation to produce negation as shown below for $\mathbf{v}_3(\neg p)$ and $\mathbf{v}_4(\neg p)$:

$$\mathbf{v}_3(\neg p) = \begin{bmatrix} g_k \\ true \\ false \end{bmatrix}, \mathbf{v}_4(\neg p) = \begin{bmatrix} g_k \\ true \\ true \end{bmatrix}$$

Again, in this situation the second criterion C_2 is completely *independent* from the first criterion C_1 and has no influence on the logic evaluation of p using C_1. In a graphic form, this situation was depicted in Fig. 10.4. If g_k evaluated p as true using the second criterion C_2 then two cases are possible:

$$\mathbf{v}_5(p) = \begin{bmatrix} g_k \\ fasle \\ true \end{bmatrix}, \mathbf{v}_6(p) = \begin{bmatrix} g_k \\ true \\ true \end{bmatrix}$$

Again, assuming here that agent g_k ignores the first criterion when negating the truth value produced by C_2 we get:

$$\mathbf{v}_5(\neg p) = \begin{bmatrix} g_k \\ false \\ false \end{bmatrix}, \mathbf{v}_6(\neg p) = \begin{bmatrix} g_k \\ true \\ false \end{bmatrix}$$

In a graphic form, it is shown in Fig. 10.5.

Now when p is true for the first criteria and false for the second criteria, we obtain

$$\mathbf{v}(p) = \begin{bmatrix} & g_k \\ C_1 & true \\ C_2 & true \end{bmatrix}, \mathbf{v}(p) = \begin{bmatrix} & g_k \\ C_1 & true \\ C_2 & false \end{bmatrix}$$

$$\mathbf{v}(\neg p) = \begin{bmatrix} & g_k \\ C_1 & true \\ C_2 & false \end{bmatrix}, \mathbf{v}(\neg p) = \begin{bmatrix} & g_k \\ C_1 & false \\ C_2 & false \end{bmatrix}$$

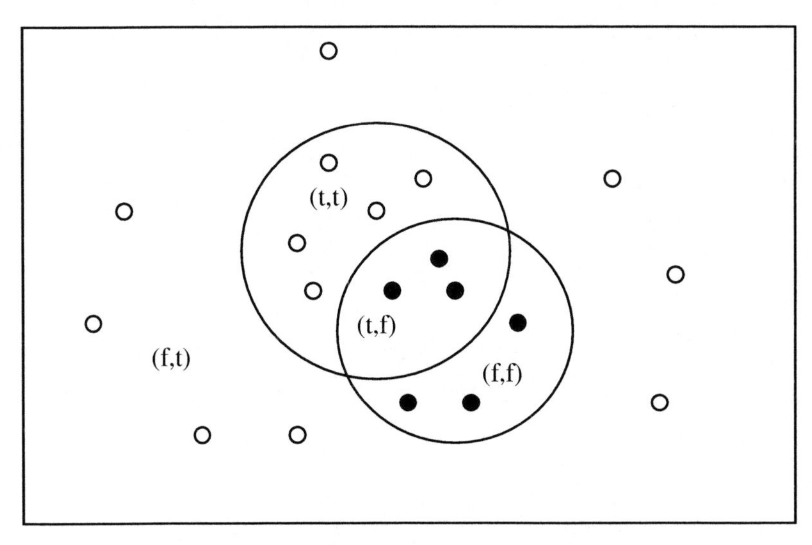

Fig. 10.5. Illustration of independent criterion C_2. White circles are agents for which p is true and black circles are agents for which p is false and ¬ p is true for criterion C_2.

The agents in self–conflict states belong to the set of agents for which p is true and p is not true. Thus,

$$\mathbf{v}(p) \wedge \mathbf{v}(\neg p) = \begin{bmatrix} & g_k \\ C_1 & true \\ C_2 & false \end{bmatrix} \wedge \begin{bmatrix} & g_k \\ C_1 & true \\ C_2 & false \end{bmatrix} = \begin{bmatrix} & g_k \\ C_1 & true \\ C_2 & false \end{bmatrix}$$

The contradiction is true for the first criterion, but when p is false for the first criterion and true for the second one, we have

$$\mathbf{v}(p) = \begin{bmatrix} & g_k \\ C_1 & false \\ C_2 & true \end{bmatrix}, \mathbf{v}(p) = \begin{bmatrix} & g_k \\ C_1 & false \\ C_2 & false \end{bmatrix}$$

$$\mathbf{v}(\neg p) = \begin{bmatrix} & g_k \\ C_1 & true \\ C_2 & true \end{bmatrix}, \mathbf{v}(\neg p) = \begin{bmatrix} & g_k \\ C_1 & false \\ C_2 & true \end{bmatrix}$$

and

$$\mathbf{v}(p) \wedge \mathbf{v}(\neg p) = \begin{bmatrix} & g_k \\ C_1 & false \\ C_2 & true \end{bmatrix} \vee \begin{bmatrix} & g_k \\ C_1 & false \\ C_2 & true \end{bmatrix} = \begin{bmatrix} & g_k \\ C_1 & false \\ C_2 & true \end{bmatrix}$$

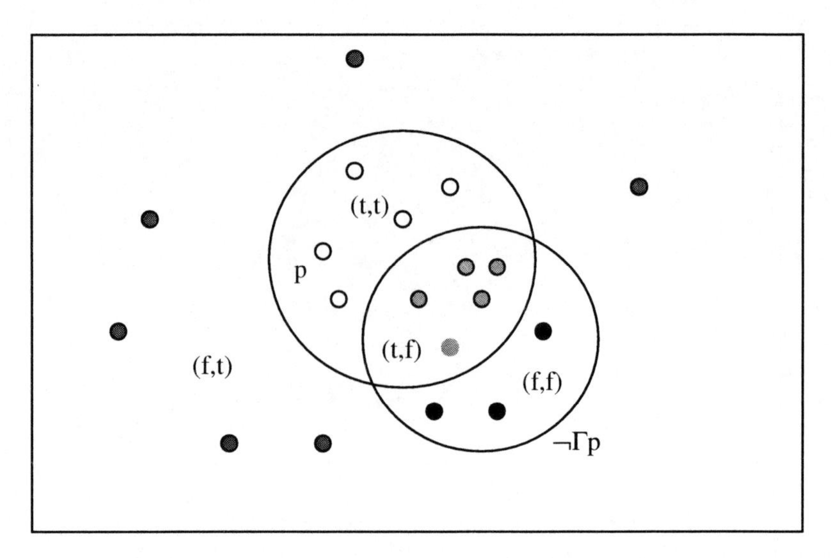

Fig. 10.6. Example with two criteria. The white circles are the agents for which p is true for both C_1 and C_2 criteria, which is denoted as (t,t). The black circles are the agents for which p is false for both C_1 and C_2 criteria, that is $\neg$ p is true for C_1 and C_2. The green agents are self-conflicting agents for which the contradiction p $\wedge\neg\Gamma$p is true for the first criterion C_1. For the red self-conflicting agents the tautology is false for the first criteria.

The tautology is false for the first criterion. Fig. 10.6 provided another example explained in its caption.

We can remark that the graph in Fig. 10.6 can be obtained by the superposition of the two criteria C_1 , C_2 as we show in Fig. 10.7.

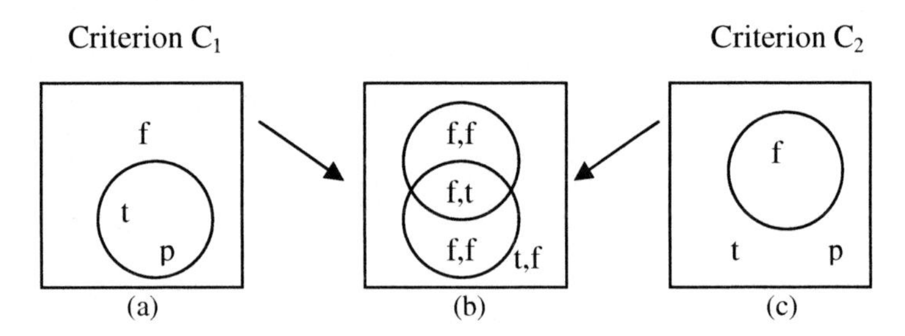

Fig. 10.7. Criteria and logic values in the second order of conflict. Case (a) presents a set where p is true for the first criterion, and case (c) presents a set where p is true for the second criterion. Case (b) shows the superposition of the two criteria.

10.4.2 Correlation in Quantum Mechanics and Second Order Conflict

Consider two interacting particles as agents. These agents are interdependent. This correlation is independent from any possible physical communication by any type of fields. We can say that the correlation is more a logic state for the particles that are dependent. Now the quantum correlation is a conflicting state because we know from quantum mechanics that there is a correlation but when we try to measure the correlation we cannot check the correlation itself. The measure destroys the correlation. So if the spin of one electron is up and another electron is down the change of the first spin when we have correlation or entanglement generate the change of the other spin instantaneously and this is manifested in a statistic correlation different from zero.

10.5 Agents and Atanassov's - Ruspini's Models

This section shows that Atanassov's intuitionistic fuzzy sets [22] and Ruspini's model [24] can gain an interpretation using agents' approach. There is a following property in [22]:

$$G(p) \cup G(\neg p) + G(1) = \text{Universe}$$

Now consider an example in Fig. 10.8, where G(p) is a set of agents in areas (2) and (3), G(¬p) is a set of agents in areas (3) and (4), and G(1) is the set of agents in area (1) with a vector of values (f,t).

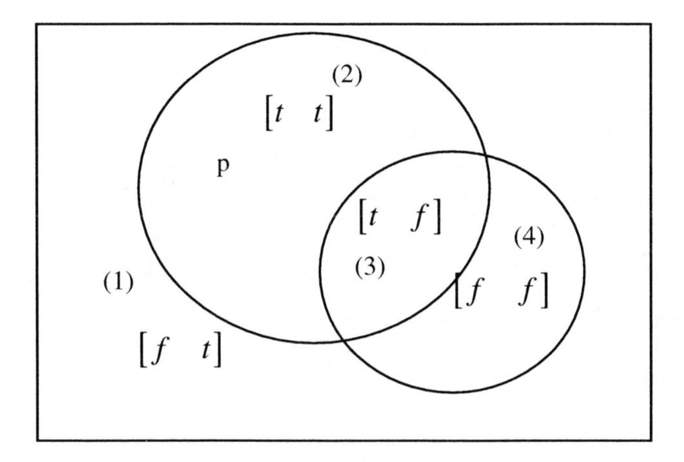

Fig. 10.8. Four classes of agents. The logic value true for p is located in (2 ,3), the criteria negation of p is located in (3,4). Set (3) and set (1) are the logically self-conflicting states.

Let set (2,3) be "Tall" and set (4,3) be "Short", that is different from the word "not tall" which is set (1,4). This consideration is in line with Atanassov's work. Now with the *distance* among set of agents that represent propositions we can define the distance between sets "Tall" and "Short". This gives agents' interpretation of the Atanassov's theory. In set 3, we have the superposition of p and negation ¬ p, where ¬ p is true for C_2. In sets 3 and 4, p is false for the criterion C_2 and p is true. In sets 2 and 3, p is true for criterion C_1. In addition, in set 3, the negation enters to the set of agents where p is true. Thus, we have

$$G(p \wedge \neg p) = G(p) \cap G(\neg p) = G(3) \neq \varnothing$$

Therefore, the contradiction is not empty. For set 1 we have

$$G(p \vee \neg p) = G(p) \cup G(\neg p) = G(2,3,4) \subseteq \text{Universe}$$

$$G(1) = G(2,3,4)^C$$

Hence, the tautology does not cover all sets of agents. If $C_1 \neq C_2$ then we may have an *ignorance* state which leads to the situation where tautology is not true. The agent can be *unaware* about presence of two different criteria, C_1 and C_2.

Thus, we can have a *contradiction* state, $g(p,C_1) \neq g(p,C_2)$ Different aspects of concepts of Ignorance or contradiction states are discussed in [15]. Using a concept of the second order of conflict we can explain the contradiction of Zadeh's rule

$$\mu(p \wedge \neg p) = \min(\mu(p), \mu(\neg p)) = \min(\mu(p), 1 - \mu(p)) \geq 0$$

to the classical logic, where $\mu(p \wedge \neg p) = 0$.

Zadeh's rule produces $\mu(p \wedge \neg p) > 0$ for the contradiction $p \wedge \neg p$. In the AUT agent approach to uncertainty, the rule of min for AND operation is valid only when $G(\neg p) \subseteq G(p)$ that is

$$|G(p \wedge \neg p)| = |G(p) \cap G(\neg p)| = \min(|G(p)|, |G(\neg p)|) = |G(\neg p)|$$

$$\text{if } G(\neg p) \subseteq G(p) \text{ then } |G(p)| \geq |G(\neg p)|$$

All agents in $G(\neg p)$ are in a state of logical self-conflict. With the introduction of the logically self-conflicting agents or contradictory agents, we solve the Zadeh's incoherence among the conjunction and the negation operations.

10.6 Sugeno Negation and Negation with Agents at the Second Order of Conflict

For the Sugeno negation,

$$\mu(\neg p) = \frac{1-\mu(p)}{1+\lambda\mu(p)} \quad \text{where } \lambda = [-1, \infty]$$

we have

$$\mu(\neg p) + \mu(p) = \frac{1-\mu(p)}{1+\lambda\mu(p)} + \mu(p) = \frac{1+\mu(p)^2}{1+\lambda\mu(p)}$$

and

$$\mu(\neg p) + \mu(p) - 1 = \lambda\mu(p)(\frac{\mu(p)-1}{1+\lambda\mu(p)})$$

Therefore,

$$\text{if } \lambda \geq 0 \text{ then } \mu(\neg p) + \mu(p) \leq 1$$

$$\text{if } \lambda < 0 \text{ then } \mu(\neg p) + \mu(p) > 1$$

Fig. 10.9 shows the sets of agents for two cases of the parameter λ

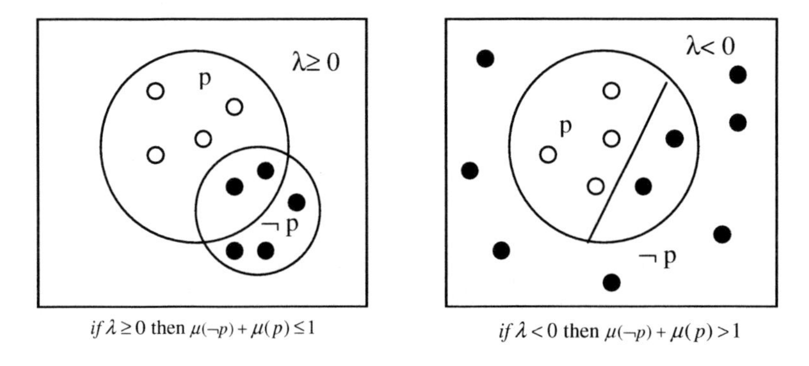

Fig. 10.9. Example for the Sugeno negation by set of agents

In Fig. 10.8 black circles show the agents for which ¬p is true and white circles show agents for which p is true for the situations with λ > 0, and λ < 0, respectively.

10.7 Evidence Theory and Agents

In the evidence theory any subset A ⊆ U of the universe U is associated with a value *m*(A) called a basic assignment probability such that

$$\sum_{k=1}^{2^N} m(A_k) = 1$$

Respectively, the belief Bel(Ω) and plausibility Pl(Ω) measures for any set Ω are defined as

$$Bel(\Omega) = \sum_{A_k \subseteq \Omega} m(A_k),\ Pl(\Omega) = \sum_{A_k \cap \Omega \neq \varnothing} m(A_k)$$

Thus, Bel measure includes all subsets that are inside Ω , The Pl measure includes all sets with non-empty intersection with Ω. In the evidence theory one and only one agent is associated with any element in the set Ω. Therefore, we have the situation shown in Fig. 10.10.

Fig. 10.10 shows set A that is overlapped with set Ω, which is divided in two parts: one inside Ω and another one outside. For the belief measure we exclude set A, thus we exclude the false state (*f,f*) and a logical self-conflicting state (*t,f*). But for the plausibility measure we accept the (*t* ,*t*) and the self-conflicting state (*f,t*).

In the belief measure, we *exclude* any possible self-conflicting states, but in the plausibility measure, we *accept* self-conflicting states.

There are two different criteria to compute the belief and plausibility measures in the evidence theory. For belief C_1 criterion is related to set Ω, thus C_1 is true for the cases inside Ω. The second criterion C_2 is related to set A, thus C_2 is false for the cases inside A.

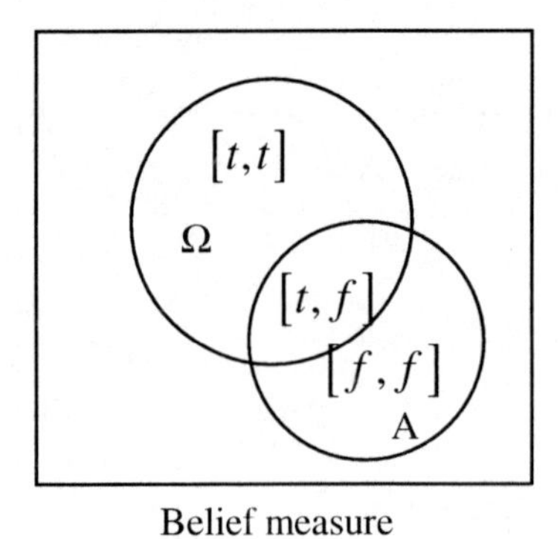

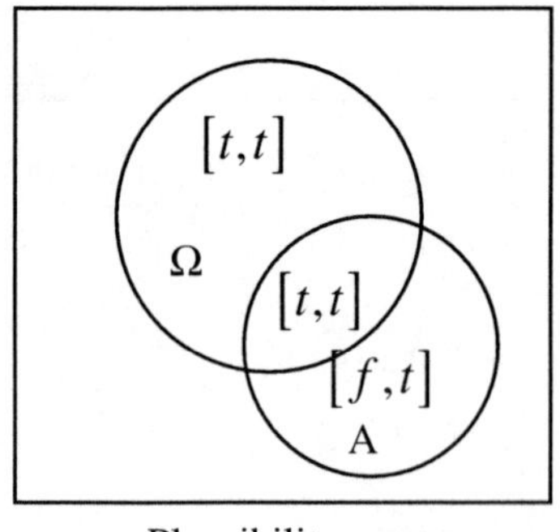

Fig. 10.10. Example of irrational agents or self-conflicting agent for the belief theory and plausible measures

Now we can put in evidence the logically self-conflicting state (t , f), where we eliminate A also if it is inside Ω. The same is applied to the plausibility, where C_2 is true for cases inside A. In this situation, we accept a logically self-conflicting situation (f,t) , where cases in Ω are false but cases in A are true.

10.8 Rough Set Theory and Agents

Below we use the logical self-conflict to study the rough set. See Fig. 10.10. In Fig. 10.11, the rough set is defined by two criteria. The first criterion C_1 assumes that p is true if x is inside the partition of the space, the second criterion assumes that p is true if x is inside the set Ω.

The rough sets have internal and external measures. When elements of the partition are accepted without any self-conflict, we get an *internal measure*. When we accept the elements of the partition having the self-conflicting agents, we obtain the *external measure*.

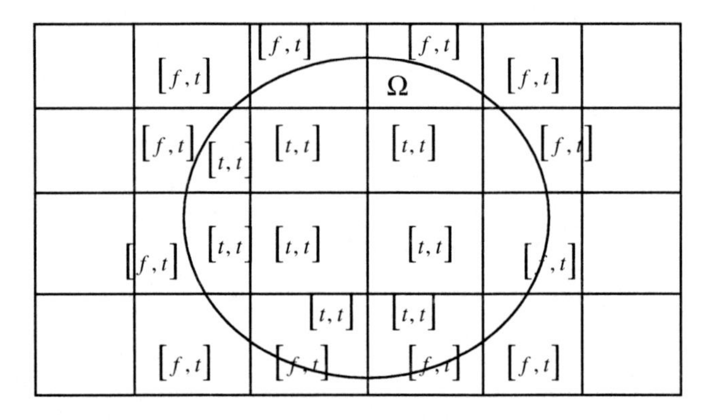

Fig. 10.11. Example of irrational agents for the Rough sets theory

10.9 AUT and Uniform Presentation of Uncertainty Theories

The agent approach to uncertainty permits to model different types of uncertainty due to: (1) ability of a set of agents to fuse conflicting logic evaluations, (2) ability of an individual agent to be self-conflicting, and (3) ability to use different negations relative to the ordinary logic evaluations. The contradiction can be true because a self-conflicting agent can evaluate p with one criterion and evaluate $\neg p$ with another criterion. Thus, p can be true for one criterion and $\neg p$ can be true for another criterion. This generates the internal irrational state. The same takes place for the tautology, p can be false for one criterion or $\neg p$ can be false for another criterion. Therefore, $p \vee \neg p$ can be false. Again, for the internal conflict we have another irrational situation for which tautology can be false.

Fusion of evaluations provided by agents and the introduction the internal structure represented by a set of criteria appear similar to the quantum interference or superposition for which conflicting positions or states can converge in one meta-position or state generated by the interference of the positions with different weights. Thus, we can say that in quantum mechanics we have a logic superposition for the same particle in different positions.

The second and higher levels of conflicts in AUT expand the concept of the agent to the concept of *self* that is an individual agent with the internal structure (presented by a set of criteria C) and/or a *fused agent* (superposition of agents, particles).

In analogy, the entangled particles create a meta-particle which itself is composed from different correlated particles. Entangled particles can be modeled by using a set of criteria C, one for any particle in the entanglement state. In other words, entanglement generates an irrational situation, where contradiction is true and tautology is false. This irrationality is represented by the instantaneous modification of the states of two particles without material communication among particles. This paradox was represented by the famous Bell inequality and by Einstein-Rosen-Podolsky or EPR paradox [26]. Here quantum mechanics violates the principle of realism for which any correlation is associated with a material communication with maximum velocity at the speed of light.

Table 10.1. Agents and uncertainty theories

Theory	*Component that uses rational agents*	*Component that uses irrational agent and self-conflict states*
Probability theory	Probability measure	No
Fuzzy logic	Membership function	Membership function
Intuitionistic fuzzy sets	Membership function	Membership function
Sugeno negation	Membership function	Membership function
Evidence theory	Belief measure	Plausibility measure
Rough Sets	Internal measure	External measure
Quantum mechanics		Particle superposition, entanglement

One of the important results of AUT is the possibility to model different types of uncertainties (probability, fuzzy sets, evidence theory evaluations, and rough set evaluations) under the same umbrella of the agents. Table 10.1 summarizes components of these theories that are modeled in AUT by rational agents and those that are modeled in AUT by irrational agents. Some components can be modeled by both types of agents. A more detailed classification of these components with different levels of conflicts and criteria using AUT is the subject of future research.

10.10 Applications

Potential application fields of AUT include verification of software agents, robotics, quantum computing, sensor networks, social modeling such as terrorism, stock market trading, marketing, web-based recommending systems that help a user to select a product, and many other fields. In all these areas agents demonstrate both rational and irrational behavior.

Different aspects of uncertainty have been considered in many types of Multi-Agent Systems (MAS). One of them is the problem of coalition formation when agents are *uncertain* about the types or *capabilities of their potential partners*. In [27] it was approached by using the probabilistic technique of a Bayesian reinforcement learning (Bayesian agents) for the situation when coalitions are formed and tasks undertaken repeatedly. Bayesian agents are rational, but their information about other agents is not complete. Here the list of types or capabilities of potential partners (other agents) are known to the agent, but it is not known which particular type is a type of the specific partner.

Rational agents operating under uncertainty also considered in [28]. These agents maximize the expected performance criterion in the form of monetary a utility payoff matrix or payoff game matrix that express preferences of agents over goods, states, or money that depend on wealth levels of agents. The wealth level of agents is considered as *private* information while the monetary payoff game matrix as *public* information that is available to each agent. The uncertainty is coming from the fact that private information is not known to other agents. Agents must negotiate to find a *stable state* using only the *public information.* The internal information about the agent that is taken into account in [28] is the *risk preference type* (risk averse, risk neutral and risk seeking). Note that from our viewpoint, even agents with risk neutral preferences can be rational or irrational, while risk averse and risk seeking agents can exhibit irrational behavior more often. This can dramatically change the negotiation result when agents attempt to reach a stable equilibrium state under uncertain games. Thus, we see that incorporation of the AUT concept of irrational agents can be beneficial in these agent negotiation tasks.

The paper [29] investigates the effect of different rationality assumptions on the performance of co-learning by multiple agents in multi-step games with close interactions between agents where an agent learns not only its own value function but also those of other agents. This paper shows complex pattern of effects resulting from different levels of rationality assumptions.

A multi-agent planning problem under temporal constraints and uncertainty of admissible operations is studied in [30]. The aim is to transform the system into one of the goal states satisfying all specified constraints with given a description of the current state of the system, a set of possible actions on the system and a description of its goal states. Here the uncertainty is associated with the set of admissible operations that are not known completely, but each agent is again rational.

The impact of uncertainty in agent-based approach in Market Engineering is considered in [31] and the uncertainty management framework for the multi-agent system is proposed in [32].

Below we outline one of the possible processes of applying the AUT:

(1) Identify a set of agents G={g};
(2) Identify a set op propositions P={p} that agents evaluate;
(3) Collect data: g(p) values for g and p in G and P. It can be a complete data set or a representative sample;
(4) Identify contradictory agents and compute agent set contradictory index (see section 10.3.1);
(5) Identify irrational agents as defined in section 10.3.1;
(6) Compute individual agent irrationality index γ(g,p) defined in section 10.3.1;
(7) Identify a set of criteria **C**={C};
(8) Collect data g(p, C_k) for g, p and C_k in G, **P** and **C** for irrational agents. It can be a complete data set or a representative sample;
(9) Identify irrational agents for a set of criteria **C** as defined in section 10.3.1;
(10) Compute individual agent irrationality index γ(g,p,**C**) defined in section 10.3.1;
(11) Test if the criteria **C** can explain irrational behavior of agents by finding criteria C_i and C_j for p such that (g(p,C_i)=True $\wedge$ g($\neg$p, C_j)=True) or (g(p,C_i)=Fase $\wedge$ g($\neg$p p, C_j)=False).
(12) Test if AND or OR operations on propositions can resolve irrational behavior of agents. The example below illustrates this situation.

Consider an example of a car selection by customer from the viewpoint of car dealers. The dealers know that customers exhibit irrational behavior quite often. They may not be able to change such irrational behavior, but can build a marketing strategy based on irrational preferences. Say, it was discovered that the customer considers car C as better than car A in spite of his/her preference A>B and B>C. Here we have a case of violation of the rational behavior (transitivity violation) by the agent. In this situation, the dealer can build an advertisement around car C not car A. On the other hand, if a dealer wants to modify preferences of the customer the dealer may add another product to be sold together with cars to make A preferable and to get rational preference of the agent. This new product can be the extended warranty for the car, which is crafted to be better for car A. The AUT vector logic with multiple criteria can model such situation by adding a new criterion. Having multiple agents and the need to construct a marketing strategy that will work for many customers, the AUT fusion (aggregation) of vector evaluations can be used. The AUT also helps to understand the reasons of the irrational behavior of the agents by the use of a set of criteria.

10.11 Conclusion

Agents and environment are major sources of uncertainty in the multi-agent systems. A set of all decision alternatives and states of the environment typically are unknown to the autonomous agent. These are fundamental *external* (to the agent) sources of uncertainty. The probability theory assumes that all alternatives are known in advance (e.g., six sides of the dice), but their occurrence is known only partially (probabilistically). Another source of uncertainty in MAS is the agent's own *internal* reasoning abilities. If the agent does not evaluate alternatives rationally and does not reason rationally, then this creates uncertainty of the agent's decisions and actions. The AUT addresses both internal and external uncertainty challenges in multi-agent systems.

The hierarchy of sets of agents relative to levels of conflicts, self-conflicts, and irrationality proposed in this paper is intended to provide a base for several areas of further studies. These studies include the foundation of probability theory, fuzzy logic, and other types of uncertainties, as well as real-world interpretation of these theories including quantum mechanics and quantum computing.

The major weakness in fuzzy logic is its ad hoc nature with operations that are not justified in advance (as it is done in the probability theory), but adjusted in many different ways (by using many different t-norms and t-conorms). The novelty of this paper is not in introduction of new formulas for initial membership function values and operations with them, but in creating a base for their *interpretation* by means of conflicting and irrational agent as an internal part of the theory. Physics provides plenty of positive examples of such interpretations that can be inspiration examples for building comprehensive logics of uncertainty for agents.

A society of agents always has a degree of conceptual conflicts and any agent can be self-conflicting. This affects agents' abilities to use resources for identifying goals and acting. In this paper, we established the fundamental logic structure that can be used for making logic decisions in a conflicting multi-agent environment. The new contributions of this paper consist of:

1) Introduced the agent uncertainty theory (AUT) to model the uncertainty in the society of agents;
2) Discovered connection among different types of uncertainties by using agents;
3) Discovered connection between classic logic and fuzzy logic using AUT that allows providing empirical, physical interpretations for them and distinction of their domains;
4) Built a new model of many-valued logic and uncertainty by using concepts of conflicts among agents and irrationality of agents;
5) Identified a way to model quantum mechanics by using agents and many-valued logic.

In the future research we plan to study more complex structures with more criteria and higher order of conflicts In fact, a set of N criteria can generate $O(2^N)$ the logically self-conflicting states and very complex conflict situations. We expect that in near future this will be an active area of new fundamental research and discoveries.

References

1. Carnap, R., Jeffrey, R.: Studies in Inductive Logics and Probability, vol. 1, pp. 35–165. University of California Press, Berkeley (1971)
2. Fagin, R., Halpern, J.: Reasoning about Knowledge and Probability. Journal of the ACM 41(2), 340–367 (1994)
3. Edmonds, B.: Review of Reasoning about Rational Agents by Michael Wooldridge. Journal of Artificial Societies and Social Simulation 5(1) (2002), `http://jasss.soc.surrey.ac.uk/5/1/reviews/edmonds.html`
4. Ferber, J.: Multi Agent Systems. Addison Wesley, Reading (1999)
5. Gigerenzer, G., Selten, R.: Bounded Rationality. MIT Press, Cambridge (2002)
6. Halpern, J.: Reasoning about uncertainty. MIT Press, Cambridge (2005)
7. Hisdal, E.: Logical Structures for Representation of Knowledge and Uncertainty. Springer, Heidelberg (1998)
8. Resconi, G., Jain, L.: Intelligent agents. Springer, Heidelberg (2004)
9. Resconi, G., Klir, G.J., Clair, U.S.: Hierarchical uncertainty metatheory based upon modal logic. International J.Gen.Systems 21, 23–50 (1992)
10. Resconi, G., Klir, G.J., Clair, U.S., Harmanec, D.: In the integration of uncertainty theories. Intern. J. Uncertainty Fuzziness knowledge-Based Systems 1, 1–18 (1993)
11. Resconi, G., Klir, G.J., Harmanec, D., Clair, U.S.: Interpretation of various uncertainty theories using models of modal logic: a summary. Fuzzy Sets and Systems 80, 7–14 (1996)
12. Harmanec, D., Resconi, G., Klir, G.J., Pan, Y.: On the computation of uncertainty measure in Dempster-Shafer theory. International Journal General System 25(2), 153–163 (1995)
13. Resconi, G., Murai, T., Shimbo, M.: Field Theory and Modal Logic by Semantic field to make Uncertainty Emerge from Information. Int. Journal General System 29(5), 737–782 (2000)
14. Resconi, G., Turksen, I.B.: Canonical Forms of Fuzzy Truthoods by Meta-Theory Based Upon Modal Logic. Information Sciences 131, 157–194 (2001)
15. Resconi, G., Kovalerchuk, B.: The Logic of Uncertainty with Irrational Agents. In: Proc. of JCIS-2006 Advances in Intelligent Systems Research. Atlantis Press, Taiwan (2006)
16. Kahneman, D.: Maps of Bounded Rationality: Psychology for Behavioral Economics. The American Economic Review 93(5), 1449–1475 (2003)
17. Kovalerchuk, B.: Analysis of Gaines' logic of uncertainty. In: Turksen, I.B. (ed.) Proceeding of NAFIPS 1990, Toronto, Canada, vol. 2, pp. 293–295 (1990)
18. Kovalerchuk, B.: Context spaces as necessary frames for correct approximate reasoning. International Journal of General Systems 25(1), 61–80 (1996)
19. Kovalerchuk, B., Vityaev, E.: Data mining in finance: advances in relational and hybrid methods. Kluwer, Dordrecht (2000)
20. Wooldridge, M.: Reasoning about Rational Agents. MIT Press, Cambridge (2000)
21. Montero, J., Gomez, D., Bustine, H.: On the relevance of some families of fuzzy sets. Fuzzy Sets and Systems 16, 2429–2442 (2007)
22. Atanassov, K.T.: Intuitionistic Fuzzy Sets. Springer, Heidelberg (1999)
23. Flament, C.: Applications of graphs theory to group structure. Prentice Hall, London (1963)
24. Ruspini, E.H.: A new approach to clustering. Information and Control 15, 22–32 (1999)
25. Priest, G., Tanaka, K.: Paraconsistent Logic. Stanford Encyclopedia of Philosophy (2004), `http://plato.stanford.edu/entries/logic-paraconsistent`

26. D'Espagnat, B.: Conceptual Foundation of Quantum mechanics, 2nd edn. Perseus Books (1999)
27. Chalkiadakis, G., Boutilier, C.: Sequential Decision Making in Repeated Coalition Formation under Uncertainty. In: Padgham, Parkes, Müller, Parsons (eds.) Proc. of 7th Int. Conf. on Autonomous Agents and Multiagent Systems (AA-MAS 2008), Estoril, Portugal, May 12-16 (2008), `http://eprints.ecs.soton.ac.uk/15174/1/BayesRLCF08.pdf`
28. Soo, V.-W.: Agent negotiation under uncertainty and risk. In: Soo, V.-W., Zhang, C. (eds.) PRIMA 2000. LNCS, vol. 1881, pp. 31–45. Springer, Heidelberg (2000)
29. Sun, R., Qi, D.: Rationality assumptions and optimality of co-learning. In: Soo, V.-W., Zhang, C. (eds.) PRIMA 2000. LNCS, vol. 1881, pp. 61–75. Springer, Heidelberg (2000)
30. Baki, B., Bouzid, M., Ligęza, A., Mouaddib, A.: A centralized planning technique with temporal constraints and uncertainty for multi-agent systems. Journal of Experimental & Theoretical Artificial Intelligence 18(3), 331–364 (2006)
31. van Dinther, C.: Adaptive Bidding in Single-Sided Auctions under Uncertainty: An Agent-based Approach in Market Engineering (Whitestein Series in Software Agent Technologies and Autonomic Computing), Birkhäuser, Basel (2007)
32. Wu, W., Ekaette, E., Far, B.H.: Uncertainty Management Frame- work for Multi-Agent System. In: Proceedings of ATS 2003 (2003), `http://www.enel.ucalgary.ca/People/far/pub/papers/2003/ATS2003-06.pdf`
33. Colyvan, M.: The Philosophical Significance of Cox's Theorem. International Journal of Approximate Reasoning 37(1), 71–85 (2004)
34. Colyvan, M.: Is Probability the Only Coherent Approach to Uncertainty? Risk Analysis 28, 645–652 (2008)

11

Agent Transportation Layer Adaptation System

Jeffrey Tweedale[1], Felix Bollenbeck[2], Lakhmi C. Jain[1], and Pierre Urlings[3]

[1] School of Electrical and Information Engineering,
Knowledge Based Intelligent Engineering Systems Centre,
University of South Australia, Mawson Lakes, SA 5095, Australia
{Jeffrey.Tweedale, Lakhmi.Jain}@unisa.edu.au

[2] Leibniz Institute of Plant Genetics and Crop Plant Research
Pattern Recognition Group, Corrensstraße 3,
D-06466 Gatersleben, Germany
feboll@gmx.de

[3] Land Operations Division,
Defence Science and Technology Organisation,
Edinburgh SA 5111, Australia
Pierre.Urlings@dsto.defence.gov.au

Abstract. Heuristic computing has consolidated into two streams of research. One that personifies software to exhibit human behaviour and an oher that provides innovative software or smart products [1]. The Turing test [2] was pivotal in providing researchers with a generally accepted method of classifying the work that now defines the major problems pursued within Artificial Intelligence (AI). Cognitive Science is one of these fields and Research in Multi-Agent System (MAS) has revealed that Agents must enter into a voluntarily trust relationship in order to collaborate, otherwise the imposed goal(s) may be aborted or fail completely [3, 4]. Current agent architectures present a finite limit to functionality when supporting one or more of these paradigms.

Discussion about a framework being developed at the University of South Australia enables individual students associated with our Knowledge-Based Intelligent Information & Engineering Systems (KES) centre to fast track the development of their research concepts via a *Plug 'n' Play* mechanism within a multi-agent blackboard architecture. This paper highlights the core architecture, we believe is required for MAS developers achieve such flexibility. The research focuses on how agents can be teamed to provide the ability to adapt and dynamically organise the required functionality to automate in a team environment. The model is conceptual and has been designed initially as a blackboard model, where each element represents a block of functionality required to automate a process in order to complete a specific task. Discussion is limited to the formative work within the foundation layers of that framework.

Keywords: Autonomy, Intelligent Agents, Teaming, Trust.

L.C. Jain, N.T. Nguyen (Eds.): Knowl. Proc. & Dec. Mak. in Agent-Based Sys., SCI 170, pp. 247–273.
springerlink.com

Acronyms

ACL Agent Communication Languages
AI Artificial Intelligence
ATLAS Agent Transportation Layer Adaption System
BA Babel Agent
BDI Beliefs, Desires, Intentions
CMA Collaboration Management Agent
CRM Cockpit Resource Management
DAI Distributed Artificial Intelligence
DARPA Defense Advanced Research Projects Agency
DSTO Defence Science and Technology Organisation
EMMA Enterprise Management and Modeling Architecture
FIPA ACL FIPA Agent Communication Languages
HCI Human Computer Interface
IA Intelligent Agent
ISO International Standards Organisation
JUTE Jack/Unreal Tournament Environment
KES Knowledge-Based Intelligent Information & Engineering Systems
KIF Knowledge Interchange Format
KQML Knowledge Query Manipulation Language
MAS Multi-Agent System
OAA Open Agent Architecture
ONR Office of Naval Research
OODA Observe Orient Decide and Act
SL Semantic Language
SME's Subject Matter Experts
SST Situation Specific Trust
TNC Trust, Negotiation, Communication
UniSA University of South Australia
UT Unreal Tournament

11.1 Introduction

Researchers have been presented many Artificial Intelligence (AI) challenges as they attempt to increase the level of personification in intelligent systems. These challenges are both technical and psychological in nature. Agent technologies, and in particular agent teaming, are increasingly being used to aid in the design of "intelligent" systems [4, 5]. In the majority of the agent-based software currently being produced, the structure of agent teams have been reliant on that defined by the programmer or software engineer.

A description of work being undertaken by the University of South Australia (UniSA) and Defence Science and Technology Organisation (DSTO) discusses the development of a model that extends the architecture capable of providing an agent with the adaptable functionality required to maintain agent teams in a Multi-Agent System (MAS). Work on this concept is based on a predictive mechanism used to manage the trust relationship between each partnership being established.

This chapter discusses why the chosen mechanism for trust is the appropriate method of how trust may be used to form successful partnerships[1]. A brief description of the communications layer of the TNC model is provided. Future work on other layers is presented to stimulate discussion and feedback.

11.2 Intelligent Agents

The capability of intelligent agents to autonomously perform simple tasks has aroused much interest. The key characteristics that make them attractive is their:

- Ability to act autonomously;
- High-level representation of behaviour[2];
- Flexibility[3];
- Real-time performance;
- Suitability for distributed applications; and
- Ability to work co-operatively in teams.

An agent system can be formed using a variety of architectures, hierarchies and communications models. A single agent in a simulation could be a process or an entity. Multiple agents can be instantiated as a team of agents that combine their effort to achieve the same task or share the load using a single platform distributed across multiple machines/networks. Different systems may be instantiated with a hierarchy, with each level performing predetermined tasks

[1] The agent architecture is based on the proposed Trust, Negotiation, Communication (TNC) model and implements trust as a bond that can be strengthened via the exchange of certified tokens gained through previous encounters (similar to Microsoft's cookies). The problems of measuring trust are two fold; the form trust should take, and the measures used need to be quantitative.

[2] A level of abstraction above object-oriented constructs.

[3] Combining pro-active and reactive behavioural characteristics.

in subordinate or supervisory roles. This association is a key to the concept of the TNC as a core element with each agent in the model.

11.2.1 Intelligent Agent Definition

Wooldridge has evolved a suitable definition of agency that describes an Intelligent Agent (IA) being: "capable of flexible autonomous action in order to meet its design objectives". Here flexible, means three things [6]:

Reactivity: intelligent agents are able to perceive their environment, and respond in a timely fashion to changes that occur in it in order to satisfy their design objectives;

Pro-activeness: intelligent agents are able to exhibit goal-directed behaviour by taking the initiative in order to satisfy their design objectives; and

Social ability: intelligent agents are capable of interacting with other agents (and possibly humans) in order to satisfy their design objectives.

11.2.2 Agent Types

A variety of agents have been created that focus on the ideal properties of an agent, for example [7]:

- Deliberative agents;
- Reactive agents;
- Interface agents (HCI); and
- Mobile agents[4].

An Agent Architecture is considered to include at least one agent that is independent or a reactive/proactive entity and is conceptually contains functions required for perception, reasoning and decision. It may also be viewed as a particular methodology for collectively connecting agents or agent sub-systems. The architecture specifies how the various parts of an agent can be assembled to accomplish a specific set of actions to achieve the systems goals.

The next major enabler required to create a new generation of agents is to incorporate "human-centric" functionality. The current evolution of agent systems has concentrated solely on agent-to-agent interaction and fail the "social ability" test [9]. Wooldridge describes social ability as "the ability to interact with other agents and possibly humans via some communication language" [10]. That language will benefit AI research, where it conforms to unified standards and communication frameworks, using the underlying protocols shown in Figure 11.1. Based on this concept, we would like to suggest that "interaction" with humans is not limited to a communication language but should be extended to include other means of interaction, such as observation and adaptation. We would also like to suggest that smart agents could work in conjunction with humans if they adapt or translate their skills (communication, learning and coordination) rather than the functions they conduct. These concepts help focus our development of

[4] Inter and intra-net.

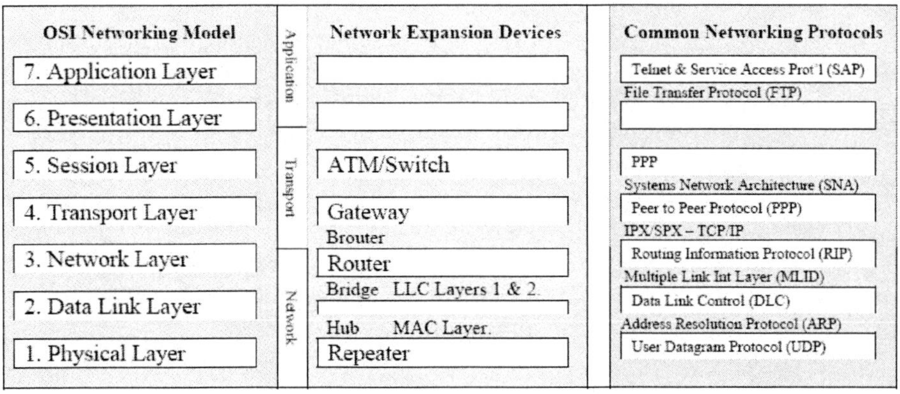

OSI Networking Model		Network Expansion Devices	Common Networking Protocols
7. Application Layer	Application		Telnet & Service Access Prot'l (SAP) File Transfer Protocol (FTP)
6. Presentation Layer			
5. Session Layer	Transport	ATM/Switch	PPP Systems Network Architecture (SNA)
4. Transport Layer		Gateway Brouter	Peer to Peer Protocol (PPP) IPX/SPX – TCP/IP
3. Network Layer	Network	Router Bridge LLC Layers 1 & 2	Routing Information Protocol (RIP) Multiple Link Int Layer (MLID)
2. Data Link Layer		Hub MAC Layer	Data Link Control (DLC) Address Resolution Protocol (ARP)
1. Physical Layer		Repeater	User Datagram Protocol (UDP)

Fig. 11.1. OSI Model Comparison, Modification of the Basic Reference Model [8]

agents adopting a human centric nature by merging one or more of the ideal attributes used in coordination, learning and autonomy.

Due to space limitation, detailed discussion about Agent Platforms, Agent Toolkits and MAS Architectures will be limited to IA in MAS teams. This limitation also applies to mobile agents classification, with the discussion limited to static agent types with deliberative, reactive and interface agents types.

11.2.3 Human-Agent Coordination

The arguement of humans retaining control of any system has endured a number of centuries. While the human is expected to be in command of the human-agent team, it is also expected that an active stream of coordination flows between the human and all team members as required. Aspects of effective coordination will include control and management, communication, self-learning, performance monitoring, warning(s), and assertive behaviour.

These aspects, in descending authority relate to, task delegation and team member assertive behaviour (which is also a critical element in the human/team training concept in many training systems[5]. Complacency and changes in communication and situational awareness were factors in crew performance and human errors resulting in many incidents and accidents since the introduction of automation in the cockpit. In an analysis of the causes of accidents carried out by Boeing almost 30 years ago, it was found that human errors on the part of the cockpit crew were the primary cause in over 73 percent of the cases investigated.

[5] In civil aviation, originally called Cockpit Resource Management (CRM), ten accidents, occurring between 1972 and 1982, have been cited as the motivation for creating CRM [11], while a following review of more than 35 accidents [12] provides the foundation leading to the official acceptance of CRM. Most of these accidents are also included in the research that allowed Billings [13, 14] to formulate his principles for a Human Centered-Automation approach. CRM has been a response to counter the effects of what Wiener [15] called "clumsy automation".

As a result of this initial investigation, the aviation world recognised that there was a need for an educational program to reduce errors and to increase the effectiveness of teams on the flight deck. CRM has been developed and successfully introduced in civil aviation for this purpose [15].

Urlings [5] concluded that since CRM has the same origin and is well-aligned with the Human Centred Automation approach, it seems logical to investigate how to translate the principles of CRM for human-human teams into requirements for constructing effective human-agent teams. It is expected that the results will not be limited to requirements and recommendations for the agent members of the team only. The principles of CRM and important skill behaviours for effective teams are well-summarised by [16] and include:

- Mission analysis. The organisation and the development of a common plan, shared by all team-members. This includes planning, strategies or contingencies for unplanned events, task assignment and continuous task prioritising.
- Assertiveness. This is a topic specifically covered by CRM programs and addresses the acceptance and expectation that all team-members raise questions when they are uncertain; or state opinions and make suggestions on decisions and procedures, even to higher ranking team-members.
- Leadership. Active leadership ensures the involvement of all members in the work of the team; prioritises and assigns tasks; reallocates tasks in dynamic situations; provides clear and organised instructions to team-members.
- Communications. The research finding is that high performing teams apply a high level of communications in abnormal situations. Information in normal situations is given when required or when asked, and communication is always acknowledged by a response.
- Situation awareness. In effective situation awareness, potential problems are identified early and other team-members are alerted. A need for action will be recognised early and an attempt will be made to understand the cause of the discrepancy in information before proceeding. Common team awareness, instead of individual awareness, is promoted.
- Decision-making. Rationales for decision-making are provided. Options are considered and alternatives generated, especially in decision-making for situations under stress.
- Adaptability/flexibility. Plans are altered when the situation demands. Other team-members are assisted when needed, and suggestions and constructive criticism are accepted.

Cross-cultural differences are often discussed in relation to CRM and may have a great impact on teaming performance and effectiveness [17]. Cultural issues are well-defined by Hofstede [18, 19] who proposed four cultural dimensions:

Power distance: The exercise and acceptance of authority.
Uncertainty avoidance: The ease to cope with novelty, ambiguity and uncertainty.
Individualism: Which addresses personal initiatives and achievements.
Masculinity: In which ambition and performance are strongly valued.

The discussion of cross-cultural differences is closely related to CRM, but falls well outside the technical scope of this chapter. But it illustrates the multi-disciplinary character and importance of social issues in effective team building. Future powerful agents will increasingly differ in important ways from the conventional software of the past. In order to design and build agents that are acceptable to their human counterparts, there will be a need to take into account the social issues, no less than the technical ones [20]. The cross-cultural differences are not covered in other sections of this article. However it is important to raise issue of the Human/Agent relationship and the effect of cultural diversity on implementing such teams. This may have a direct bearing on classification and architecture.

11.2.4 Agent Classification

Although [21] completed an exhaustive search in 1996 to derive a definition for an agent, we know from experience that trust and cultural diversity cause this definition fail due to the personified functionality required in the real world. Bratman [22], Russel and Norvig [23], Jennings and Wooldridge [24] and Nwana [25] all contributed to enhancing this definition, although many of the links need to be enunciated. The features discussed include: autonomy (decisions/actions), reactivity (perception, behaviour or beliefs), pro-activeness (purposefulness/goals/ intentions) and social ability (ability to personify, communicate/coordinate/ collaborate, or interact). Each has been debated, for instance Frankcik and Fabian [26] classify gestures (written or spoken) as the ability to stimulate or *act*. Reynolds [27] and Tu [28] compare perception with the paradigm of *behaviour* and Bratman [22] attempts to tie this together into an architecture called Beliefs, Desires, Intentions (BDI). He describes beliefs as data structures that represent *knowledge*, desires as *goals* and intentions as *plans*. Evans [29] labels this approach to agency as *orthodoxy of agent technology*, where intentions can be mapped to *opinions* and actions (re-actions) a consequence of *intentions* as implemented by Roa and Georgeff [30] in their BDI architecture.

The question of how to classify agents has been debated since the first scripts originated. Some researchers preferred functionality, while others used utility or topology. Agreement on categories has not been rationalised, however three classes dominate. These classifications include: mobility, reasoning (such as reactive- deliberative), attributed models (such as those used in autonomy), planning, learning, and cooperative/collaborative/communicative agents [3]. Alternatively, Nwana chooses to classify agents using mobility, reasoning, learning, information and data fusion. Noting these differences, pressure is also emanating from within the Distributed Artificial Intelligence (DAI) community to include interaction within the BDI agent paradigm with interaction into this definition. BDI can be used to achieve human like intelligence, however such research cannot be fitted into the above definition of truly "smart agent", as most applications still lack the main ideal characteristics of "coordination and learning". To be acknowledged as the next interruption along the AI paradigm continuum, intelligence with interaction must be demonstrated as the major impediment of next generation AI applications by incorporating agents capable of learning and self managed group coordination (teaming).

11.2.5 Agent Architectures

A real-world system takes inputs as sensors and react appropriately by modifying the outputs as necessary. Simulation models rely on the same approach. They monitor sensors that stimulate decision making that may cause changes to outputs. Three architectures have been dominant in AI research:

1. Blackboard Systems;[6]
2. Contract Nets;[7] and
3. Frameworks.[8]

11.2.6 Building Blocks

We have chosen to use the JACK BDI agent architecture and intend to demonstrate that it is the paradigm of choice for complex reactive systems [32]. We believe that the architecture needs some refinement and the development environment extended in order to suit most systems where human involvement is the most crucial factor. Examples of such systems could readily be used in the cockpit of an airplane and in complex reactive systems involving critical human decision making in real time, such as surveillance and control operators building situation awareness. MAS capable of coordination and learning could conceivably be incorporated to automate or assist operators in conducting these roles.

In more precise terms, a blackboard may be thought of as a componentised system, where each box could function as a database, series of pigeon holes or behave with an unknown black box concepts that represent the specified aspects of a system or sub-systems engaging the problem. This happens in an environment where experts are modular software subsystems, called knowledge sources, that are capable of representing different points of view, strategies, and knowledge formats, required to solve a problem. These problem-solving paradigms include:

- Bayesian networks
- Genetic algorithms
- Rule-based systems
- Case-based systems
- Neural networks
- Fuzzy logic systems
- Legacy (traditional procedural) software systems
- Hybrid systems

[6] See the Evolution of Blackboard Control Architectures [31].

[7] Architectures based on “Contract Net” where task allocation is done through a bidding system.

[8] Architectures constructed using a single agent “Framework” that constructs a series of detailed plans implemented by a collection of those agent types.

11.2.7 Multi-Agent System (MAS)

A multi-agent system may be regarded as a group of ideal, rational, real-time agents interacting with one another to collectively achieve their goals. To achieve their goals, each one of these individual agents needs to be equipped to reason not only about the environment, but also about the behaviour of other agents. Based on this reasoning, agents need to generate a sequence of actions and execute them. Multi-agent systems are mostly targeted to be an ideal solution for problems in heterogeneous environments. One of the classic example of heterogeneous environment is a game of soccer. In a heterogeneous environment (e.g. Soccer), each agent needs to have a strategy to solve a particular problem(s). In this context, a strategy refers to a decision-making mechanism that provides a long term consideration for selecting actions to achieve specific goals. Each strategy differs in the way it tackles the solution space of the problem. The presence of multiple agents necessitates the need for a different treatment. There is a coordination mechanism that handles the interaction between the agents. This mechanism is responsible for the implementation of the agents' actions and also the planning process that goes into the implementation. Traditionally, depending on their approach to solving a problem of this nature, agents have been divided into three main categories [25]:

- Deliberative
- Reactive
- Hybrid

11.2.8 Multi-Agent System (MAS) Teams

To extend the concept of agents, teams, systems and systems of systems, we could use the soccer example, based on MAS teams, one offensive and the other defensive. This problem escalates as the problems associated with co-ordination, especially in a dynamic environment, where the teams can switch roles in any given instance.

11.3 Forming Agent Teams

Forming agent teams generally requires prior knowledge of the resources and skills required by teams to complete a goal or set of decomposed tasks. This will be determined by the maturity and composition of the team, its members and any associated team[9]. The maturity could be measured across each phase of *Team Development*' as described by Tuckman [33].

[9] Each Agent may with other agents or group of agents (a Team). That agent may also interact or collaborate with other systems, within or across multiple environments. Agents can collaborate freely with other agents, although *Teams* are forced to communicate using traditional hierarchical methodologies. This structure needs to become more flexible in order to deliver efficiencies.

Personified characteristics/functions are possible in BDI agent architectures, although each consumes resources that may only be used by relatively few agents, many sporadically during its life. To reduce these overheads, only the resources/functionality required for a specified period would be instantiated and then released some time after that function is no longer required. Based on this premise, interactions between agents within teams and between teams can generally be catalogued.

There are three underpinning issues to consider in order to effectively form agent teams, these are *Communication*, *Negotiation* and *Trust*. Communication is concerned with the means of communication between agents such that they can understand each other. In-fact, early agent development relied on the idea that intelligence is an emergent property of complex interactions between many simple agents. For example: The *Open Agent Architecture (OAA)* [35] defines agents as any software process that meets the conventions of the OAA society, communication between agents is managed by facilitators, which are responsible for matching requests with the capabilities of different agents. *Aglets* [36] are Java objects that can move from one host on the network to another, hence they are also called mobile agents. Finally, *Swarm* [37] provides a hierarchical structure that defines a top level *observer swarm* and a number of *model swarm* are then created and managed in the level below it. As discussed later, the concept of communications is complex. The modes of transmission are shown in Figure 11.2 and include: simplex (like a radio or Television), half duplex (like an older push-to-talk intercom or radio telephone/CB radio) and full duplex transmissions (a telephone). Communication is also described based on the connectivity classification, such as; information push or pull, broadcast, messaging an interpersonal communications. Thus a message can be passed using all three modes between two or more parties, where interpersonal communication would generally use a full duplex mode with scale being limited by Metcalfe's Law [38] with base^n network connectivity. Mode and connectivity classifications can be mixed to suite specific application(s) within any system or number of systems being run in any environment.

The second problem is Negotiation, this problem is concerned with how to form teams. There are many aspects to consider in regards to negotiation. Generally, development of teams involves separating the requirements of a team from the requirements of individual agents. This includes assigning goals to a team as a whole, and then letting the team figure out how to achieve it autonomously. A team is constructed by the definition of a number of roles that are required in order to achieve the goals of the team. Additionally, agents can be specifically developed to perform one or more roles. An important feature of this approach is that agents are assigned with roles at runtime and can also change roles dynamically when required. Hence, one agent may need to perform one or more roles during its operation. Examples of agent development platforms that follow this approach are: *MadKit* [39] is a multi-agent platform written built upon the an organizational model called *Agent/Group/Role* and agents may be developed in many third party languages. JACK Teams [40] is an extension to the JACK

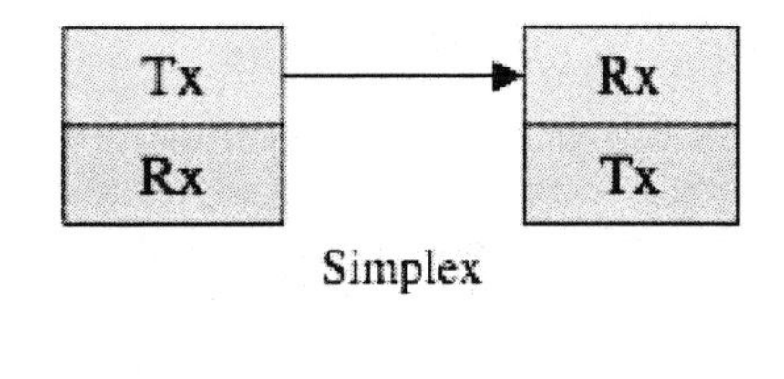

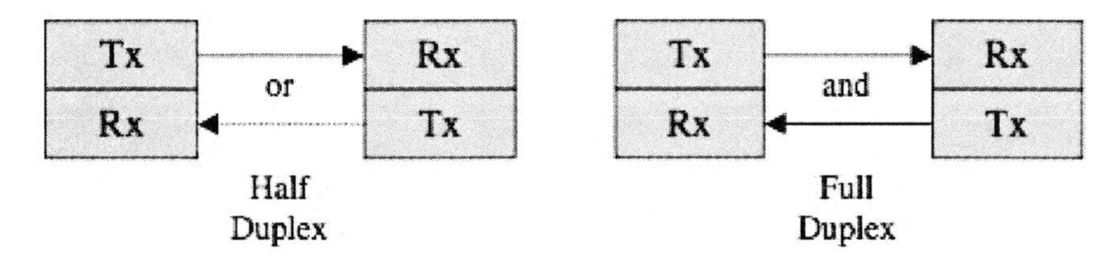

Fig. 11.2. Transmission Modes [34]

Intelligent Agents platform that provides a team-oriented modeling framework, specifically this allows the designer to specify features such as team functionality, roles, activities, shared knowledge and possible scenarios.

The third problem is Trust, it involves deciding how an agent should handle trust in regards to other agents. For example, should an agent trust information given by another agent, or trust another agent to perform a particular task. The level of trust is not easily measured, although loyalty can be used to weight information and consequently the strength of bond that is created. The fragility of that bond reflects on the frequency and level of monitoring required for the team to complete the related portion of a task. For further details on trust, the reader may refer to [41].

We must communicate with clarity and accuracy, using an appropriate level of timely feedback [42]. The interface must be intuitive enough to show sensitivity or retain focused attention on the goal or decomposed task(s). The interface must also cater for ad-hoc relationships. It would personify the processes that humans use to form friendships (usually by establishing mutual bonds), encoding the metaphor of man-machine communications/trust. Suchman [43] describes the lack of mutual understanding as the fundamental cause of miscommunication errors, such as "false alarm" and "the garden path". He concludes by describing an effective Human Computer Interface (HCI), as one that must include an understanding of situational context relating to real world knowledge with broadband communications (visual, verbal, and non-verbal). Moray [44] on the other hand suggests designers need to discover the user's functional mental model and provide information as and when expected. Chambers and Nagel [45] also completed a study based on automated pilots and concluded that the human or controller is ultimately responsible and therefore cannot be left out-of-the-loop.

We expect IA will retain their architectural foundations, but that the availability of more appropriate reasoning models and better design methodologies will assist them being used increasingly in mainstream software development. Furthermore, better support for human-agent teams will see the development of a new class of intelligent decision support applications. Autonomous connectivity using one of the existing communication modes will be required to assist with the higher level interoperability. The most common languages currently include:

- KIF,
- ACL,
- KQML, and
- FIPA/ACL.

11.3.1 Knowledge Interchange Format (KIF)

The KIF language [23] was designed as a solution to the translational problem. It is a logic language used to describe things within computer systems such as expert systems, intelligent agents and so on. KIF was specifically designed as a mediator to translate between two disparate languages. KIF is a prefix version of first order predicate calculus with extensions to support non-monotonic reasoning and definitions. The language description includes both a specification for its syntax and one for its semantics.

11.3.2 Agent Communication Languages (ACL)

Agent communication languages have been used for years in proprietary MAS. Yet agents from different vendors or even different research projects cannot communicate with each other [46]. Components will be added dynamically and will be autonomous (serve different users or providers and fulfill different goals) and heterogeneous (be built in different ways) in MAS. To fulfill the concept of *interoperability* and *autonomy* [47] agents must be able to talk to each other to decide what information to retrieve or what physical action to take, such as shutting down an assembly line or avoiding a collision with another robot. The mechanism for this exchange is the agent communication language.

Theoretically, an ACL should enable heterogeneous agents communicate. However many ACLs are being modified to suite specific implementations of multi-agent applications, creating a mix of proprietary and non-proprietary protocols. This results in agent systems with incompatible communications generally causing crashes or data corruption. Fortunately most ACLs now conform to industry standards and each project/agent system now interoperate successfully. ACL was examined because its *Knowledge Sharing Initiative*'s complement the work done by DARPA on *KIF*.

11.3.3 Knowledge Query Manipulation Language (KQML)

The first inter-project ACL was the KQML, proposed as part of the US Defense Advanced Research Projects Agencys Knowledge-Sharing effort in the late 1980's [48]. Queries in KQML are referred as *performatives*, a term adapted from speech theory[10] KQML statements syntax is a subset of LISP prefix notation grammar, although parameters in performatives are indexed by keywords rather than position. A statement is composed of a reserved performative keyword. This can be unilateral communication request like *reply, ask-one* or broadcast operations *broker-all, ask-all*, followed by the parameters of the performative. Parameters contain information on the communication act itself and contents of the message (parameters *language* and *ontology*).

The sample query in Figure 11.3 shows that the different communication layers in KQML performatives are well separated. Since KMQL is modular and uncomplex semantic descriptions can be used with KQML [49].

11.3.4 FIPA Agent Communication Languages (FIPA ACL)

In the early 1990s, France Tèlècom developed *Arcol* which includes a smaller set of primitives than KQML. The primitives are all assertives or directives, but unlike KQML they can be composed. Arcol has a formal semantics, which presupposes that agents have beliefs and intentions, and can represent their uncertainty about various facts. Arcol gives performance conditions, which define when an agent may perform a specific communication.

```
(achieve :language KIF :ontology motors
 :reply-with q1
 :content (=(val (torque motor1)(sim-time 5))(scalar 2kgf))
```

Fig. 11.3. Sample Query in KQML Format [49]

A FIPA ACL message contains a set of one or more message parameters. The only parameter that is mandatory in all FIPA ACL messages is the performative, although it is expected that most ACL messages will also contain sender/receiver and content parameters. As mentioned above, if an agent does not recognize or is unable to process one or more of the parameters or parameter values, it can reply with the appropriate not-understood message. Some parameters of messages might be omitted in FIPA ACL when their value can be deduced by the context of the conversation as shown in Figure 11.4. However, FIPA ACL does not specify any mechanism to handle such conditions, therefore those implementations that omit some message parameters are not guaranteed to interoperate with each other.

[10] An utterance is a performative when its function is to perform the action mentioned rather than to express a proposition about it (as a constative does).

```
(inform :sender agent1 :receiver
 hlp-auction-server :content(price (big good02) 150 )
 :in-reply-to round4 reply-with bid4 :language sl
 :ontology hlp-auction)
```

Fig. 11.4. FIPA ACL sample query [50]

11.3.5 FIPA Versus KQML

FIPA ACL is superficially similar to KQML. Its syntax is identical to that used by KQML , except for the different names for some reserved primitives. KQML separates the outer language that defines the intended meaning of the message and the inner language, or content language that denotes the expression represented by the beliefs, desires and intentions of the interlocutors, as described by the meaning of the communication primitive, apply. The FIPA ACL specification document claims that FIPA ACL (like KQML) does not make any commitment to a particular content language. Although such a claim holds true for most primitives, there are FIPA ACL primitives for which some understanding of the language Semantic Language (SL) is necessary for the receiving agent to understand and process the primitive [50]. Agent communication architectures apart from KQML and FIPA ACL are either ad hoc JATLite, JAFMAS or not fully documented as in proprietary systems.

11.3.6 Blackboard-Based Architectures

The blackboard architecture concept was conceived by researchers in the field of Artificial Intelligence more than a decade ago. Originally designed for the HEARSAY-II speech-understanding system [38], the blackboard architecture is one of the most widely used architectures in symbolic MAS. The intent of this research was to address issues of information sharing among multiple heterogeneous problem-solving agents.

The blackboard architecture is not very specific and can be seen as a 'meta-architecture', for implementing other architectures. The name, blackboard architecture, was chosen to evoke a metaphor in which a group of experts gathers around a blackboard to collaboratively solve a complex problem. The blackboard model is based on a division into independent modules that do not communicate data directly, but interact by sharing data using a common framework, such as; ACL and message passing systems. Blackboards form a good way of communication between heterogeneous information sources. The sources do not need to know about each other, the only thing they see is the blackboard. The model chosen to wrap our projects ensures that it has structure and is able to comply with the rigor of model agency. This framework is shown in the blackboard model of Figure 11.5.

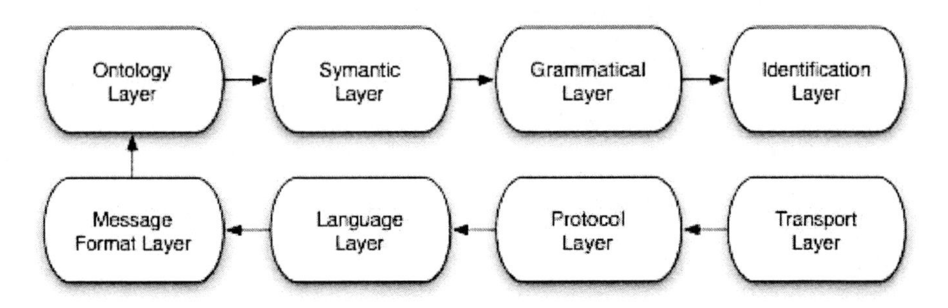

Fig. 11.5. ATLAS Blackboard Model [7]

11.3.7 Examples Subject to Current Research

The Enterprise Management and Modeling Architecture (EMMA)[11] is a hybrid human-computer decision making system that consists of a "Problem Solving" subsystem using a well maintained corporate knowledge base generated from the inputs supplied by Subject Matter Experts (SME's). Its architectural design has six layers, with each layer reliant on the previous layers functionality. These layers include: the network, data, information, organisational, coordination and marketing layers. As described these layers concentrate on the information management system concept and omit the more traditional International Standards Organisation (ISO) layer approach, although the concepts and relationships are visible.

Multi-layering and heterogeneity are chosen as desirable principles in many structures, including those using communications and learning capability. It is natural to implement autonomous characters in multi-media as autonomous agents due to the need to personify the entity and achieve autonomous responses. This concept is generally achieved using multi-layered teams of agents, each with a unique scope of functionality/responsibility. Multi-layering and heterogeneity are desirable principles of autonomous embedded agents. FreeWill+ is an example of a structured agent used to produce animated graphics that communicate and present the ability to learn [51]. Like all *animated* creatures, these agents need to reflect the capacity of possessing an *artificial mind*. Given the difficulty of this task, the FreeWill+ designers choose to embed intelligence based on augmented levels of automation. Such features could include environmental awareness, purposefulness, attitude and the illusion of life. Using a fusion of agents in a team hierarchy, a holistic, autonomous being can be generated, based on the real physical constraints of the environment being created. The overriding supervision does stifle a level individual autonomy at the expense of the team and its goals (such as flocking or crowd dynamics). *CHRIS* [52] embodies elements of an agent reasoning and learning framework and was initially developed as an extension into research on JACK agents at the University of South Australia [53].

[11] EMMA is based on research conducted under Defense Advanced Research Projects Agency (DARPA) contract #F30602-91-C-0016 and the Office of Naval Research (ONR) under contract #N00014-92-J-1298.

11.3.8 Related Examples

CHRIS equips a JACK agent with the ability to learn from actions that it takes within its environment, it segments the agent reasoning process into five stages based on Boyd's Observe Orient Decide and Act (OODA) loop [54], Rasmussen's Decision Ladder [55] and the BDI model. The collaboration module itself has been implemented within the Orientation stage, specifically between the State and the Identification operation.

In the Jack/Unreal Tournament Environment (JUTE), negotiation is performed using an authoritative Collaboration Management Agent (CMA). Subordinate agents simply need to be able to perform the `Cooperation` role. The current implementation only supports goal-based collaboration relationships, where an agent negotiates for another agent to achieve a particular goal. Finally, an event called `RequestCollaboration` is used to ask the CMA for collaboration.

A demonstration program was written that provides limited human-agent collaboration. It uses two agents, the first agent called `Troop` which connects to a computer game called Unreal Tournament (UT) using UtJackInterface [56] and controls a player within the game. The second agent is called `HumanManager` and is used to facilitate the communication with humans encountered within the game. The demonstration shows that the `Troop` agent is given the goal hierarchy. The Defend and Attack goals are executed in parallel, also, considering that it is not possible for the `Troop` agent to perform both the `Defend` and `Attack` goals, it decides to handle the `Attack` goal and then asks the CMA to organise for any friendly human player seen in the game to take responsibility for the Defend goal. The sequence of operations for the demonstrations is:

Create the agents: Create `Troop`, CMA and the `HumanManager` agents and use the `scenario.def` file to construct a team with the `Cooperation` role between the CMA and the `HumanManager`.

Set the `Win` goal: Activate the goals of `Win`, `Defend` and `Attack`, so that subgoals are subsequently activated automatically in parallel. `Attack` is handled by the `Troop` agent which subsequently attacks any enemy that comes within the field of view. For demonstration purposes, the `Attack` goal succeeds after the agent attacks five enemy players.

Monitor the Message Queue: A `RequestCollaboration` message is sent to the CMA for the `Defend` goal. The CMA then executes an `@team_achieve` for any sub-teams that perform the `Cooperation` role. The `HumanManager` agent then negotiates and performs assessment with the human in order to satisfy the `Defend` goal.

Intelligent agent technology is at an interesting point in its development [57]. Commercial strength agent applications are increasingly being developed in domains as diverse as meteorology, manufacturing, war gaming, capability assessment and UAV mission management. Furthermore commercially supported development environments are available and design methodologies, reference architectures and standards are beginning to appear. These are all strong indicators of a mature technology. However, the uptake of the technology is not as rapid or as pervasive as its advocates have expected. It has been touted as becoming the paradigm of choice for the development of complex distributed systems and as the natural progression to object oriented programming. Agent technology is still in need of the killer application and more mature tools to promote its acceptance in commercial circles. There are also major technological reasons, such as; the architecture (Harvard and Von Neuman microprocessors), parallelism, speed and capacity. Although there are current impediments, the future is positive and agent technology will become another disruption in the technology continuum.

The development of intelligent agent applications using current generation agents is not yet routine. Certainly providing more intuitive reasoning models and better support frameworks will help, but we see behaviour acquisition as a major impediment to the widespread application of the intelligent agent paradigm. The distinguishing feature of the paradigm is that an agent can have autonomy over its execution - an intelligent agent has the ability to determine how it should respond to requests for its services. This is to be contrasted with the object paradigm, where there is no notion of autonomy and objects directly invoke the services that they require from other objects. Depending on the application, acquiring the behaviours necessary to achieve the required degree of autonomous operation can be a major undertaking and one for which there is little in the way of support. The problem can be likened to the knowledge acquisition bottleneck that beset the expert systems of the 1980's. There is a need for principled approaches to behaviour acquisition, particularly when agents are to be deployed in behaviour rich applications such as enterprise management. Cognitive Work Analysis has shown promise in this regard, but further studies are required.

Alternatively, the requirement for autonomous operation can be weakened and a requirement for human interaction introduced. Rather than having purely agent-based applications, we then have cooperative applications involving teams of agents and humans. Agent-based advisory systems can be seen as a special case of cooperative applications, but we see the interaction operating in both directions - the agent advises the human, but the human also directs and influences the reasoning processes of the agent. Existing architectures provide little in the way of support for this two way interaction. What is required is that the goals and intentions of both the human and the agent are explicitly represented and accessible, as well as the beliefs that they have relating to the situation. This approach provides a convenient way to address the difficulties associated with the behaviour acquisition of any autonomous operation. By making visible the agent's longer term goals and intentions, as well as the rationale behind its immediate recommendation, this approach also provides a mechanism for building trust between humans and agents. It should also be noted that in many applications, such as cockpit automation and military decision making, full autonomy is not desirable - an agent can provide advice, but a human must actually make the decision. Because of these reasons, we expect to see an increasing number of applications designed specifically for human/agent teams.

Learning has an important role to play in both cooperative and autonomous systems. However the reality is that it is extremely difficult to achieve in a general and efficient way, particularly when dealing with behaviours. The alternative is to provide the agent with predefined behaviours based on a priori knowledge of the system and modified manually from experience gained with the system. This has worked well in practice and we expect that it will remain the status quo for the immediate future, hence the goal is to produce a re-useable, *Plug-n-Play*, dynamic demonstrator model.

11.4 The Demonstrator Concept

The concept of the Babel Agent (BA) is similar to that of the babel fish[12]. It has been designed to accept a number of agent language format and translate it into another as required. To prove the concept, the following agent languages have been implemented, although it is feasible to expand this list given sufficient information about the language to be embodied. Figure 11.6 shows a number of disparate agents from various sources and with a number of the listed languages or dialects. Agents seeking a translation or guidance/supervisor needs to communicate with the BA. The BA is capable of translating between a specified combination of languages, or capable of conducting a series of specified tasks, such as add, subtract, divide and multiply. Where the BA is unable to complete a request, it will attempt to contact another agent within the system known to have the required skill. Once contacted, the BA will forward the task to the known resource and return the processed result to the originator. If busy or approaching an overloaded condition it may of load that task and monitor the transaction or completely transfer the task given it finds/spawns another BA.

11.4.1 The Atomic Concept of the TNC Model

The concept of the TNC model is shown in Figure 11.7, which details the type of interaction possible between single agents [25, 59, 60]. The model works for a hierarchy (team) of autonomic agents and by loose association for another agent team[13] within a complex system. The goal of the model being to provide a flexible structure that enables agents to team together without prior configuration (adaptation). Trust is the centre attribute used to form and maintain he partnership(s) and is required for agents joining or already within a team.

[12] The babel fish was originally a biblical reference of the Tower of Babel which led to God confusing the languages of Man in order to prevent the Tower's construction. It is depicted as a small, yellow, leech like fish, reported to be the oddest thing in the Universe. The babel fish was more recently seen on the television series called "Hitch Hiker's Guide to the Galaxy" by Douglas Adams. It feeds on brainwave energy received from its carrier and all those around it. It absorbs all unconscious mental frequencies from this brainwave energy to nourish itself with. It provides a symbiotic relationship by constructing a telepathic matrix in the brain that combines conscious thoughts and speech signals. The practical upshot of all this is that if you stick a babel fish in your ear you can instantly understand anything said in any language [58]".

[13] Or human beings in future systems.

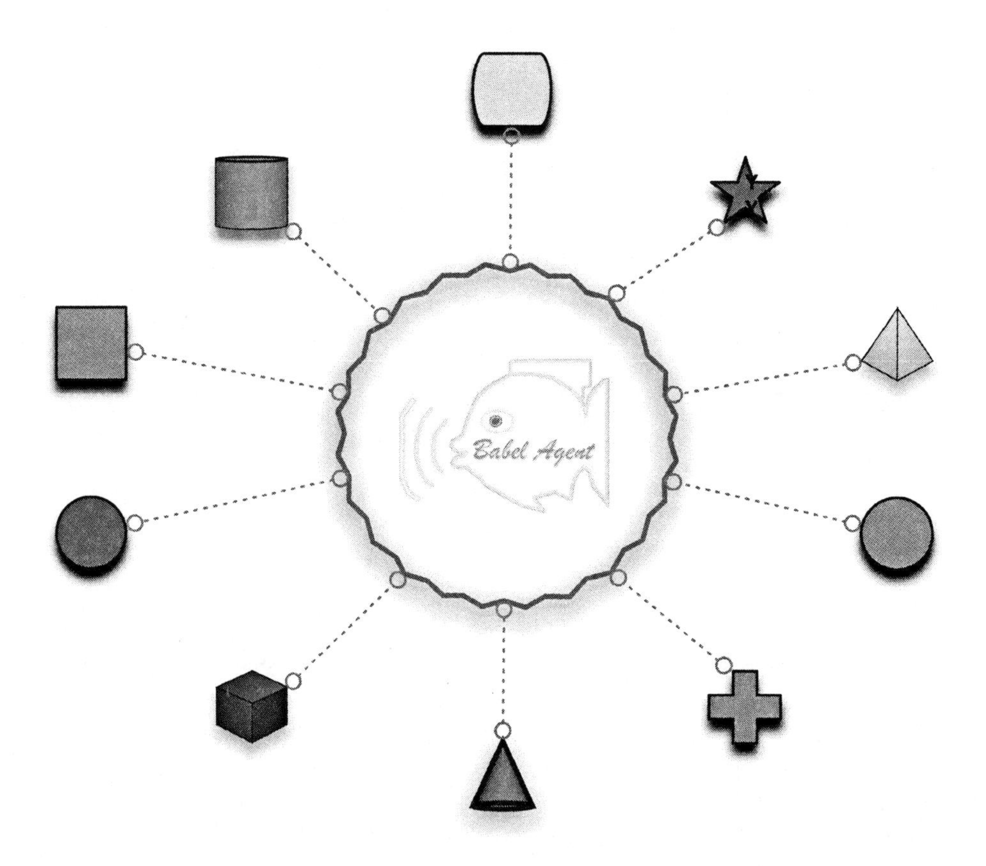

Fig. 11.6. The Proposed Babel Fish Concept

The model is effectively a series of wrappers for agents, that extend the communication ability inherent in agents to incorporate an ability to 'independently' negotiate partnerships. The model provides a communications interface, negotiation mechanism and a trust monitoring capacity.

The application of agents implemented in the form of the model would be as such. An environment would be built with predefined resources and skills which may be context dependent. When a resource is called upon, by another agent or environment, a team (comprising of one or many agents in the form of the proposed model) would be instantiated by the resource manager. Each team would have a hierarchy, where each agent would fulfill a specific skill. The partnerships within the team would be coordinated by a controller agent and the level of control based on loyalty through the bond established[14].

[14] The agents within the team would also have the ability to seek partnerships outside of the direct team when they require or are able to provide additional resources. The partnerships established would be based on trust, which would strengthen or wain over time.

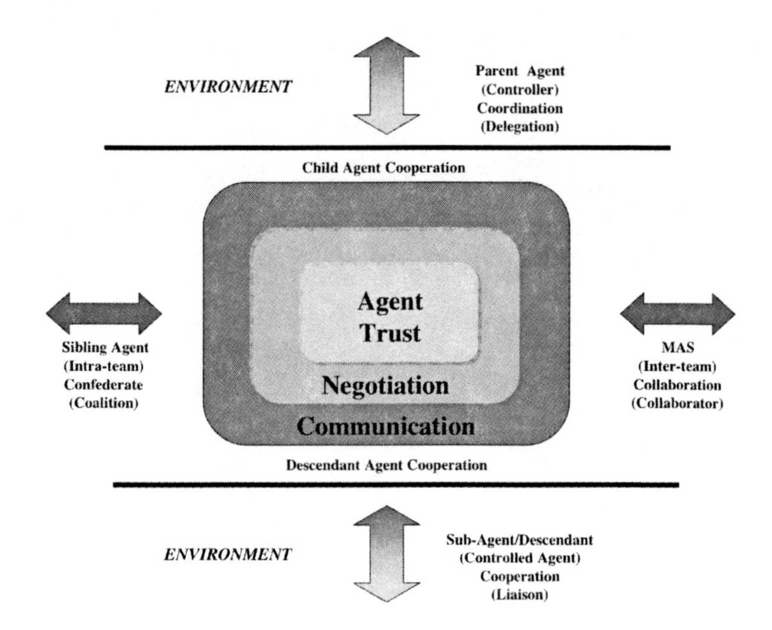

Fig. 11.7. The Proposed Trust, Negotiation and Communication Agent Model

The approach is analogous to that used by humans. When Human decision makers are presented a problem they are not able to solve themselves, they may choose to form a team of people they trust to assist in making that decision. The team is likely to have a hierarchical structure of some form in order to effectively manage the project in terms of scheduling and tasking. The structure may include multiple levels. The TNC model is based on a similar concept. When decisions are required to be made within an environment, a team would be assembled with the current resources and skills considered necessary to achieve the goal. If either resources or skills are lacking within the environment, the necessary skills or resources may be obtained from another environment or source.

Once a team has been instantiated, agents within the team at each level are optionally able to collaborate with other agents in the team at the same level or with agents at a similar level in another team. Ordinarily, agents are controlled using team hierarchy, with commands being issued from controller agents, at a level above, or delegated to agents at a level below (see Figure 11.7).

11.4.2 The Agent Transportation Layer Adaption System

Discussion in this section will be confined to the complete outer ring of the TNC model. This communication component has been labeled the Agent Transportation Layer Adaption System (ATLAS) model. This concept is based on the Basic OSI Model [8]. Significant effort on this topic has resulted in the design and specification of many syntactical (agent communication languages) and semantical (knowledge interchange formats) architectures in order to facilitate distributed

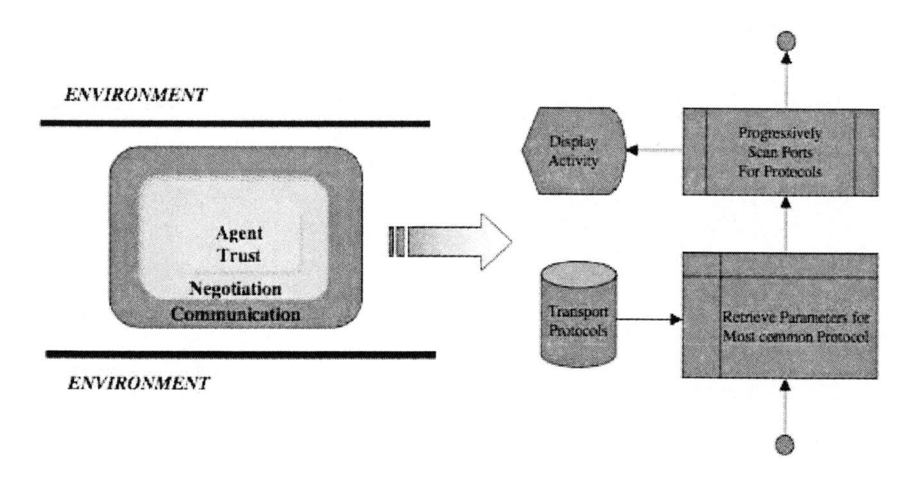

Fig. 11.8. Port Scanner/Polling Layer

work collaboration. Nevertheless little work has been undertaken on reflection, or the evaluation of the communication process(es) within system(s), in order to include the communication process in the environment that the agent is constrained within. Most languages used to co-ordinate and collaborate within their hierarchy.

Trust may be obtained from another environment or source. Hence, using the closed-world-assumption that the original model or system has little or no prior knowledge of any new environment(s) it is proposed that agents be used to cross these system boundaries and to establish new trust relationships. Agents may use different communication languages or knowledge interchange formats, ontologies and semantics to effectively communicate with the new environment. The trust model comes into play in order to establish a relationship, built on available information or adapted over time using an ongoing communication process compatible with that new environment. Such adaptation requires time for both systems to reflection and evaluation the consequences of any past communication performed while operating in that environment (possibly by introducing a reinforcement learning system).

11.4.3 The Communications Protocol Layer

A flowchart of the ATLAS protocol handler is shown in Figure 11.9. This is based on the ISO model of communications and is used to establish/determine the actual protocol being used to establish the transmission across a communications channel. The ATLAS model concentrates' on using an electrical signal along a network medium. It scans the environment to establish the presence of a signal and when detected, attempts to decode its format. In this case; UDP, TCP/IP and FTP have been implemented as known standards. Figure 11.10 displays the basic concept of polling ports to test for new communication events (This represents the second step of the ATLAS process).

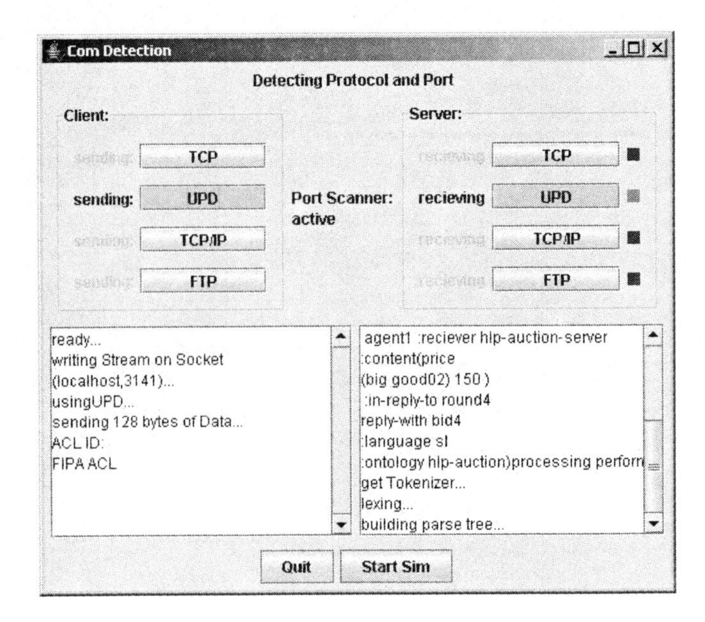

Fig. 11.9. The Prototype Port Scanner GUI [8]

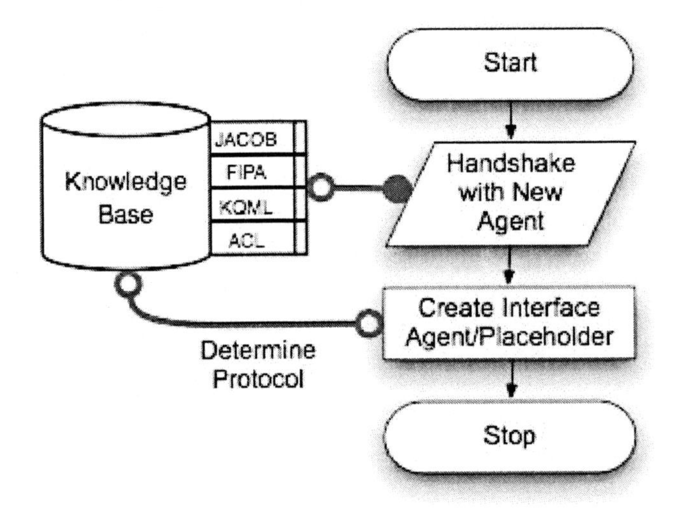

Fig. 11.10. The Proposed Communications Protocol Layer

11.4.4 Portscanner Prototype

Establishing communication between heterogeneous agent systems is crucial for building trust relationships between agent instances in these systems. Interchanging messages between different agent systems situated in different environments possibly affects all layers of the communication model, including transport protocol/system ports, ACL, semantic layers and Ontologies. To establish trust relationships with alien agent instances, the systems architecture must provide sufficient means to each agent instance, in order for agents to be *multilingual.*

Any implementation of the TNC layer model within an agent system must provide the capacity for handling multiple signal sources simultaneously. While these signals may originate from different transport language protocols, they must be capable of communicating the intended operation or method of how the knowledge is encoded. Incoming stimuli[15] must be recognized using sensor networks and processed based on the message format and content. In a multi-agent environment, agents must be capable of handling several incoming and outgoing communication request concurrently, while avoiding deadlocks or corrupting data being feed to the knowledge base. Multi-threaded systems are well suited to meeting these requirements. They can also be used to implement data integrity strategies required by database management systems. Such implementations consists of a multi-threaded client/server architecture, where instances of function are simultaneously instantiated as either clients or servers in the communication process.

Incoming communication requests can be forked to autonomous threads capable of handling specific protocols and further processing and translation. To build a *proof of concept* for the communication layer of the TNC model we implemented a multi threaded client/server system to simulate the processing of multiple signal sources. These input stimuli are described by clients connecting to a server using a random protocol in our function model. The server thread models the communication interface of the receiving agent. The servers master thread accepts connection and forks it into a new thread. This thread is specific for the determined protocol type and the received information is preprocessed before being committed to the knowledge base. The communication prototype is written in Java using Sun's TM system library for multi-threading and networking. The function model is a simple Java application implementing the above architecture and visualizing the on-going communication.

The choice of protocols for this conceptual model is limited to common protocols in web applications. Nevertheless these protocols can meet MAS specific needs: UDP as non connecting protocol for broadcast communication, TCP as a connecting protocol for address specific communication. FTP for transferring large amounts of data can be useful when exchanging full knowledge bases or the parameters required for existing AI connectivity (especially between agents and their sub-systems). The implementation of the described architecture in the *portscanner* prototype serves not only as a proof of concept for the proposed multilingual agent. The client/server infrastructure can also be used in further development of a fully functional system, which includes the detection, processing and translation of different formats, ACLs, semantics and ontologies.

---

[15] Such as acoustic, ultrasonic, radio, microwave, infrared and even light or network based traffic.

11.5 Future Research

This research raises many issues and attempts to merge a number of pre-existing concepts. A concept demonstrator has been developed to enhance the research and enable further study into the connectivity and auto negotiations of agents, systems or sub-systems, prior to establishing trust. The TNC model is based on *Trust* being *Negotiated* through a common framework of *Communication*. Presently the model is only at a conceptual stage and is expected to be refined over the next few years. However, work will commence on the implementation of the model to provide the end goal of agents having the capability to autonomously implement the administrative aspects of team formation (including human operators) and task/resource management of any agent in an MAS.

For the formation of a bond to occur between two agents the communications layers of both agents must jointly facilitate establishment of the protocol to be used. Establishing and implementing this process across a finite set of protocols will constitute the first phase of the implementation of the model.

The second phase will be to establish and implement methods of measuring the loyalty to establish the strength of the trust within a bond. Loyalty can be used as a weight that can be applied to determine the fragility of the collaborative bond between siblings or inter-team entities. Cohen [61] describes a Situation Specific Trust (SST) model that is comprised of a qualitative (informal core) that describes the mental attributes and includes a quantitative (prescriptive extension). This could be used to refine trust estimates formed during the negotiation phase. These estimates must be presented or exchanged using a flexible taxonomy that is easy to interpret and be used in the formative stages prior to the agents establishing bonds between themselves.

The third phase will see the implementation of agents with the adaptive abilities required to establish the bonds and/or initiate the process(es) of providing agents the ability to autonomously form team bonds that solve specific tasks within defined environments.

Acknowledgments

We wish to express our greatest appreciation to the reviewers and referees for their visionary feedback, which were used to enhance the quality of this chapter. Initial results of this research appear in the Journal of Intelligent and Fuzzy Systems, Vol. 19(6), 2008 [62].

References

1. McCorduck, P.: Machines who think, pp. 1–375. Freeman, San Francisco (1979)
2. Moor, J.H.: Briefly noted: the turing test: The elusive standard of artificial intelligence. Comput. Linguist. 30(1), 115–116 (2004)
3. d'Inverno, M., Luck, M.: Understanding agent systems. Springer, New York (2001)
4. Wooldridge, M., Jennings, N.R.: The cooperative problem-solving process. Journal of Logic and Computation 9(4), 563–592 (1999)

5. Urlings, P.: Teaming Human and Machine. PhD thesis, School of Electrical and Information Engineering, University of South Australia, University of South Australia (2003)
6. Wooldridge, M.J.: An Introduction to Multiagent Systems. Wiley and Sons, New York (2002)
7. Ehlert, P., Rothkrantz, L.: Intelligent agents in an adaptive cockpit environment. Technical Report DKE01-01, Version 0.2, Delft University of Technology, the Netherlands (2001)
8. ISO:IEC: Information Technology - Basic Reference Model. ISO 7498-1. International Standards Organisation, Geneve, Switzerland (1982)
9. Heinze, C., Goss, S., Josefsonn, T., Bennett, K., Waugh, S., Lloyd, I., Murray, G., Oldfield, J.: Interchanging agents and humans in military simulation. AI Magazine 23(2), 37–48 (2002)
10. Wooldridge, M., Jennings, N.R.: Theories, architectures, and languages: A survey, intelligent agents. In: Wooldridge, M.J., Jennings, N.R. (eds.) ECAI 1994 and ATAL 1994. LNCS, vol. 890, pp. 1–39. Springer, Heidelberg (1995)
11. Satchell, P.M.: Cockpit Monitoring and Alerting Systems. Ashgate Publishing Ltd., Aldershot (1993)
12. Kayten, P.J.: The accident investigatorís perspective. In: Wiener, E.L., Kanki, B.G., Helmreich, R.L. (eds.) Cockpit Resource Management, vol. 10, pp. 283–314. Academic Press, San Diego (1993)
13. Billings, C.E.: Human-centered aircraft automation: A concept and guidelines. August Technical Memorandum 103885, NASA-Ames Research Center, Moffett Field, CA (1991)
14. Billings, C.E.: Aviation Automation: The Search for a Human-Centred Approach. Lawrence Erlbaum Associates, Inc., Mahwah (1997)
15. Wiener, E.L., Curry, R.: Flight-deck automation: Promises and problems. Technical Report NASA TM 81206, NASA, Moffett Field, CA (1982)
16. Prince, C., Salas, E.: Training and research for teamwork in the military crew. In: Wiener, E.L., Kanki, B.G., Helmreich, R.L. (eds.) Cockpit Resource Management, vol. 12, pp. 336–337 (1993)
17. Johnston, N.: CRM: Cross-Cultural Perspectives. In: Wiener, E.L., Kanki, B.G., Helmreich, R.L. (eds.) Cockpit Resource Management, vol. 13, pp. 367–398. Academic Press, San Diego (1993)
18. Hofstede, G.: Cultureís consequences: international differences in work-related values. Sage, Beverly Hills (1980)
19. Hofstede, G.: Cultures and Organisations: Software of the mind. McGraw-Hill, Maidenhaid (1991)
20. Bradshaw, J.M., Sierhuis, M., Acquisti, A., Feltovich, P., Hoffman, R., Jeffers, R., Prescott, D., Suri, N., Uszok, A., Van-Hoof, R.: Adjustable autonomy and human-agent teamwork in practice: An interim report on space applications. In: Hexmoor, H., Facone, R., Castelfranchi, C. (eds.) Agent Autonomy, pp. 243–280. Kluwer, Dordrecht (2002)
21. Franklin, S., Graesser, A.: Is it an agent, or just a program?: A taxonomy for autonomous agents. In: Proceedings of the Third International Workshop on Agent Theories, Architectures and Languages. Budapest, Hungary, Budapest, Hungary, pp. 193–206 (1996)
22. Bratman, M.E.: Intentions Plans and Practical Reason. Center for the Study of Language and Information (1999)
23. Russel, S., Norvig, P.: Artificial Intelligence: A Modern Approach, 2nd edn. Prentice-Hall, Inc., Englewood Cliffs (2003)

24. Jennings, N.R., Wooldridge, M.: Applications of intelligent agents. In: Agent technology: foundations, applications, and markets, Secaucus, NJ, USA. Springer, New York (1998)
25. Nwana, H.S.: Software agents: An overview. In: McBurney, P. (ed.) The Knowledge Engineering Review, vol. 11(3), pp. 205–244 (1996)
26. Francik, J., Fabian, P.: Animating sign language in the real time. In: Applied Informatics. In: 20th IASTED International Multi-Conference, Innsbruck, Austria, pp. 276–281 (2002)
27. Reynolds, C.W.: Flocks, herds, and schools: A distributed behavioral model. Computer Graphics 21(4), 25–34 (1987)
28. Tu, X., Terzopoulos, D.: Artificial fishes: Physics, locomotion, perception, behavior. Computer Graphics 28, 43–50 (1994)
29. Evans, R.: Varieties of learning. In: Rabin, E. (ed.) n AI Game Programming Wisdom, Charles River Media, Hingham MA, vol. 2 (2002)
30. Rao, A., Georgeff, M.: BDI Agents: From theory to practice. In: Proceedings for the 1st International Conference on Multi-Agent Systems (ICMAS 1995), pp. 312–319. AAAI Press, California (1995)
31. Carver, N., Lesser, V.: The evolution of blackboard control architectures. Technical Report UM-CS-1992-071, University of Massachusetts (1992)
32. Kinny, D., Georgeff, M., Rao, A.: A methodology and modelling techniques for systems of bdi agents. In: Perram, J., Van de Velde, W. (eds.) MAAMAW 1996. LNCS, vol. 1038, pp. 56–71. Springer, Heidelberg (1996)
33. Mann, R.: Interpersonal styles and group development. American Journal of Psychology 81, 137–140 (1970)
34. Held, G.: Basic Telegraph and Telephone Operations, 3rd edn., pp. 23–45. John Wiley and Sons, Georgia (2001)
35. Cheyer, A., Martin, D.: The open agent architecture. Journal of Autonomous Agents and Multi-Agent Systems 4(1), 143–148 (2001)
36. Lange, D.B.: Java aglet application programming interface (J-AAPI) white paper - draft 2. Technical report, IBM Tokyo Research Laboratory (1997)
37. Group, S.D.: Documentation set for swarm 2.2. (accessed August 25, 2005), `http://www.swarm.org/swarmdocs-2.2/set/set.html`
38. Gilder, G.: Metcalfes law and legacy. Technical report, Forbes (1993)
39. Ferber, J., Gutkecht, O., Michel, F.: MadKit development guide (accessed August 25, 2005), `http://www.madkit.org/madkit/doc/devguide/devguide.html`
40. AOS: JACK intelligent agents teams manual 4.1 (accessed March 2, 2006), `http://www.agent-software.com.au/shared/resources/index.html`
41. Tweedale, J., Cutler, P.: Human-Computer Trust in Multi-Agent Systems. In: Proceedings of the 10th International Conference on Knowledge Based Intelligent Information and Engineering Systems (KES 2006), October 9-11, Bournemouth, England (2006)
42. Rasmussen, J.: Diagnostic reasoning in action. IEEE Transactions on Systems, Man, and Cybernetics 23(4), 981–992 (1993)
43. Suchman, L.A.: Plans and Situated Actions: The Problem of Human-machine Communication. Cambridge University Press, Cambridge (1987)
44. Moray, N.: Intelligent aids, mental models, and the theory of machines. Int. J. Man-Mach. Stud. 27(5-6), 619–629 (1987)
45. Chambers, A.B., Nagel, D.C.: Pilots of the future: Human or computer? Communications of the ACM 28(11), 1187–1199 (1985)

46. Singh, M.P.: Agent communication languages: Rethinking the principles, Carolina State Uni. (1998)
47. Steels, L.: Adaption multi-agent learning. In: Alonso, E., Kudenko, D., Kazakov, D. (eds.) AAMAS 2000 and AAMAS 2002. LNCS, vol. 2636, pp. 559–572. Springer, Heidelberg (2003)
48. Finin, T., Fritzon, R., McKay, D., McEntire, R.: KQML as an Agent Communication Language. In: Adam, N., Bhargaa, B., Yesha, Y. (eds.) Proceeding of the 3rd international Conference on Information and Knowledge Managment (CIKM 1994), pp. 456–463. ACM Press, New York (1994)
49. Wooldridge, M.: Verifying that agents implement a communication language. In: Joint Sixteenth National Conference on Artificial Intelligence (AAI 1999) and Eleventh Innovative Applications of Artificial Intelligence Conference (IAAI 1999), Orlando, FL, USA, pp. 52–57 (1999)
50. Labrou, Y., Finin, T., Peng, Y.: The current landscape in agent communication languages. IEEE Intelligent Systems 14(2), 1–11 (1999)
51. Szarowicz, A., Francik, J., Mittmann, M., Remagnino, P.: Layering and heterogeneity as design principles for animated embedded agents. Inf. Sci. Inf. Comput. Sci. 171(4), 355–376 (2005)
52. Sioutis, C., Ichalkaranje, N.: Cognitive hybrid reasoning intelligent agent system. In: Khosla, R., Howlett, R.J., Jain, L.C. (eds.) KES 2005. LNCS, vol. 3682, pp. 838–843. Springer, Heidelberg (2005)
53. Sioutis, C.: Reasoning and Learning for Intelligent Agents. PhD thesis, School of Electrical and Information Engineering, University of South Australia (2006)
54. Hammond, G.T.: The Mind of War: John Boyd and American Security. Smithsonian Institution Press, Washington (2004)
55. Rasmussen, J., Pejtersen, A., Goodstein, L.: Cognitive Systems Engineering. Wiley and Sons, New York (1994)
56. Sioutis, C., Ichalkaranje, N., Jain, L.: A framework for interfacing BDI agents to a real-time simulated environment. In: Abraham, A., Koppen, M., Franke, K. (eds.) Design and Application of Hybrid Intelligent Systems. Frontiers in Artificial Intelligence and Applications, pp. 743–748. IOS Press, Amsterdam (2003)
57. Tweedale, J., Ichalkaranje, N., Sioutis, C., Jarvis, B., Consoli, A., Phillips-Wren, G.: Innovations in multi-agent systems. Journal of Network Computing Applications 30(3), 1089–1115 (2007)
58. Adams, D.: Hitchhikerś Gude to the Galaxy. Ballantine Publishing Group, New York (1995)
59. Wooldridge, M., Jennings, N.: Agent theories, architectures, and languages: A survey. In: Wooldridge, M., Jennings, N.R. (eds.) Intelligent Agents - Theories, Architectures, and Languages, Proceedings of ECAI 1994 Workshop on Agent Theories, Architectures, vol. 890, pp. 403–442. Springer, Heidelberg (1995)
60. Consoli, A., Tweedale, J.W., Jain, L.: The link between agent coordination and cooperation. In: 10th International Conference on Kowledge Based Intelligent Information and Engineering Systems, Bournemouth, England. LNCS. Springer, Heidelberg (2006)
61. Cohen, M.S.: A situation specific model of trust to decision aids. Cognitive Technologies (2000)
62. Tweedale, J., Jain., L.C.: Interoperability with multi-agent systems. Journal of Intelligent and Fuzzy Systems 19(6) (2008)

12

Software Agents to Enable Service Composition through Negotiation

Claudia Di Napoli

Istituto di Cibernetica "E. Caianiello" - C.N.R.
C.Dinapoli@cib.na.cnr.it

Abstract. The management of computational resources is becoming a crucial aspect in new generation distributed computing systems like service-oriented ones because of the decentralized, heterogeneous and autonomous nature of these resources. As such they cannot be managed by adopting a centralized approach, but more sophisticated computing methodologies are necessary. In this paper we propose to use software agent negotiation as a means to compose services necessary to provide a new composite service can be provided. In particular, we propose an automated negotiation mechanism to select the service providers that meet the requirements of service consumers on the provision of multiple interconnected services. The negotiation mechanism allows for the evaluation of dependent issues that are negotiated upon when multiple interconnected services are required, and it relies on an iterative process so to improve the possibility of reaching an agreement by letting both service consumers and providers to exchange more proposals and counter-proposals in order to accommodate to the dynamic and changing nature of service-oriented environments.

Keyword: Software Agents, Negotiation, Service–Oriented Computing.

12.1 Introduction

Service-Oriented Computing (*SOC*) is the new promise in the field of distributed computing. The technological advances in the recent years and the pervasiveness of the Internet in the everyday life shifted the distributed computing paradigm from the possibility of exploiting the computing power offered by a collection of computational resources distributed over the network, towards the possibility of exploiting any type of software and hardware commodity available on the network.

In order to make service-oriented systems a viable approach to extensively provide computational capabilities to end users to solve large scale problems, middleware mechanisms are necessary to enable the sharing, selection, and aggregation of computational resources distributed across multiple administrative domains, depending on their availability, capability, performance, cost and users' quality-of-service requirements.

In traditional computing systems the coordinated access to resources that need to be consumed in an aggregated manner is guaranteed by resource

L.C. Jain, N.T. Nguyen (Eds.): Knowl. Proc. & Dec. Mak. in Agent-Based Sys., SCI 170, pp. 275–296.
springerlink.com

management systems designed to operate under the assumption that they have complete control of all necessary resources and thus they implement mechanisms and policies necessary for the effective use of those resources in isolation. In service-oriented environments this assumption cannot be made and more sophisticated computational methodologies for managing the composition of resources that are heterogeneous, located across separately administrated domains, and that inevitably adopt different policies for their use without relying on a centralized control are necessary.

Another aspect that impacts the aggregation of resources in service-oriented computing is the possibility that their provision will be regulated by market-based mechanisms, i.e. subject to a sort of payment. In this scenario, it becomes necessary to adopt methodologies that can take into account both users and providers requirements when accessing resources and not only the optimization of the resource usage [1].

We propose that the middleware mechanism to manage coordinated access to distributed resources/services is based on *software agent negotiation* both to guarantee the autonomy of resources, their coordination, and the satisfaction of both users and providers requirements in respectively requiring and providing them.

The paper is organised as follows. A brief overview of service-oriented computing paradigm together with the opportunity of using software agents as an enabling technology for SOC-based systems are discussed in sections 12.2 and 12.3; section 12.4 describes the representation adopted for requests of composition of services, while section 12.5 discusses the reason to adopt negotiation as the mechanism to select the services necessary for a composition of services; sections 12.6, 12.7, and 12.8 report respectively the negotiation mechanism proposed in the present work, and the behaviour of the participants in the negotiation; section 12.9 discusses the first experiments carried out to test the proposed mechanism; finally section 12.10 reports some related work to better position the contribution of the proposed work, and conclusions close the paper in section 12.11.

12.2 Service-Oriented Computing

Service Oriented Computing (SOC) is the new computing paradigm for distributed systems. The basic idea consists in representing any computational resource as a network-accessible commodity and then using these resources as building blocks to develop distributed applications [2].

One of the main technical challenges to provide service-based distributed applications is to automate the composition of resources once their interoperability is guaranteed. A service–oriented approach addresses the interoperability problem by adopting a uniform way to represent resources in order to encapsulate as much as possible their intrinsic heterogeneous nature. According to this approach a resource is represented as a *service*, i.e. a computational functionality defined through a set of well-defined interfaces, and a set of standard protocols used to invoke the service from those interfaces [3]. A service may be information, or a virtual representation of some physical good or activity, and it has to be identified, published, allocated, and scheduled [4]. Services are autonomous, platform-independent computational entities that can be described, published, discovered, and assembled on-demand to provide added-value applications.

Web services are currently the most promising SOC-based technology that uses Internet as a communication medium and Internet-based standards (such as SOAP [5], WSDL [6], and others) to support service integration. Web service technology allows to separate service access interface definition from service implementation details making services reusable for creating distributed applications combining software components residing in different administrative domains.

It is natural to view service-oriented systems in terms of the entities providing and consuming services: services are offered by *service providers* as a commodity to be consumed by *service consumers* under particular conditions. Nevertheless, most research efforts in SOC-based technology did not focus on the conceptual representation of providers and consumers that play a crucial role when focusing on mechanisms to allow the provision and the composition of services in complex, evolving environments where there might be competition of resources, or conflicts may occur when demand exceeds supply. Higher-level mechanisms than the ones provided by web service layers are necessary to support collaboration or cooperation and to refine service planning or provisioning at run-time [7]. These mechanisms are more related to the role of service providers and consumers than the functionality represented by a service, so services need to be equipped with middleware technology able to model the behaviour of providers and consumers.

12.3 Software Agents for Service-Oriented Computing

Software agents are a natural way to represent service providers and consumers and their defining characteristics are essential to realize the full potential of service-oriented systems [8].

Software agents are autonomous problem solving entities, situated in an environment, able to reach their own objectives and to respond to the uncertainty of the environment they operate in, and they are equipped with flexible decision making capabilities [9]. The above characteristics make software agents a useful computational paradigm to model respectively providers that offer services (which they have total control on) at given conditions, and consumers that require services at other, sometimes conflicting, conditions. Providers and consumers interoperate according to specified protocols and interfaces and establish their own conditions to provide or consume services, and finally they can adopt different decision making mechanisms to accommodate to the dynamic and changing nature of the open environment in which they operate in.

Furthermore software agents are able to cooperate by providing their capabilities in an aggregated and coordinated manner so that more sophisticated capabilities can be provided to service consumers when required. This aspect of agency allows to model the provision of a composition of services as the possibility of sharing computers, software, data, and other resources according to well established sharing rules. So, aspects of agency become even more useful when composition of services are required, i.e. when service consumers cannot interact just with a single service provider for what they need. In fact, in such a case service providers need to coordinate the provision of their services and also to accommodate to the needs of consumers that are interested in the provision of the whole composition and not of the individual components.

Software agents characteristics allow to represent:

- the distributed nature of the provided services through the location of service provider agents in different control domains,
- the different behaviours service providers may have in providing their services through the possibility for the different service providers to adopt different and autonomous decision making mechanisms ,
- the available services through *agent capabilities* representing what they can potentially do,
- the provision of services through *agent actions* representing what they actually provide,
- the request of services through *agent interactions*, i.e. the possibility of exchanging request/reply messages,
- the quality-of-services as *agent preferences*, i.e. parameters that specify providers and consumers expectations on non-functional characteristics of a service they respectively provide or require.

In the present work two types of agents are introduced: *Service Agents* (SAs) that are the providers of services, and *Service Market Agents* (SMAs) that represent service consumers, or more specifically they act on behalf of service consumers. These agents are modelled as self-interested software agents since they are autonomous and independent business entities that do not usually have common objectives, but they are more likely to have conflicting interests. The Service Market Agent can play the role of a Service Agent and viceversa.

A *service* is described by a set of *metadata*:

- the *service identifier*, i.e. a unique identifier for the service,
- the *service description*, i.e. a description of the service functionality represented by a *service type*,
- the *service provider identifier*, i.e. a unique identifier for the service provider,
- the *service attributes*, i.e. a set of parameters that are related to the non-functional characteristics of service execution, or more generally its quality-of-service in a broad sense.

A *composite service* is a service composed of a set of capabilities that cannot be provided as one service by a single Service Agent. A composite service $cs = \{s_1 \mapsto s_2 \mapsto \ldots \mapsto s_n\}$ is a workflow of n services, each one provided by a service provider, so it corresponds to an aggregation of Service Agents that provide the components of the composition of services according to established conditions they agreed upon. The aggregation of Service Agents can be seen as a *Virtual Organisation* [10] formed to provide a single reference provider for the entire composition.

12.4 The User Request: An Abstract Workflow

The user request is in the form of an *abstract workflow* (AW). An abstract workflow is a directed acyclic graph $AW = (ST, P)$ where $ST = ST_1, \ldots ST_n$ is a set of nodes, and P is a set of directed arcs (see Figure 12.1).

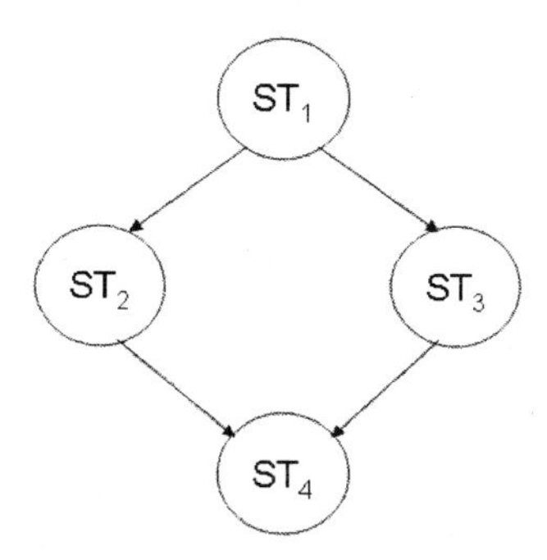

Fig. 12.1. A composite service

Each node represents an *abstract service*, i.e. a service description that specifies the functionality required. It is assumed that the description of a service is given according to an ontology referring to the considered application domain. Each directed arc that connects two nodes represents a *precedence relation* among the corresponding abstract services. A precedence relation $p = (ST_i, ST_j)$ of P implies that an instance of the abstract service ST_j cannot start its execution until an instance of the abstract service ST_i finishes its execution due to a dependence relation between ST_i and ST_j. Usually a precedence relation occurs among service instances, but it can be assumed that it is possible to specify an execution order according to which the required functionality should be delivered. In the present work it is assumed that the precedence relations are provided together with the user request. According to the precedence relations in an abstract workflow, there are *sequential abstract workflows* whose abstract services, once instantiated, have to be executed in a sequential order, and *concurrent abstract workflows* characterized by the possibility of executing some abstract services, once instantiated, in a concurrent way because there are no dependencies among them.

Once a consumer submit a request, a discovery process takes place resulting in a *Partially Instantiated Abstract Workflow* ($PIAW$), i.e. an abstract workflow specifying for each required abstract service the set of potential service instances that are advertised when the request is issued. A PIAW is a graph structure

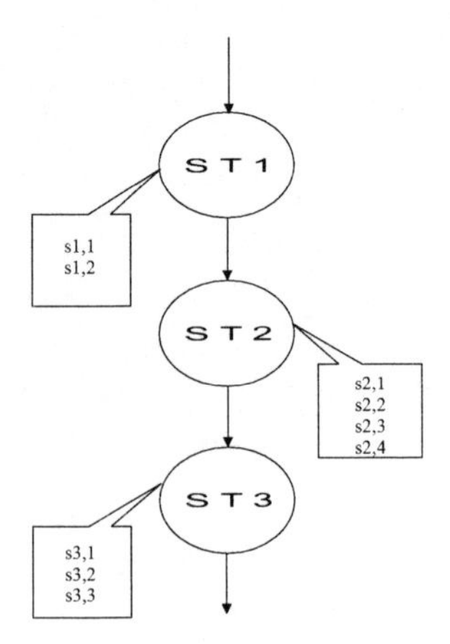

Fig. 12.2. A partially instantiated sequential abstract workflow

composed of as many nodes as the number of required abstract services, and each node refers to the potential set of available services providing the required functionality. Figure 12.2 shows a PIAW for a sequential workflow composed of 3 abstract services. It should be noted that there is a bijective mapping between a service instance and the corresponding service provider, so in the notation used in Figure 12.2, the service $s_{1,1}$ is an instance for the abstract service of type ST_1 provided by the Service Agent SA_1.

In order for a user request to be fulfilled, one service instance for each required service type needs to be selected. Once the selection process successfully end the PIAW becomes an *Instantiated Workflow* (*IW*) is a workflow in which only one service instance is specified for each service type required for the composite service.

12.5 Negotiation for Service Composition

In service-oriented scenarios it is necessary to organise compositions of services on demand in response to dynamic requirements and circumstances. It is likely that in service-oriented systems more service providers can provide the same service, i.e. the same functionality, but at different conditions, or even the same provider can provide the same service at different conditions according to its own policies that can dynamically change. These conditions may refer to non-functional characteristics of the provided service, or other characteristics like price, time to deliver, and so on. These characteristics can be generically referred as a quality-of-service in its broader sense: they are not static conditions, but they can dynamically change and as such they cannot be advertised together with the service description. Furthermore, service providers and service consumers have typically conflicting objectives, so it is unlikely that the conditions required by a user match the ones required by the providers.

A computational mechanism suitable to attempt to reach an agreement on conflicting objectives is *automated negotiation*, i.e. "the process by which a group of agents come to a mutually acceptable agreement on some matter" [11].

In the present work, we propose to use automated negotiation as the mechanism to select the providers that can successfully take part in the provision of a composite service, i.e. whose services can be provided under the conditions necessary to fulfill the user request.

It is assumed that these conditions come from the fact that when a user requires a composite service *cs* it specifies the type of services necessary for the composition, the precedence relations occurring among them, and the deadline ($T_{deadline}$) by when the composite service should be delivered as its preference. In our approach, the dependencies occurring among the service components allow to specify the execution order in which services need to be delivered and as such they are regarded by the SMA as time constraints on the delivery of each service. So, in our scenario, the issue to be negotiated upon is the service time to deliver, i.e. the $time_{start}$ and the $time_{end}$ of each service representing respectively the time when the service is expected to start its execution and the time when the service is expected to end its execution, taking into account the total time for executing the composite service should be within the specified deadline.

The negotiation process allows to select the providers that are able to provide the required services under time conditions that can meet both the deadline required by the user, and also the precedence relations specified by the abstract workflow. Due to these precedence relations, settling the time for the provision of one service cannot be done without considering settling the times for the provision of the other services in the composition. So negotiating foe compositions of services means that only when all SAs agree on the time issues, the composite service can be successfully delivered to the end user according to its preferences. This type of negotiation requires new protocols that are flexible enough to accommodate for both service dependencies and the dynamic nature of the environment in which providers and consumers operate in.

12.6 The Multi-phase Multi-iteration Negotiation Mechanism

In dynamic and changing environments like service-oriented ones, it is not advisable to adopt a fixed negotiation protocol completely determined at design time because it cannot take into account service attributes that are not static features of a service, and also the changing conditions of the environment like the number of available providers, or their characteristics that can also dynamically change in time.

For this reason, we propose a *Multi-phase Multi-Iteration negotiation mechanism* able to deal with the uncertainty about the availability of the service providers, with the multiple issues that need to be negotiated upon, with the

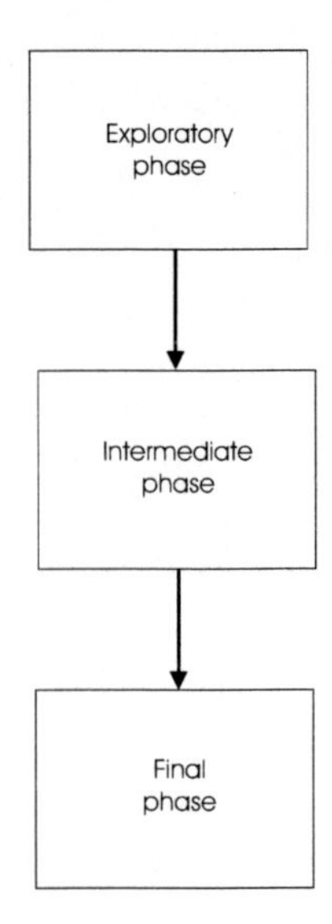

Fig. 12.3. Multi-phase negotiation mechanism

dependences among these issues, and also with the necessity to complete the negotiation process according to time constraints related to the actual execution of the composite service.

The negotiation mechanism we propose consists of 3 phases (see Figure 12.3):

1. *Exploratory phase* to find out the number of Service Agents available to enter negotiation, and their preference over the issues to be negotiated upon,
2. *Intermediate phase* to iterate the process of exchanging offers and counter-offers among the negotiation participants so as to increase the chances of reaching a common agreement,
3. *Final phase* to end the negotiation either with a success leading to a contractual agreement, or with a failure.

The protocol of the negotiation mechanism is *asynchronous* in that messages sent respectively by the SMA and the SAs do not block the sender until a reply is received. In such a way the sender can carry on with its own processing activity after sending the message. The asynchrony of the protocol is necessary to allow for a future distributed implementation of the protocol in which concurrent negotiations will take place at the same time so agents should be able to receive messages coming from agents taking part in different negotiations. In addition, the asynchrony of the protocol allows the involved agents to receive also messages that are not related to the negotiations they are involved in (e.g. messages requiring the actual execution of a service).

12.6.1 The Exploratory Phase

The Exploratory phase is aimed at collecting information on the availability of services required for the composition, and so on the level of competition offered by the environment when a request for a composite service is issued. The information collected is essentially the number of potential SAs available to provide their services for delivering a specific composite service to the user, and the conditions they offer for the provision of their services. The protocol adopted for the Exploratory phase is a variation of the Contract Net Protocol (CNP) introduced by Smith [12]. The main difference is that the SMA (which corresponds to the manager in CNP) does not award a definite contract to the potential contractors after it receives their replies back. This is because the proposed protocol allows for a re-submission of the proposals from the involved SAs, once the SMA checked whether these proposals can meet the time dependence and user constraints after they are assembled together.

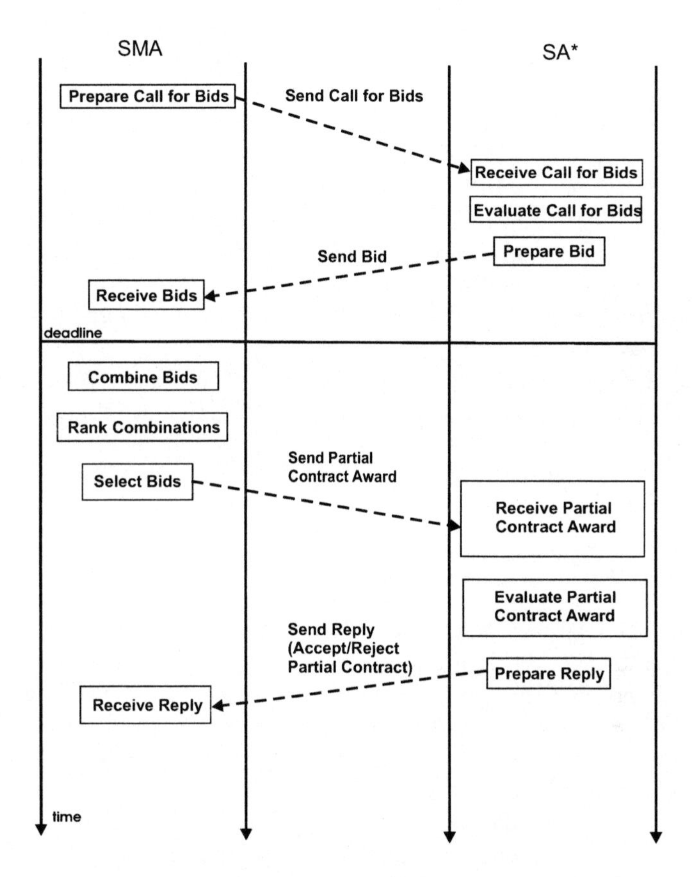

Fig. 12.4. The Exploratory phase negotiation protocol

The Exploratory phase is initiated by the SMA when it receives a request of a composite service from a user. So, the initial state of this phase is characterized by the reception of a user request.

The Figure 12.4 describes the protocol of the Exploratory phase, in terms of the interactions that take place among the participants, and the actions that are performed by the participants. The notation SA^* means that more than one SA is involved at the same time in the negotiation since messages are sent by the SMA to all potential providers. The interactions are described by the dashed arrows in Figure 12.4 representing the following exchanged messages:

- *Call for Bids sending*: the SMA prepares a call for bids for each service type required in the abstract workflow, and it sends the messages to the SAs providing the required types of service, specifying the deadline by when a reply should be received;
- *Bid message sending*: the SAs that received the call for bids may reply with a bid (if they decide to do so, since the protocol allows them also to also ignore the message they received) containing the identifier of the service instance they provide, together with the values of the service attributes required in the call for bids, i.e. the delivery time for the service they provide by assuming that an estimated execution time for delivering the service is known to the SA;
- *Partial Contract Award message sending*: if the SMA received bids for all the abstract services required in the abstract workflow, then it evaluates the bids, and for the acceptable ones, it awards a *Partial Contract* to the respective SAs; otherwise if no bids were received for at least one service type, it declares a *Failure* to all SAs because it is not possible to provide the required composite service;
- *Accept/Reject message*: the SAs that have been selected by the SMA reply to the Partial Contract Award either accepting or rejecting it.

The Exploratory phase terminates once the SMA evaluated the received bids and selected accordingly the SAs that take part in the next phase.

12.6.2 The Intermediate Phase

The Intermediate phase defines the *flexibility* of the protocol by allowing the SMA to iterate the Exploratory phase protocol with a number of SAs for a variable number of times: this means to iterate the number of times the SMA is allowed to send again call for bids to the SAs that have already replied with bids, asking for new bids.

It should be noted that it is the Exploratory protocol that is iterated (i.e. the interactions that take place), and not the Exploratory phase. In fact, the Exploratory phase includes also individual actions performed by the involved agents, and these actions may be different depending on the number of iterations that have already taken place (as explained later on).

12.6.3 The Final Phase

The Final phase defines the completion of the negotiation, and it takes place after the Exploratory phase protocol has been iterated for the number of times fixed by the considered protocol in the Intermediate phase.

The protocol of the Final phase consists of the following exchanged messages:

- *Failure message sending*: the SMA declares a failure of the negotiation to the participants in case of unsuccessful negotiations, i.e. when it is not possible to find a suitable combination of services in terms of times;
- *Final Contract Award message sending*: the SMA awards Final contracts to the successful negotiators, i.e. to SAs selected to provide the service instances for the required composite service according to the agreed upon time constraints.

12.6.4 The Protocol Variants

The proposed negotiation protocol is flexible since the number of times the SMA and the SAs are allowed to send respectively call for bids and bids may vary according to the number of iterations that are specified in the Intermediate phase, and this number can be decided by the SMA at the beginning of the negotiation according to different evaluation criteria. This means that the proposed protocol is deployed as a *set of protocols*, where each protocol is referred to as the *i–P* protocol with i representing the number of iterations taking place during the Intermediate phase. So, the following negotiation protocols are defined by the set of protocols:

- *0–P* composed of the Exploratory phase and the Final phase;
- *1–P* composed of the Exploratory phase, the Intermediate phase in which the Exploratory protocol is iterated once, and the Final phase;
- *N–P* composed of the Exploratory phase, the Intermediate phase in which the Exploratory protocol is iterated N times, and the Final phase.

The choice of adopting one particular protocol may depend on the time that can be spent in negotiation, and on the desired outcome of the negotiation in terms of number of possible solutions, or a better time to deliver for a solution, and so on.

12.7 The Service Market Agent Behaviour

The SMA life-cycle consists in performing *interaction actions* with the environment it operates in, i.e. with the user and the SAs, by receiving and sending messages, and *individual actions* consisting in preparing call for bids, in evaluating the messages received from the SAs, and in preparing the replies (reported in Figure 12.5).

Depending on the phase of the negotiation protocol, and the iterations that have taken place during the Intermediate phase, the replies consist of sending either new call for bids, or a partial contract award, or a final contract award.

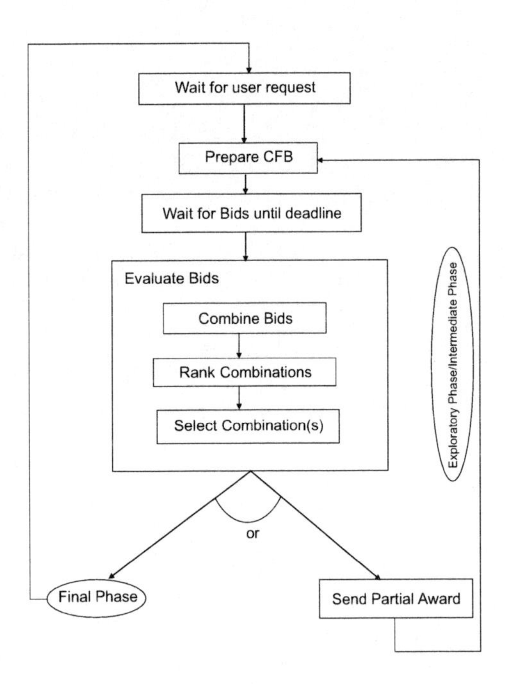

Fig. 12.5. The SMA actions during negotiation

For the time being, it is assumed that at each iteration all the participants SAs are selected for the successive iterations.

When the negotiation starts, the SMA prepares as many call for bids as the number of service types required in the abstract workflow containing the type of service required for the composition, and the time to deliver for that service, i.e. its $time_{start}$ and $time_{end}$. For the first call for bids these times are calculated by dividing the time interval required by the user for the execution of the composite service, $[0, MaxDeadline]$, for the number of the required service types, and assigning the time slots corresponding to the position of the service type in the abstract workflow. For example, for a user request expressed by the abstract workflow reported in Figure 12.2 with a $MaxDeadline = 100$, the contents of the initial `CallForBids` for the service type $ST1$, $ST2$, $ST3$ are respectively:

```
CallForBid = (ST1, servid?, 0, 32)
CallForBid = (ST2, servid?, 33, 66)
CallForBid = (ST3, servid?, 67, 100)
```

The notation `servid?` means that the SMA requires the identifier of the service instance providing the functionality of type STi.

In the successive phase, i.e. the Intermediate phase, the SMA sends always as many `CallForBids` as the number of the required service types, but only to the selected SAs. The process of sending `CallForBids` is iterated for the number of times required by the particular chosen protocol.

Every time the SMA receives all the Bids after the fixed deadline expired, it collects and evaluates them. All received Bids cannot be evaluated separately from the others because the SMA needs to check whether they can be combined so that they meet the user requirements in terms of time, but also the constraints coming from the precedence relations of the composite service. So the SMA combines the received Bids and evaluates the combinations according to a utility function. The number of possible combinations is given by:

$NumComb = \prod_{i=1}^{N} b_i$

where b_i is the number of bids received for the abstract service of type ST_i and N is the number of abstract services required in the abstract workflow. Of course, by increasing the number of SAs and the number of services required in the composition, the number of possible combinations increases exponentially. For this reason, the SMA could limit the number of combinations it considers for evaluation, i.e. it can consider only a subset of possible combinations in the next stages of the negotiation. In fact, it is foreseen (but not yet included in the proposed negotiation mechanism) that the SMA may adopt previously collected information on SAs it is negotiating with to rank them and to select only a subset to be considered in the Intermediate phase.

The evaluation of the Bid combinations is carried out by checking first if there are *feasible* combinations, i.e. combinations for which the time constraints required by the abstract workflow structure are met. For example for the abstract workflow reported in Figure 12.2, the following constraints are to be checked for each combination:

$$(et(ST2) \leq bt(ST3)) \wedge ((et(ST1) \leq bt(ST2))$$

In order to verify that the constraints are met, the SMA calculates for each combination the *time overlap* and the *time slack* among consecutive services, i.e. among services for which a dependence relations occurs according to the structure of the abstract workflow. The time overlap and the time slack for a sequential abstract workflow are given by the following expressions:

$$TimeOverlap(Comb_i) = \sum_{j=1}^{N-1} \lambda * (bt_{j+1} - et_j) \ (1)$$

where $\lambda = 1$ if $(bt_{j+1} - et_j) < 0$ (i.e. there is an overlap in time) while $\lambda = 0$ if $(bt_{j+1} - et_j) > 0$ (i.e. there is a time slack).

$$TimeSlack(Comb_i) = \sum_{j=1}^{N-1} \lambda * (bt_{j+1} - et_j) \ (2)$$

where $\lambda = 0$ if $(bt_{j+1} - et_j) < 0$ (i.e. there is an overlap in time) while $\lambda = 1$ if $(bt_{j+1} - et_j) > 0$ (i.e. there is a time slack).

If $TimeOverlap(Comb_i) = 0$, then $Comb_i$ is a feasible combination, otherwise the combination is unfeasible.

The SMA ranks combinations using the following function:

$$Rank(Comb_i) = \alpha TimeSlack(Comb_i) + \beta TimeOverlap(Comb_i)$$

where the constants α and β are used to differentiate the impact of the time overlap and the time slack on the rank value of a combination. In fact, the time overlap is a measure of the unfeasibility degree of the combination (given the dependence constraints to be met among the services in the composition for the considered sequential workflows), while the time slack is a measure of how close to the required deadline the composition can be delivered, and as such it cannot impact the ranking value in the same way as the time slack.

Feasible combinations are ranked according to an utility function given by:

$$Util(Comb_i) = \frac{MaxDeadline - Duration(Comb_i)}{MaxDeadline}$$

i.e., the combination of services that provides the quickest delivery time is considered the best feasible combination.

12.7.1 The SMA Decision Making

Now, according to the chosen protocol and the results of ranking bid combinations, the Service Market Agent can make different decisions.

0-P Protocol. If there are feasible `Bid` combinations, the SMA selects the highest ranked combination and it starts the Final phase of the protocol where final contracts are awarded to the corresponding SAs. If there are not feasible combinations, the Final phase takes place where a failure message is sent to the SAs involved in the Exploratory Phase.

1-P Protocol. If all `Bid` combinations are feasible, the Final phase takes place by awarding final contracts to the SAs that provided the highest ranked feasible combination. If there are unfeasible `Bid` combinations, the SMA chooses the highest ranked unfeasible combination. Then, the Intermediate phase takes place in which new `CallForBids` are sent to the SAs to search for more feasible combinations. The SMA may select a subset of SAs that take part in the Intermediate phase and a new set of `CallForBids` are formulated. The new time values sent in the `CallForBids` are determined according to the time overlap and the time slack values of the selected best unfeasible combination and the last `Bids` received for that combination in the following way:

- the number of services whose provided time slots need to be moved is calculated (according to the time limits imposed by the minimum start time that is 0, and the maximum end time that is $MaxDeadline$);
- the total overlap is split among the services whose provided time slots need to be moved according to the time limits, by consecutively calculating new time slots for each service according to their length and their precedence relations;
- if the new set cannot be built using this heuristics because the time slack cannot compensate for the time overlap, then the remaining time until $MaxDeadline$ is considered to calculate the new `CallForBids`;
- if the new set of time slots referring to the new `CallForBids` is equal to the set provided in the previous negotiation iteration, then the calculated values are varied adding to all of them a percentage of the length of each considered service in the composition.

This is just an ad hoc heuristic developed to start experimenting with the proposed protocol, but a deeper study on the type of algorithms that can be adopted for these types of problems need to be carried out.

If the SMA receives `Bids` different from the ones previously selected, it checks if the obtained combinations are feasible, and it ranks them again. Then the Final phase takes place. If feasible combinations are found, the SMA awards final contracts to the SAs of the combination with the highest ranking value. Otherwise, it sends a failure message to all SAs involved in the negotiation.

N-P Protocol. The steps are the same as for the 1-P protocol, with the difference that N additional new `CallForBid` messages are sent by the SMA during the Intermediate phase if a feasible solution is not found, or more feasible solutions are searched for.

12.8 The SA Behaviour

The SAs life–cycle consists of:

- waiting for incoming messages,
- processing the received messages according to the order in which they are received,
- preparing a reply to each received message,
- sending the reply to the sender of the received message.

Messages concerning the actual invocation of a service are not considered.

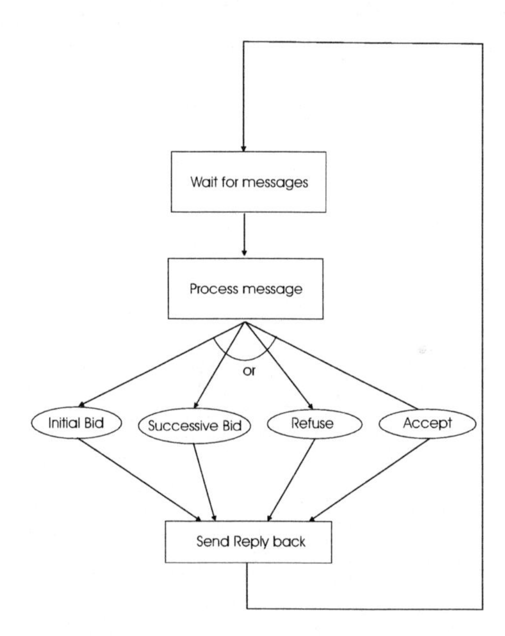

Fig. 12.6. The SA life–cycle for each negotiation phase

During negotiation, the SA performs *interaction actions* (to send and receive messages), and *individual actions* consisting in evaluating the messages it receives from the SMA and in preparing a reply.

The Figure 12.6 describes the SAs life–cycle loop and the messages it is allowed to send during negotiation. The loop is iterated for the number of times the SA receives negotiation messages from the SMA, and different individual actions will take place depending on the phase of the negotiation.

Since SAs are meant to deliver services, we assume they are also committed to serve previously received requests. For this reason, we associate to each SA a *profile*, not known to the SMA or to the other SAs, that represents the SA's expected workload in a time interval.

A time interval is expressed in discrete *time units* (tus), where $1tu$ is considered the smallest unit. The profile is expressed in terms of the number of time slots it has already reserved in that interval to execute the services it provides and it is given by a piecewise constant function $p(t)$ defined on discrete time

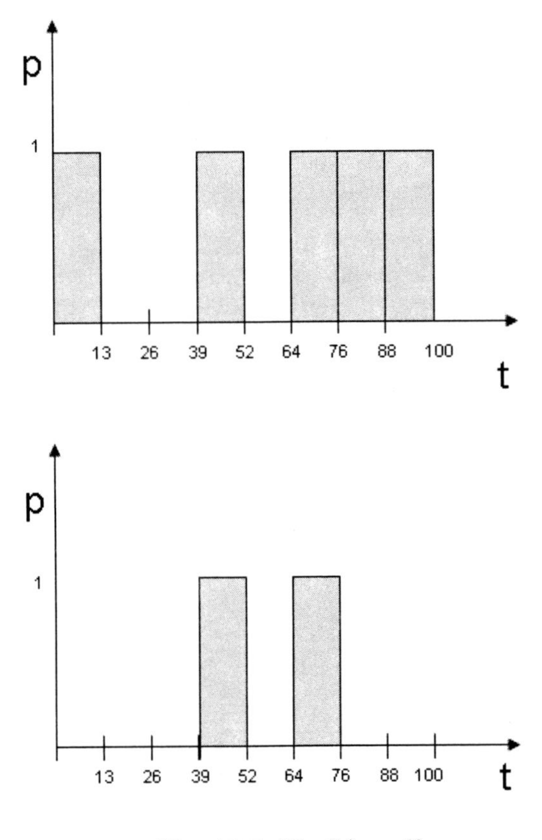

Fig. 12.7. The SA profiles

intervals of fixed length referred just to as *time slots* (TS). The function is equal to 1 or 0 in the TS intervals. If $p(TS) = 1$ then TS is a busy time slot, i.e. the SA cannot provide the service in that time slot; if $p(TS) = 0$ then TS is a free time slot, i.e. it can potentially provide the service in that time slot.

For the time being, the length of a time slot is equal to the the estimated execution time of the service provided by the Service Agent, so $TS = servlength$. Services of the same type are assumed to have the same estimated execution time.

According to the distribution of the workload, two types of SAs are considered:

1. *Busy*: i.e. less than half of the available time slots are free,
2. *Free*: i.e. more than half of the available time slots are free.

The profile impacts the decisions the SA makes during negotiation.

Different profiles for a time interval [0, 100] of a Service Agent providing a service of length 12 are shown in Figure 12.7, respectively for a Busy and Free type of SA.

12.8.1 The SA Decision Making

When the SA evaluates a `CallForBid` sent by the SMA, it tries to meet the request. In fact, the SA is aware that the SMA is trying to find a feasible combination of services, and it is in the interest of the SA to contribute to find a combination. The rationale of this assumption is that in this way the SA has more chances to be the one selected to take part in the Virtual Organisation and hence to make a profit by charging for the service it provides. This is why, when receiving a `CallForBid`, the SA evaluates the time slot proposed for the execution of the service it provides against its workload, according to the following *acceptability rule*:

> A proposed time slot is acceptable if it is included in a free time slot, i.e. a time slot for which the workload is 0. Otherwise the time slot is not acceptable.

If the time slot proposed in the `CallForBid` by the SMA is acceptable, the SA prepares a reply `Bid` containing the same values for the service start time and end time attributes as the ones included in the `CallForBid`. If the time slot is not acceptable, the SA applies the following *closest time slot rule*:

> If there are free time slots in the considered time interval, the TS closest in time to the one proposed by the SMA is selected, i.e. the time slot with the maximum time overlap with the one proposed is selected for the reply.

If there are no time slots available, the SA replies with a reject message.

12.9 The First Experimental Set-Up

The initial experimentation is aimed at collecting information on the performance of the different examined protocols. As pointed out earlier, the protocols are different in the number of times the SMA is allowed to send new `CallForBids` to the SAs, and the protocols considered for the first set of experiments are:

1. *0–P* composed of the Exploratory phase and the Final phase;
2. *1–P* composed of the Exploratory phase, the Intermediate phase with one iteration, and the Final phase;
3. *2–P* composed of the Exploratory phase, the Intermediate phase with two iterations, and the Final phase.

The performance of the protocol is measured as the percentage of feasible combinations of services that is obtained after negotiation with respect to the total number of possible combinations of services. A feasible combination is an Instantiated Workflow, i.e. a set of instantiated services that comply with the dependence relations specified by the Abstract Workflow, and with the time constraints specified by the user request. The total number of combinations depends on the number of `Bids` that the SMA receives from the available SAs.

In the experiments the number of available SAs ranges from 2 to 6 for each service type required for a sequential composite service of 3 services. So the total number of available SAs ranges from 8 to 18, and accordingly, the number of possible combinations is 8, 27, 64, 125, and 216.

For the first set of experiments, the expected length of the required services, and the maximum deadline are fixed, and the workload for each SA is distributed in the time interval $[0, MaxDeadline]$, where $MaxDeadline$ is the time set by the SMA as the time by when the user who issued the request wants to obtain the result of the execution of the composition of services.

For each set of experiments, all SAs are assumed to be of the same type, i.e. they are all either *free* or *busy* providers. The profile of each SA is generated when it is initialised, and it remains constant during the negotiation. A busy time slot cannot be proposed in a `Bid` because it represents a time interval for which the SA has already committed to do something else, and this is why its workload is 100%. Full timeslots are randomly generated within the considered time interval. It is also assumed that all SAs have always a number of free time slots, so they can always reply to a received `CallForBid`.

12.9.1 The Obtained Results

The results shown in Figure 12.8 are obtained by running the simulation program 50 times for each set of available SAs. Each run calculates the percentage of feasible combinations obtained at the end of the negotiation with respect to the number of all possible combinations. The obtained percentage is averaged on the 50 runs. The simulation results show that the percentage of feasible combinations increases when more iterations of the protocol are allowed (i.e. ranging from 0-P to 2-P protocol). This is an expected result since with more iterations the possibility of obtaining feasible combinations should increase because SAs are allowed to propose different time slots that could meet the time constraints.

Unfortunately, within one protocol the increase in the number of SAs does not affect the number of feasible combinations, as it would have been expected.

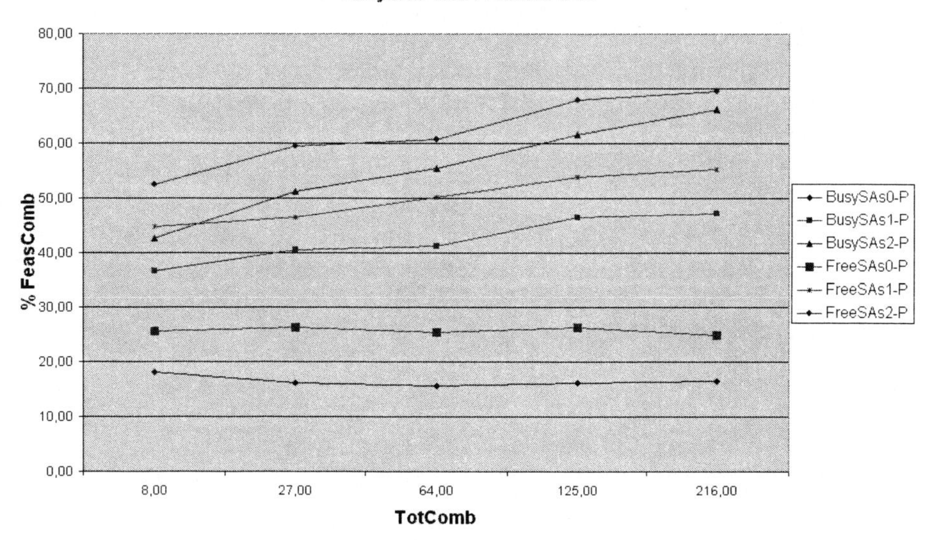

Fig. 12.8. Experimental results

The main reason for this behaviour is that with a randomly generated profile for each SA, only by increasing the number of SAs could increase the number of obtained feasible combinations. Furthermore, it is foreseen that the number of available SAs could affect the results in a more evident way when there are busy SAs whose number of free time slots is low.

12.10 Related Works

It is well recognised that in service-oriented systems where computational capabilities can be rented over the Internet, composed to deliver a required outcome, and then discarded, an agreement between the providers of such capabilities and their consumers is necessary. Such an agreement, known in the literature as Service Level Agreement (SLA) [13], [14] represents a form of guarantee for the provision of a service both from the consumer and the provider side.

As pointed out earlier, an agreement become even more necessary when composition of services are required since the conditions under which a service has to be delivered may depend on the conditions at which another service in the composition is delivered.

The possibility of using software negotiation to manage the provision of computational resources has been widely explored both in the service-oriented research area [15], [16], [17], and in the agent-oriented research area [18], [19], [20].

However, most of the research efforts concentrated on the provision of single services, so that usually negotiation occurs between a provider and a consumer upon a given set of negotiation issues. If the negotiation is successful the agreed upon values for the negotiated issues represent the delivery conditions specified in the SLA. In fact, in the service-oriented area, the conditions under which single services should be delivered for a composition of services are usually manually determined and configured resulting in systems that are not flexible in the case of unavailability of some services in the composition and also not fully automated. On the other hand, in the agent-oriented area, the conditions under which single services should be delivered for a composition of services are the results of aggregating one-to-one and one-to-many negotiations for the provision of single services [21].

The contribution of the present work is the attempt to use negotiation for establishing the quality-of-service constraints for each service by globally evaluating the proposals of each provider, and then to select the providers suitable for delivering the entire composition accordingly.

12.11 Conclusions

It is widely recognized that the ultimate success of service-oriented environments relies on the possibility of selling and buying services in the same way as goods and services are bought and sold in the real world economy [22]. Furthermore, in service-oriented systems the possibility of providing compositions of services to end users in a transparent manner, so that added-value applications can be built aggregating specialized services offered by different providers is crucial. In fact, consumers usually require high level services coming from the composition of simpler functionalities offered by different providers. It is likely that there are dependencies among the services in the composition, so an execution order is determined and it should be respected for the successful delivery of the whole composition. In this scenario it is necessary to coordinate service provision before service execution because the unavailability of one service at the right time causes the delivery of the whole composition to fail. In fact, if a composite service composed of two services provided by different providers should be provided to a user available to pay some money to obtain the composite service, what happens if one service is successfully delivered and the other one is not? The user does not want to pay for something that he/she did not get, but at the same time the provider that successfully delivered the service cannot end up providing something for free! Nevertheless, in highly dynamic environments where the forces of demand and supply are continuously changing, fixing all the aspects linked to the provision of a service may not be possible. So, high level computational mechanisms are necessary to automate the process of reaching agreements taking into account on one hand consumer and provider needs, and on the other hand the possible interdependencies among the services in a composition. In such cases the possibility of negotiating some service characteristics at the time of the request is crucial.

The novelty of the present work is to adopt a negotiation mechanism not to reach an agreement between two parties in a space of already known solutions, but as a means to select collectively the services that can better meet the requirements coming from both the user preferences and the constraints of the required composite service (in terms of dependencies among its components). In fact, services are selected as a result of a successful negotiation process before the actual execution takes place.

In order to reach this objective services need to be equipped with middleware mechanisms that allow for automated negotiation, in particular for a flexible negotiation mechanism that accounts for the dynamic nature of service-oriented systems. The proposed negotiation mechanism allows to deal with:

- imperfect knowledge typical of open environments like the SOC-based ones where negotiators are unlikely to have perfect knowledge about each other;
- criticality of time typical of on demand negotiations where the time spent in negotiation can be dynamically adjusted according to the conditions related to the current requests;
- environment volatility in terms of uncertainty about the actual participants in a negotiation since service providers and consumers freely join and leave the environment;
- the interdependencies among the issues to be negotiated upon occurring when composition of services are required.

It is planned to further experiment with the proposed family of protocols by increasing the number of SAs, and by varying their profile in order to better specify decision making strategies for both the SAs and the SMAs. The main purpose of the experimentation will be to collect information on when it is useful to iterate the negotiation, and according to which parameters whose values cannot be statically determined. The collected information could help the SMA to decide which protocol to adopt in different situations.

References

1. Buyya, R., Abramson, D., Giddy, J.: An economy driven resource management architecture for global computational power grids. In: Proceedings of The 2000 International Conference on Parallel and Distributed Processing Techniques and Applications (PDPTA 2000), Las Vegas, USA (2000)
2. Papazoglou, M.P., Georgakopoulos, D.: Service–oriented computing. Communications of the ACM 46(10), 24–28 (2003)
3. Foster, I., Kesselman, C., Nick, J., Tuecke, S.: The physiology of the grid: An open grid service architecture for distributed system integration. Technical report Open Grid Service Infrastructure WG (2002)
4. De Roure, D., Jennings, N.R., Shadbolt, N.: The Semantic Grid: A future e–Science infrastructure, pp. 437–470. Wiley, Chichester (2003)
5. Mitra, N., Lafon, Y.: Soap version 1.2 part 0 (April 2007), `http://www.w3.org/TR/soap12-part0/`
6. Booth, D., Liu, C.K.: Web services description language (wsdl) version 2.0 part 0 (June 2007), `http://www.w3.org/TR/wsdl20-primer`
7. Payne, T.R.: Web services from an agent perspective. IEEE Intelligent Systems 23(2), 11–14 (2008)

8. Foster, I., Jennings, N.R., Kesselman, C.: Brain meets brawn: Why grid and agents need each other. In: Proc. 3rd AAMAS, pp. 8–15 (2004)
9. Jennings, N.: An agent–based approach for building complex software systems. Communication of the ACM 44(4), 35–41 (2001)
10. Foster, I., Kesselman, C., Tuecke, S.: The anatomy of the grid: Enabling scalable virtual organizations. The International Journal of High Performance Computing Applications 15(3), 200–222 (2001)
11. Jennings, N.R., Faratin, P., Lomuscio, A.R., Parsons, S., Sierra, C., Wooldridge, M.: Automated negotiation: prospects, methods and challenges. Int. Journal of Group Decision and Negotiation 10(2), 199–215 (2001)
12. Smith, R.G.: The contract net protocol: High–level communication and control in a distributed problem solver. IEEE Trans. on Computers 29(12), 1104–1113 (1980)
13. Keller, A., Ludwig, H.: The wsla framework: Specifying and monitoring service level agreements for web services. Journal of Network and Systems Management, Special Issue on E-Business Management 11(1) (2003)
14. Czajkowski, K., Foster, I., Kesselman, C., Sander, V., Tuecke, S.: Snap: A protocol for negotiating service level agreements and coordinating resource management in distributed systems. In: Feitelson, D.G., Rudolph, L., Schwiegelshohn, U. (eds.) JSSPP 2002. LNCS, vol. 2537, pp. 153–183. Springer, Heidelberg (2002)
15. Ouelhadj, D., Garibaldi, J., MacLaren, J., Sakellariou, R., Krishnakumar, K.: A multi-agent infrastructure and a service level agreement negotiation protocol for robust scheduling in grid computing. In: Sloot, P.M.A., Hoekstra, A.G., Priol, T., Reinefeld, A., Bubak, M. (eds.) EGC 2005. LNCS, vol. 3470, pp. 651–660. Springer, Heidelberg (2005)
16. Li, J., Yahyapour, R.: Negotiation strategies for grid scheduling. In: Chung, Y.-C., Moreira, J.E. (eds.) GPC 2006. LNCS, vol. 3947, pp. 42–52. Springer, Heidelberg (2006)
17. Nassif, L.N., Nogueira, J.M., de Andrade, F.V.: Distributed resource selection in grid using decision theory. In: Seventh IEEE International Symposium on Cluster Computing and the Grid (CCGrid 2007), pp. 327–334. IEEE, Los Alamitos (2007)
18. Collins, J., Tsvetovat, M., Mobasher, B., Gini, M.: Magnet: A multi-agent contracting system for plan execution. In: Proceedings of Workshop on Artificial Intelligence and Manufacturing: State of the Art and State of Practice, pp. 63–68 (1998)
19. Collins, J., Ketter, W., Gini, M.: A multi-agent negotiation testbed for contracting tasks with temporal and precedence constraints. International Journal of Electronic Commerce 7(1), 35–57 (2002)
20. Norman, T.J., Preece, A., Chalmers, S., Jennings, N.R., Luck, M., Dang, V.D., Nguyen, T.D., Deora, V., Shao, J., Gray, A., Fiddian, N.: Conoise: Agent-based formation of virtual organisations. In: 23rd SGAI Int. Conf. on Innovative Techniques and Applications of AI, pp. 353–366 (2003)
21. Stein, S., Payne, T.R., Jennings, N.R.: Flexible service provisioning with advance agreements. In: Proc. of Seventh International Conference on Autonomous Agents and Multi-Agent Systems (AAMAS 2008). IEEE, Los Alamitos (in press, 2008)
22. Wooldridge, M.: Engineering the computational economy. In: IST 2000: Proceedings of the Information Society Technologies Conference, Nice, France (2000)

13

Advanced Technology towards Developing Decentralized Autonomous Flexible Manufacturing Systems

Hidehiko Yamamoto

Dept. of Human and Information Systems, Faculty of Engineering, Gifu University
1-1, Yanagido, Gifu, Japan
yam-h@gifu-u.ac.jp
http://www1.gifu-u.ac.jp/~yamlab/

Abstract. One of the advanced technology for controlling production systems is a decentralized autonomous Flexible Manufacturing System (DA-FMS). The autonomous decentralized FMS aims at high production efficiency by giving self-control or decentralizing the plan, design and operation of FMS. This paper discusses the intelligent real-time decision making necessary for realizing an autonomous decentralized FMS with Automatic Guided Vehicles (AGVs) and Machining Centers (MCs). This research develops a real-time production control method called reasoning to anticipate the future (RAF) based on the predictions that forecast not only current production situations but also anticipate future ones. The paper also describes a method to rank competing hypotheses by oblivion and memory (ranking by oblivion and memory; ROM) for improving the reasoning efficiency of RAF.

13.1 Introduction

With the development of network technology and information communication technology, the basic technology of a new production system has been developed. One of the directions it can take is a decentralized autonomous Flexible Manufacturing System (DA-FMS). The DA-FMS aims at high production efficiency by giving self-control or decentralizing the plan, design and operation of FMS. This chapter discusses the advanced technology for realizing DA-FMS with Automatic Guided Vehicles (AGVs) and Machining Centers (MCs). As the advanced technology, the two stories, the real-time production control method called reasoning to anticipate the future (RAF) and the method to control DA-FMS by using a memory, are described.

In general, if it is possible to make plans by considering near future trends and information, it is considered wiser than acting blindly. As the AGV actions' decision, the first story introduces this idea that RAF carries out a real-time decision making for AGV actions based on the predictions that forecast not only current production situations but also anticipate future ones. Also, the developed decision making method is applied to a DA-FMS. Because of the results, it can be seen that multi-production that keeps the target production ratio is possible even though neither AGV actions' plans nor parts input schedules are given beforehand. Especially, it can be shown that the

L.C. Jain, N.T. Nguyen (Eds.): Knowl. Proc. & Dec. Mak. in Agent-Based Sys., SCI 170, pp. 297–322.
springerlink.com

method will operate a FMS without influencing the production ratio even when unpredicted troubles happen, which is often seen in an actual factory.

This RAF applies hypothetical reasoning to the number of next actions that can be considered for the AGV (competing hypotheses). However, if the number of agents included in the hypothetical reasoning process in the RAF is increased, the number of next actions that are considered as competing hypotheses also increases. As a result, the replacement of true and false hypotheses and number of repetitions of discrete production simulations produced by these replacements are increased and the increase will give rise to the problem of decreased reasoning efficiency of the RAF. The second story reports a method to solve these problems. The reported method is called Ranking by Oblivion and Memory (ROM). This ROM differs from conventional methods, and seeks to make optimum use of the characteristics of DA-FMSs.

13.2 Decentralized Autonomous FMS Model

The construction of a DA-FMS that this chapter deals with is shown in Fig. 13.1. It shows a Parts Warehouse that supplies parts for a factory, a Products Warehouse for finished parts from MCs, some AGVs that carry parts and some MCs are arranged. Each AGV carries one part. AGVs move on the lattice lines of the figure at a uniform

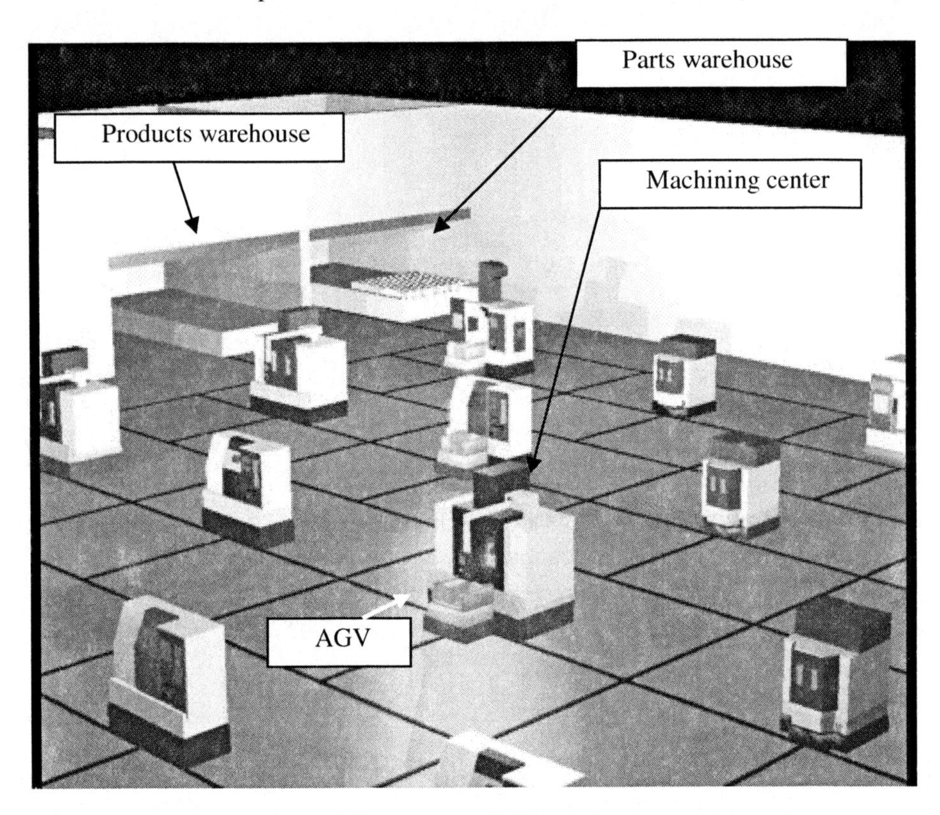

Fig. 13.1. DA-FMS model

velocity. MCs can work several kinds of parts and each of the parts has decided manufacturing processes and manufacturing time. Some MCs do the same kinds of work processes. The set of the same type MC is called a Group MC and is distinguished by describing subscripts, for example, ${}_{g}MC_1$, ${}_{g}MC_2$, … . Each MC in the same Group MC is distinguished by attaching a hyphen and figures after the name of Group MC, for example, $MC_{1\text{-}1}$, $MC_{1\text{-}2}$, .

Note that later sentences (section 13.2 ~ section 13.8) uses the term " parts ". The parts meaning is not limited to the same variety but includes different varieties.

The contents of information exchanges and cooperative actions between each agent in a DA-FMS is basically the following. The Parts Warehouse sends the information on the names of parts that are in the Parts Warehouse. The AGV sends both the information on the name of the part that the AGV currently has and the information on its next destination. The MC sends both the information on the name of the part it is currently manufacturing and the information about the remaining manufacturing time. When necessary, an agent uses the received information to make the agent movement decisions.

There are a few researches for a decentralized autonomous FMS [1-4]. They do not consider what will happen in FMS. This research's characteristic is to forecast the near future situations of FMS. In this point, the research [5] is different from the ordinal researches.

13.3 Reasoning to Anticipate Future

13.3.1 Outline of Future Forecast Reasoning

One of FMS's characteristics is to realize efficient production by jointly sharing manufacturing operations among the MCs in a Group MC that has the same manufacturing processes. The parts delivers by AGVs are responsible for the sharing operation. Because the AGV moving distance and time and the waiting time in front of each MC bay are inexplicably linked with FMS operating efficiency. According to which parts the AGV will deliver to which MC and which part the AGV will take, the FMS operating efficiency can change much.

With the predetermined scheduling for AGVs' movement, which is the ordinary method, it is difficult to deal with unpredicted troubles such as manufacturing delay and machine breakdown. If the ordinary method is used, once this kind of unpredicted trouble occurs, re-consideration of the production schedule must be necessary. Moreover, in a FMS that uses many AGVs and MCs, it is difficult to make a predetermined schedule for efficient AGV movement order and parts delivering order.

In order to solve the problems, the first story topic shows the following process procedure for carrying out DA-FMS: ① use each agent information, ② forecast several future steps of probable AGV actions, ③ forecast probable FMS operating situations. The reasoning to anticipate the future (**RAF**) to decide the AGV's next moving actions based on the prediction results of ③ is presented and solves the above mentioned problems. RAF resembles a chess strategy that moves a piece after anticipating the several alternatives for one move. The decisions by RAF are both where the AGV moves next and which part it carries next. In this way, predetermined parts delivering schedule is not needed.

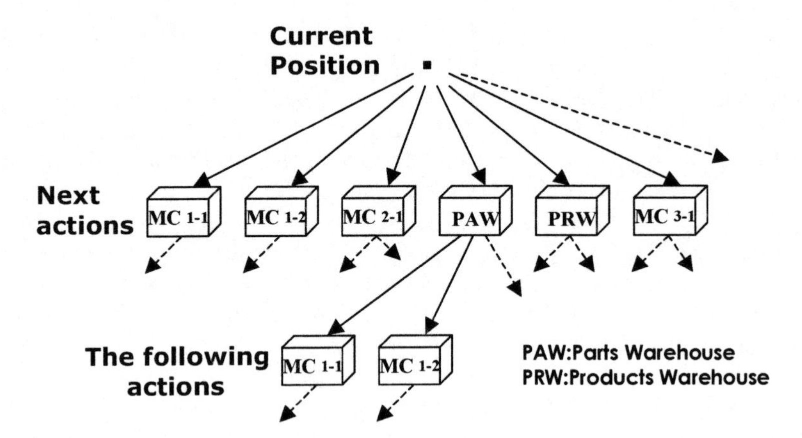

Fig. 13.2. Reasoning tree construction

The probable action (**Next Action**) that a AGV will take next is not decided as a single action but as many actions because there are some MCs that are doing the same manufacturing process jointly, maybe the AGV transfer finished parts to the Products Warehouse, or maybe deliver a new part into the FMS from the Parts Warehouse, and there is a possibility that another AGVs will make the same action. Hypothetically, if an AGV chooses one of the above actions, in an actual FMS, each agent in the FMS keeps doing its chosen operation. When an AGV needs the choice of Next Action again, it chooses a single Next Action from among the possible choices again. In this way, the operating situation of a FMS is expressed as the choice process of unending cycle of AGV Next Actions. That is, it is expressed as a tree construction which includes nodes corresponding to possible AGVs Next Actions. The tree construction can be extended infinitely, as shown in Fig. 13.2. The strategy of RAF considers the possible Next Actions that the AGV will be able to take locally a few steps ahead as a forecastable range, as well as globally forecasting phenomena happening in the FMS in a near future and, then going back to the present, decides which choice should be chosen at present. In order to do RAF, by the hypothetical reasoning which considers the choices that the AGV will be able to take as competitive hypotheses, the decision process is controlled.

13.3.2 Future Forecast and Competitive Hypotheses

Hypothetical reasoning regards events that can happen simultaneously as competitive hypotheses, classifies each hypothesis among them into a true hypothesis and the rest false hypotheses, then hypothetically continues to reason with the true hypothesis and follows the true hypothesis till a contradiction occurs[6].

Based on hypothetical reasoning, RAF tentatively decides AGV Next Actions a few steps ahead locally and forecasts FMS operating situations globally. In this situation, what are established as competitive hypotheses strongly depends upon RAF executions. RAF establishes competitive hypotheses in the following way.

Considering the Next Action that an AGV will be able to take from a standpoint of the AGV, two kinds of actions are possible, ① where it will move next, ② which part it will take next. This classification is reflected in the two kinds of competitive : competitive hypotheses for moving places (**C-hypotheses-move**) and competitive hypotheses for parts (**C-hypotheses-parts**). C-hypotheses-move may be analyzed into three types : (1) move to MC bay to exchange parts, (2) move to the Parts Warehouse to input new parts , (3) move to the Products Warehouse to deposit parts when all manufacturing processes are finished. As the elements of C-hypotheses-move, each MC ($MC_{1\text{-}1}$, $MC_{1\text{-}2}$, …), the Parts Warehouse and the Products Warehouse are established. As the elements of C-hypotheses-parts, each part (P_1,P_2,P_3, …) is established.

Now, the functions of RAF will be described. When an actual FMS is operating, RAF beforehand forecasts near future DA-FMS operating situations. This means the thinking process called RAF begins just before an AGV in the actual world has to choose its Next Action. When RAF begins, the first step is to search all possible next actions that the AGV could take. The actions are called **Next Action Set**. The next step is to choose a single Next Action from among Next Action Set. The third step is to simulate with all agents what would happen if the Next Action is chosen as the AGV's next action till a certain AGV needs to choose a Next Action. This simulation is called **Simulation in Hypotheses (SiH)**. In carrying out SiH, the situation comes that an optional AGV in the simulation searches Next Action Set. In other word, this situation is that an AGV in SiH has to make a decision which place the AGV moves next as an AGV in the real world has to. At this time, the Next Action Set is re-searched, one Next Action is chosen from among the Next Action Set and SiH is carried out again. By the results of the SiH that is followed by the choice of a Next Action, DA-FMS future operating situations can be seen again. That is, RAF can be shown as tree construction where the three layers repeatedly lie one upon another : ① the action choice that an AGV in the actual world took last, ② the Next Action Set of the AGV in the actual world, ③ Next Action Set of an AGV in SiH, as shown in Fig. 13.3.

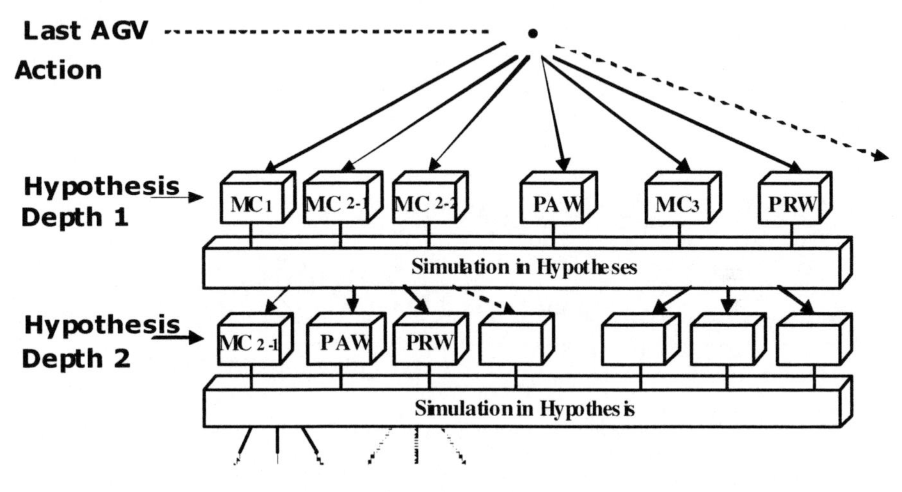

Fig. 13.3. RAF tree construction

In Fig. 13.3, the layer corresponding to one depth from the root of the tree construction is called one step hypotheses depth (D_{hy}=1), the node set belonging to hypotheses depth is called forecasting actions one step ahead (Forecast[1]). Forecasting actions located below each node of the forecasting actions of one step ahead corresponds to hypotheses depth 2 (D_{hy}=2) and are called as forecasting action of two step ahead (Forecast[2]). In the same way, forecasting actions corresponding to hypotheses depth are expressed as three step ahead, four step ahead, … . Then, RAF considers forecastable actions an optional n step ahead, Forecast[n], as a forecastable range, regards forecastable actions in each layer as competitive hypotheses and carries out the reasoning control by regarding one optional forecastable action among each layer of competitive hypotheses as a true hypothesis and the remaining forecastable actions as false hypotheses. By judging DA-FMS operating situations n steps ahead, the actual AGV Next Action is decided.

The algorithm to carry out RAF is described as below. First, the terms used in the algorithm are defined.

[Definition] Standard to Judge True and False: Standard to judge the contradictions in hypothetical reasoning with the results of DA-FMS total operating efficiency gained by the executions of SiH. The standard has ***s*** kinds of standards' range. If the standard is not satisfied, it is judged that a contradiction occurs. Some concrete examples are :

- ***if*** DA-FMS total operating efficiency is over $\boldsymbol{e}(1)$%, it is true,; ***if*** under, false.
- ***if*** DA-FMS total operating efficiency is over $\boldsymbol{e}(2)$%, it is true,; ***if*** under, false.
- ***if*** DA-FMS total operating efficiency is over $\boldsymbol{e}(3)$%, it is true,; ***if*** under, false.

…

- ***if*** DA-FMS total operating efficiency is over $\boldsymbol{e}(\boldsymbol{s})$%, it is true,; ***if*** under, false.

as $\boldsymbol{e}(1) > \boldsymbol{e}(2) > \boldsymbol{e}(3) > \ldots > \boldsymbol{e}(\boldsymbol{s})$

[Definition] Group MC Selection Priority Value, *Fm*: The values indicates the aim how long each Group MC will operate and is expressed as Equation (1). The Group MC that has the large Priority Value is considered to have many jobs left and the priority ranking that is selected as a true hypothesis is given a high ranking.

$$Fm(gMc.N) = \sum_{Pn=1}^{Pn=Pn'} gMc.process.time(Pn) \times \left\{ \frac{production.rate(Pn)}{100} - \left(\frac{finished.parts.N(Pn) + \dfrac{inprocess.parts.N(Pn)}{2}}{all.finished.parts.N + \dfrac{all.inprocess.parts.N}{2}} \right) \right\} \tag{13.1}$$

gMc: name of Group MC

Pn:parts variety （*Pn*=1 ~ *Pn'* ）

gMc.process.time(*Pn*): time that Group MC needs to manufacture parts *Pn production.rate*(*Pn*): target production ratio (%) of parts *Pn*

finished.parts.N(*Pn*): number of parts *Pn* when all processes are finished (number of parts *Pn* in Products Warehouse)

inprocess.parts.N(*Pn*): number of parts *Pn* that are in process or in being transferred (number of parts *Pn* that AGV or MC has)

all.finished.parts.N: number of all parts where all processes are finished (number of all parts in Product Warehouse)

all.inprocess.parts.N: number of all parts that are in process or being transferred (number of all parts that AGV or MC has)

[Definition] Parts Warehouse Selection Priority Value, *Fp*: The value indicates how many parts are in process or in being transferred and is expressed as Equation (13.2). This value becomes the priority ranking for competitive hypotheses. The value is an integer after being rounded off.

$$F_p = \left(1 - \frac{\max.parts.N - all.inprocess.parts.N}{\max.parts.N}\right) \times destination.N \quad (13.2)$$

max.parts.N : possible maximum parts input number (sum of AGVs and MCs)

destination.N : number of parts destinations (sum of AGVs, MCs, Parts Warehouse, and Products Warehouse)

[Definition] Products Warehouse Selection Priority Value, F_f: The value indicates the parts condition of the Products Warehouse and is expressed as Equation (13.3). The value corresponds to the priority ranking in competitive hypotheses.

$$F_f = destination.N - F_p \quad (13.3)$$

[Definition] Parts Selection Priority Value, *V(Pn)* :The value indicates how many parts are still waiting to be input into the production line and is expressed as Equation (13.4). The part that has a large Priority Value has a high rank order of priority that is taken from the Parts Warehouse by AGVs and is input into the production line.

$$V(Pn) = all.process.time(Pn) \times \left\{ \frac{production.rate(Pn)}{100} - \left(\frac{finished.parts.N(Pn) + \frac{inprocess.parts.N(Pn)}{2}}{all.finished.parts.N + \frac{all.inprocess.parts.N}{2}} \right) \right\} \quad (13.4)$$

all.process.time(*Pn*); total manufacturing time for parts *Pn*

[Definition] Job Variance Value, F_d(MC.N): In the case of a Group MC that has the same type MC, it is necessary to keep the job equality among MCs, that is, each MC should be doing an equal amount of work. In order to do this, the Job Variance Value F_d(MC.N) expressed with Equation (13.5) is adopted. The equation is based on MC operating efficiency. The MC whose operating efficiency is low becomes a large Job Variance Value and this MC is likely to be chosen for work.

$$F_d(Mc.N) = \frac{100}{Mc.efficiency} \tag{13.5}$$

Mc.efficiency ; operating efficiency of Mc.N (%)

[Algorithm of RAF]

Step1: Establish hypotheses depth D_{hy}=1 and the Standard to Judge True and False s =1.

Step2: Search all Next Actions (Forecast[D_{hy}]) that can be forecastn and classify them into competitive hypotheses elements of C-hypotheses-move and C-hypotheses-parts, as shown in Equations (13.6) and (13.7). The elements are established so that the left side element in the parentheses has a high priority. At this stage, the ranking is tentative.

C-hypotheses-move= {Parts Warehouse, Products Warehouse, $MC_{1\text{-}1}$, $MC_{2\text{-}1}$,…} (13.6)

C-hypotheses-parts= {P_1, P_2, P_3, …} (13.7)

Step3:Confirm the current position of an AGV that needs to decide its next action and carry out the following rule.

If { the AGV location is in the Parts Warehouse and the AGV does not have parts}
Then{Go to Step5}
Else{Go to Step4}

Step4: Carry out Hypothetical Reasoning for Moving Decisions from Step4-1~Step4-9.

Step4-1: Replace MC, the element among competitive hypotheses C-hypotheses-move, with Group MC that the MC belongs to, as shown in Equation (13.8). At this time, repeated Group MCs are integrated into one Group MC. The resulting Group MC is called a **Competitive Group MC**.

C-hypotheses-move = {Parts Warehouse, Products Warehouse, ${}_gMC_1$, ${}_gMC_2$, ${}_gMC_3$, …} (13.8)

Step4-2: Find Group MC Selection Priority Value *Fm* for each Competitive Group MC among competitive hypotheses C-hypotheses-move and renew the elements' row of competitive hypotheses C-hypotheses-move by changing the Group MC row with the highest Group MC Selection Priority Value *Fm* first.

Step4-3: Find Parts Warehouse Selection Priority Value, *Fp* and Products Warehouse Selection Priority, F_f and renew the elements' row of competitive hypotheses C-hypotheses-move by inserting Parts Warehouse and Products Warehouse in the priority ranking positions among competitive hypotheses C-hypotheses-move, whose positions correspond to acquired Parts Warehouse Selection Priority Value, *Fp* and Products Warehouse Selection Priority, F_f .

Step4-4: Randomly Select an optional competitive Group MC, ${}_gMC_\alpha$ from among competitive hypotheses C-hypotheses-move.

***Step4-4-1*:** Search all MCs belonging to ${}_{g}MC_{\alpha}$, call it $MC_{\alpha\text{-}\beta}$and find their MC's Job Variance Values, $F_d(MC_{\alpha\text{-}\beta})$.

***Step4-4-2*:** Compare each value of Job Variance Values, $F_d(MC_{\alpha\text{-}\beta})$ and make a list called MC_{α}-List such that MCs form a queue according to the Job Variance Value with the highest value first, like Equation (13.9).

$$MC_{\alpha}\text{-List} = \{MC_{\alpha\text{-}1}, MC_{\alpha\text{-}2}, MC_{\alpha\text{-}3}, \cdots\} \quad (13.9)$$
$$\text{as } F_d(MC_{\alpha\text{-}1}) \geqq F_d(MC_{\alpha\text{-}2}) \geqq F_d(MC_{\alpha\text{-}3}) \geqq \cdots$$

***Step4-4-3*:** Renew the elements of competitive hypotheses by replacing ${}_{g}MC_{\alpha}$ with MC_{α}-List.

***Step4-4-4*:** Renew the elements of competitive hypotheses C-hypotheses-move by giving the remaining competitive Group MC the repeated processes from *Step4-4-1* to *Step4-4-3* .

***Step4-5*:** Select the action whose priority ranking is No.1 from among the elements of competitive hypotheses C-hypotheses-move, corresponding to the left end element, establish it as a true hypothesis and establish the remaining elements as false hypotheses.

Step4-6*:** By using a true hypothesis, carry out SiH till an AGV must make its Next Action choice and calculate FMS total operating efficiency ***E at the time when SiH stops.

***Step4-7*:** Perform the following rule.

If{ $e(s) \leqq E$ }, **Then**{ Go to Step6}

Else{ Admit that a contradiction has occurred and go to Step4-8}

***Step4-8*:** Perform the following rule.

If { An element that has not been chosen as a true hypothesis among the competitive hypotheses C-hypotheses-move still exists }

Then { Replace a true hypothesis with a false hypothesis, select the next priority ranking element among competitive hypotheses C-hypotheses-move as a true hypothesis and return to step4-6 }

Else {Go to Step4-9}

***Step4-9*:** Perform the following rule.

If { Hypothesis depth D_{hy}=1}, **Then** { Establish $s \leftarrow s+1$ and return to Step2}

Else{ Backtrack after establishing $D_{hy} \leftarrow D_{hy} - 1$ and select the next priority ranking hypothesis element among competitive hypotheses of the layer D_{hy} as a true hypothesis. If the selected hypothesis belongs to C-hypotheses-move, return to Step4-6. If not, go to Step5.}

***Step 5*:** Carry out Hypothetical Reasoning for Parts Decisions from Step5-1 to Step5-6.

Step5-1*:** Calculate Parts Selection Priority Value *V(Pn)* for ***n kinds of parts *Pn* among competitive hypotheses C-hypotheses-parts and renew the elements in competitive hypotheses C-hypotheses-parts by changing the parts row with a large Parts Selection Priority Value *V(Pn)* .

***Step5-2*:** Select the part whose priority ranking is No.1 from among the elements of competitive hypotheses C-hypotheses-parts, corresponding to the left end element, establish it as a true hypothesis and establish the remaining elements as false hypotheses.

Step5-3*:** By using that true hypothesis, carry out Simulation in Hypotheses till the AGV is forced to make its Next Action choice and calculate FMS total operating efficiency ***E at the time when Simulation in Hypotheses stops.

***Step5-4*:** Perform the following rule.

If { $e(s) \leqq E$ }, **Then**{ Go to Step6}

Else { Admit that a contradiction occurs and go to Step5-5}

***Step5-5*:** Perform the following rule.

If { An element that has not been chosen as a true hypothesis among competitive hypotheses C-hypotheses-parts still exists }

Then { Replace a true hypothesis with a false hypothesis, select the next priority ranking element among competitive hypotheses C-hypotheses-parts as a true hypothesis and return to step5-3 }

Else {Go to Step5-6}

***Step5-6*:** Perform the following rule.

If { Hypothesis depth $D_{hy}=1$}, **Then** { Establish $s \leftarrow s+1$ and return to Step2}

Else{ Carry out backtracking after establishing $D_{hy} \leftarrow D_{hy} - 1$ and select the next priority ranking hypothesis element among competitive hypotheses of the layer D_{hy} as a true hypothesis. If the selected hypothesis belongs to C-hypotheses-move, return to *Step4-6*. If not, go to Step5.}

***Step6*:** Perform the following rule.

If { $D_{hy} < n$}, **Then** { Establish $s \leftarrow s+1$ and return to Step2}

Else {Go to Step7}

***Step7*:** Select a true hypothesis in hypotheses depth $D_{hy}=1$ as the next action of an actual FMS and execute the actual FMS.

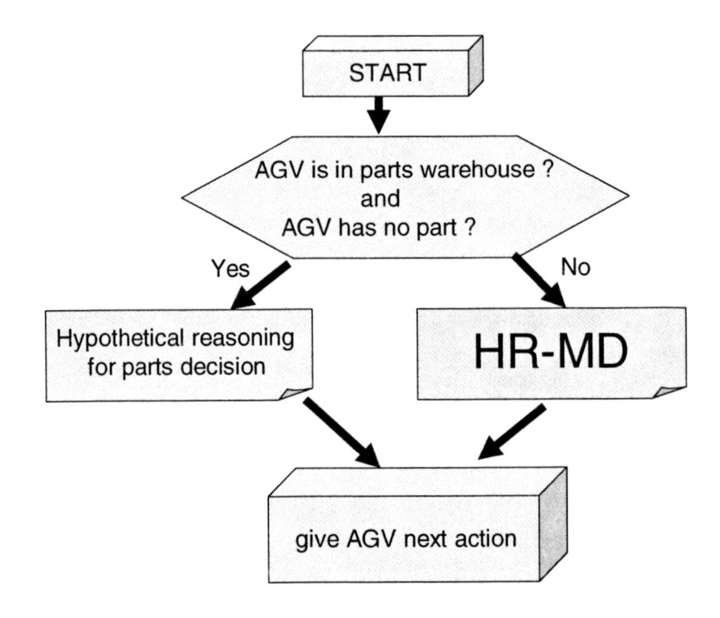

Fig. 13.4. Algorithm of RAF

Fig. 13.4 shows the chart of an RAF algorithm. RAF has two kinds of parallel reasoning : 1) Hypothetical Reasoning for Moving Decisions (HR-MD) that decides the moving destination of an AGV and 2) Hypothetical Reasoning for Parts Decisions that decides input parts. The reasoning backtracking can be made by a checking contradiction occurrence and a hypotheses depth.

HR-MD considers the practicable next moving destinations of an AGV as competitive hypotheses C-hypotheses-move, proceeds in its reasoning by dividing each hypothesis among the competitive hypotheses C-hypotheses-move into a true hypothesis and false hypotheses and finally decides the AGV moving destinations. Although a conventional hypothetical reasoning optionally divides hypotheses into a true hypothesis and false hypotheses, HR-MD makes ranking the possibility to be selected as a true hypothesis from among competitive hypotheses by using Group MC Selection Priority Value, Parts Warehouse Selection Priority Value, Products Warehouse Selection Priority and Job Variance Value and adopts the hypotheses generating method based on the priority ranking to make true and false hypotheses.

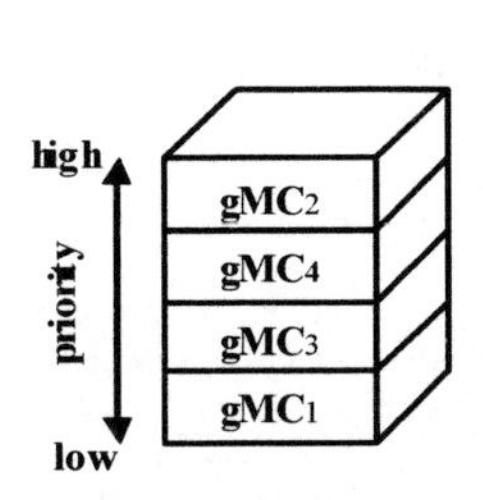

Fig. 13.5. Example 1

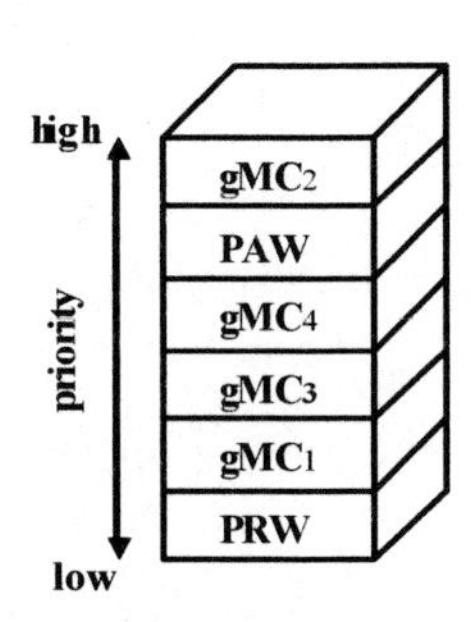

Fig. 13.6. Example 2

Let me describe the hypotheses generating method based on the priority ranking. The method has three processes : ① by using Group MC Selection Priority Value, the moving actions for each Group MC that has the same manufacturing processes are given a priority ranking and are listed according to the ranking : ② by using Parts Selection Priority Value and Products Warehouse Selection Priority, the moving actions for Parts Warehouse and Products Warehouse are given the priority ranking and are listed according to the ranking : ③ by using Job Variance Value, the selection ranking for each MC among a Group MC are ranked and listed according to the ranking. For example, let's consider the case that ${}_{g}MC_1$ ~ ${}_{g}MC_3$ has one MC for each, ${}_{g}MC_4$ has three MCs, $MC_{4\text{-}1}$ ~ $MC_{4\text{-}3}$. When ${}_{g}MC_2$, ${}_{g}MC_4$, ${}_{g}MC_3$ and ${}_{g}MC_1$ in turn are generated in process ①, as shown in Fig. 13.5 and each of Parts Warehouse Selection Priority Value and Products Warehouse Selection Priority value is calculated as 2 and 6 in process ②, the competitive hypotheses elements from the ranking 1 to the ranking 6 are decided as shown in Fig. 13.6. Because ${}_{g}MC_4$ has four MCs, the priority selection for the three is carried out in process ③. In the case where the priority ranking for the three MCs is decided as $MC_{4\text{-}2}$,$MC_{4\text{-}1}$ and $MC_{4\text{-}3}$, ${}_{g}MC_4$ is replaced with the ranking list. As a result, the final priority ranking for competitive hypotheses elements is decided as shown in Fig. 13.7.

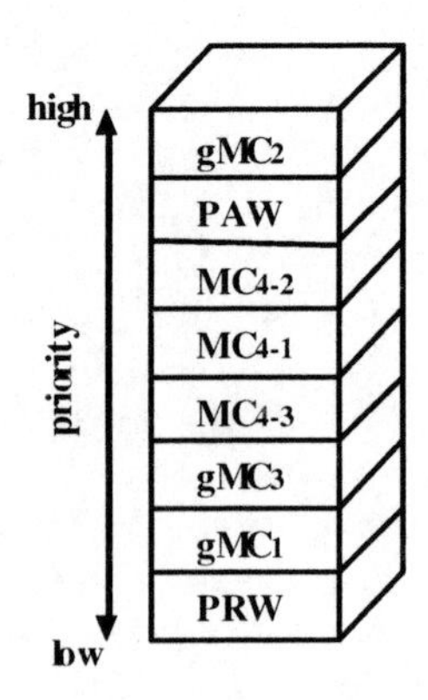

Fig. 13.7. Example 3

Hypothetical Reasoning for Parts Decisions is the reasoning that decides which part an AGV takes when the AGV arrives at the bay of the Parts Warehouse. Because of the reasoning, it is not necessary to have a prior parts input scheduling system.

13.3.3 Application Examples of RAF

The RAF described above was applied to the operations of a DA-FMS. As there is no actual DA-FMS production system, nine kinds of DA-FMSs are constructed in a computer and some numerical experiments are carried out. The nine DA-FMSs are the production systems whose number of parts subject to manufacturing, MCs, Group MCs and AGVs are different, as shown in Table 13.1. That is, <**Type 1**> parts number 3, Group MC number 3, MC number for each Group MC 1 and AGV number 3, <**Type 2**> parts number 3, Group MC number 3, MC number for each Group MC 1,2,1 and AGV number 3, <**Type 3**> parts number 3, Group MC number 3, MC number for each Group MC 2 and AGV number 3, <**Type 4**> parts number 6, Group MC number 6, MC number for each Group MC 1 and AGV number 5, <**Type 5**> parts number 6, Group MC number 6, MC number for each Group MC 2 and AGV number 5, <**Type 6**> parts number 6, Group MC number 6, MC number for each Group MC 3 and AGV number 5, <**Type 7**> parts number 9, Group MC number 8, MC number for each Group MC 1 and AGV number 5, <**Type 8**> parts number 9, Group MC number 8, MC number for each Group MC 2 and AGV number 5 and <**Type 9**> parts number 9, Group MC number 8, MC number for each Group MC 3 and AGV number 5. The factory layout of Type9 is shown in Fig. 13.1. Other Types' layouts are the ones that MCs disappear according to MCs number. The manufacturing time for each part is different. For example, in a case where Type 1, the manufacturing time of parts P_1,P_2 and P_3 are established as Table 13.2. The target production ratios for each of the parts are also different as follows. They are : P_1:P_2:P_3 =5:6:2 from Type 1 to Type 3 ; P_1:P_2:P_3:P_4:P_5:P_6=5:6:3:3:2:1from Type 4 to Type 6 ; P_1:P_2:P_3:P_4:P_5:P_6:P_7:P_8:P_9 = 5:6:3:3:2:1:4:5:2 from Type 7 to Type 9.

Table 13.1. DA-FMS variations

Styles	Parts kinds	gMC(MC)	AGV
1	3	3(1,1,1)	3
2	3	3(1,2,1)	3
3	3	3(2,2,2)	3
4	6	6(1,1,1,1,1,1)	5
5	6	6(2,2,2,2,2,2)	5
6	6	6(3,3,3,3,3,3)	5
7	9	8(1,1,1,1,1,1,1,1)	5
8	9	8(2,2,2,2,2,2,2,2)	5
9	9	8(3,3,3,3,3,3,3,3)	5

Table 13.2. Examples of machining time

Parts	P_1	P_2	P_3
Machining Time (seconds)	gMC_1 180	gMC_2 120	gMC_3 180
	gMC_2 120	gMC_1 60	gMC_2 150
	gMC_3 120		

Allowing for unpredicted troubles that happen where an actual DA-FMS operating, the numerical experiments adopt four operating conditions : <**Condition 1**> there are not any unpredicted troubles : <**Condition 2**> each AGV randomly breaksdown three times a day (24 hours) and its breakdown time is five minutes : <**Condition 3**> parts manufacturing time at each MC is randomly extended 10 % : <**Condition 4**> both unpredicted troubles of Condition 2 and Condition 3 happen. In Condition 2, Condition 3 and Condition 4, ten kinds of happening time for manufacturing time extensions and breakdowns are established as unpredicted troubles' random conditions by adopting ten random series. As a result, Type 1 executed one numerical experiment under Condition 1 and executed ten numerical experiments under each other Condition. A numerical experiment time 24 hours is adopted and n of forecastable actions range (Forecast[n]) in hypothetical reasoning is established as 3.

One result of the numerical experiments is shown in Table 13.3. Table 13.3 indicates the production outputs for each of the four Conditions in Type 5. As a comparison, the numerical experiment of the case that n of forecastable actions range (Forecast[n]) is 1 was carried out. Judging from the result, the outputs of the case n=3 were bigger than that of the case n=1 under every Condition. In other Types, the same results were obtained. Fig. 13.8 shows the output ratio for each of the parts of Type 5. All four Conditions could get the ratio very close to the target production ratio, $P_1:P_2:P_3:P_4:P_5:P_6=5:6:3:3:2:1$ even though a conventional prior parts input scheduling system was not used. In other Types, the same results were also obtained. For example, the production ratio of Condition 4 in Type 9 is 5.049: 6.028:3.049:3.049: 2.042: 1.000:4.049:5.092:2.049 and its target ratio is $P_1:P_2:P_3:P_4:P_5:P_6:P_7:P_8:P_9$ = 5:6:3:3:2:1:4:5:2.

Table 13.3. Simulation results

A \ B	Condition1	Condition2	Condition3	Condition4
1		573 (550)	562 (551)	566 (538)
2		572 (550)	556 (552)	543 (545)
3		576 (558)	556 (550)	564 (537)
4		574 (560)	554 (546)	560 (552)
5		580 (557)	560 (547)	558 (543)
6		575 (555)	568 (545)	568 (544)
7		576 (550)	561 (555)	559 (546)
8		574 (558)	565 (550)	556 (556)
9		565 (553)	566 (545)	559 (546)
10		573 (558)	563 (555)	553 (541)
Average	578 (559)	573.8 (554.9)	561.1 (549.6)	558.6 (544.8)

A: **Random Numbers** B: **Conditions**

In consequence, it was ascertained that the DA-FMS using RAF can keep a target production ratio even if unpredicted troubles happen.

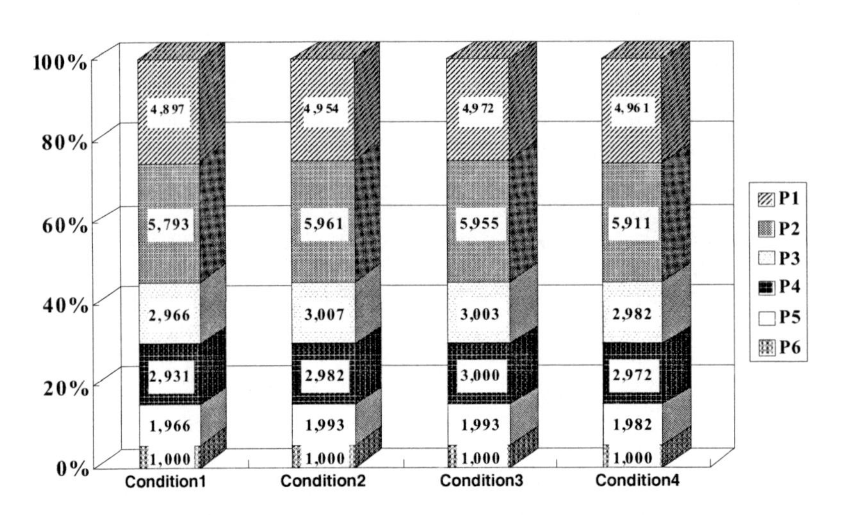

Fig. 13.8. Production ratio of type 5 DA-FMS

13.4 Background of Memory Necessity in DA-FMS

In the DA-FMS in this chapter, production is carried out with each AGV determining its own actions through RAF. RAF can anticipate up to several future steps available to the AGV. Then, through predictions based on discrete production simulations of phenomena that may occur within these several future steps, RAF works backward to the present to decide which of the options the AGV should choose at the present point in time. This process is controlled by the application of hypothetical reasoning to the several competing options (hypotheses) available to the AGV.

Fig. 13.4 shows the flow of RAF. RAF includes two parallel reasoning processes: HR-MD and hypothetical reasoning for parts decisions. Occurrence of a contradiction causes the reasoning to backtrack. The HR-MD that was applied in this chapter is described in the following.

To decide the next action of an AGV, HR-MD considers competing hypotheses for each destination MC, parts warehouse, and products warehouse, separates the competing hypotheses into true and false hypotheses, and finally decides the destination of that AGV. Specifically, it (1) sets a priority ranking for destinations of MC groups that perform the same machining process, (2) sets a priority ranking for parts warehouse and product warehouse destinations, and (3) sets a priority ranking for each MC within a group. It then sets a priority ranking for destinations available to the AGV in each of the competing hypothesis, in the order of likelihood of being adopted as a true hypothesis. This ranking is called a moving priority ranking (MP-ranking). Then, from among the competing hypotheses ranked by priority, one is selected as a true hypothesis from among the hypotheses with high priority rankings, and the remaining are considered false hypotheses. The hypothetical reasoning [7-9] then begins.

13.5 Memory Contents

In RAF, the reasoning and simulations are repeated continuously as true and false hypotheses are exchanged until contradictions are no longer found. Because of this, a problem occurs in that the number of exchanges of true and false hypotheses and number of production simulations that must be re-executed because of these exchanges, increase with greater numbers of MCs or AGVs. In the present study, therefore, good reasoning efficiency was defined as fewer exchanges or production simulations. To solve this problem of excessive exchanges, a method to rank competing hypotheses by oblivion and memory (ranking by oblivion and memory; ROM) is presented to improve the reasoning efficiency of RAF. The ROM system is based on the idea that when a production situation occurs that is the same as one in the past, the same destination as in the past is more likely to be selected; that is, it has a high probability of being selected as the true hypothesis.

An important point in ROM is which memories are retained and how these data are used. In the following, therefore, we will first consider the contents of memory.

Humans often remember the past in the form of situation-phenomena; for example, “when… (situation), we were… (phenomena).” The memory contents used in ROM are similarly expressed in this form combining situation and phenomena. Specifically, a past production situation and the AGV destination selected at that time. Thus, the memory expresses 4 items: the 3 production situations of (a) moving priority ranking, (b) the location of the AGV, and (c) the location of the parts, together with (d) the AGV destination decided at that time (determined action). Next, the contents of these 4 items, as well as 3 terms using these 4 items, are defined as follows.

[Definition] (a) Moving priority ranking: the priority ranking determined for each destination available to the AGVs as reasoned by HR-MD.

[Definition] (b) AGV positions: the current position of each AGV or the destination it is moving toward. This corresponds to the parts warehouse, Product warehouse and each MC.

[Definition] (c) Parts positions: the place where each part existing in the AD-FMS is located. This corresponds to the parts warehouse, product warehouse, each MC and each AGV.

[Definition] (d) Decided action: the final destination selected for the AGV under the production situations of (a)~(c).

[Definition] Single memory: the memory set of (a), (b), (c) and (d).

[Definition] Whole memory: the set of all single memories, M_{number}.

[Definition] Oblivion degree: the values attached to all elements of whole memory, from 1 to M_{number}. The values are decided using a carry up method. For example, if a new memory corresponding to a single memory is assigned oblivion degree M_{number}, the single memory whose oblivion degree had been M_{number} is given a new oblivion degree of $M_{number} - 1$. This carry up method is carried out with other single memories in whole memory. As a result, the single memory whose oblivion degree had been 1 up to that point will be deleted from whole memory. The smaller the oblivion degree is, the sooner the memory is forgotten.

According to the above definitions, ROM can have many memories. Like humans, ROM has memories that are both easier and more difficult to forget.

The relation between whole memory and single memory is shown in Fig. 13.9. Examples of ROM memory contents (a) ~ (c) are shown in Table 13.4 to Table 13.6 and the signs used in these tables are described below.

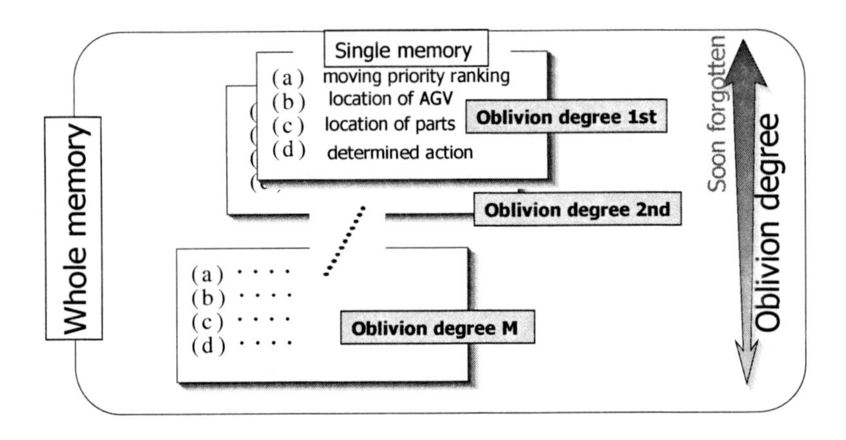

Fig. 13.9. Whole memory and single memory

$G(x)$: the currently available destinations for each AGV. The values of x are 1, 2,..., m'. $G(x)$ is also considered to be the following:

$$G(x) \in \{Parts\_warehouse, Products\_warehouse, MC_{1-1}, MC_{1-2}; \cdots\}$$

$P'(x)$: MP-ranking f $G(x)$. The values of $P'(x)$ are 1, 2, 3, 4,...

$_{AGV}L(m)$: the current location or destination of AGV_m. The values of m are 1, 2,..., m'. $_{AGV}L(m) \in \{Parts\_warehouse, Products\_warehouse, MC_{1-1}, MC_{1-2}; \cdots\}$

$_{Parts}L(s)$: the current place of part P_s. The values of s are 1, 2,..., s'.

$$_{AGV}L(s) \in \{Parts\_warehouse, Products\_warehouse, MC_{1-1}, MC_{1-2}, \cdots, AGV_1, AGV_2, \cdots\}$$

Table 13.4. Memory (a) example

x	1	2	3	4	5	6	7
$G(x)$	MC_{1-1}	MC_{1-2}	MC_{1-3}	MC_{2-1}	MC_{2-2}	Parts Warehouse	Products Warehouse
$P'(x)$	4	1	3	2	5	7	6

Table 13.5. Memory (b) example

m	1	2	3	4
AGV_s	AGV_1	AGV_2	AGV_3	AGV_4
$_{AGV}L(m)$	Parts Warehouse	MC_{1-2}	MC_{2-2}	MC_{2-1}

Table 13.6. Memory (c) example

s	1	2	3	4	5
Parts	P_1	P_2	P_3	P_4	P_5
$_{Parts}L(s)$	MC_{1-1}	MC_{1-2}	AGV_1	MC_{2-1}	MC_{2-2}

Table 13.4 shows the contents of memory (a). The table indicates that the MP-rankings, $P'(x)$, for the destinations of the seven competing hypotheses, MC_{1-1}, MC_{1-2}, MC_{1-3}, MC_{2-1}, MC_{2-2}, parts warehouse and products warehouse are 4, 1, 3, 2, 5, 7, 6, respectively. Table 13.5 shows the contents of memory (b). It indicates that $_{AGV}L(m)$, the location of each AGV, is parts warehouse, MC_{1-2}, MC_{2-2} and MC_{2-1} respectively. Table 13.6 shows the contents of memory (c). It indicates that $_{Parts}L(s)$, the location of each part, is MC_{1-2}, AGV_1, MC_{2-1} and MC_{2-2} respectively.

13.6 ROM Function

First, this section describes how the memories in ROM are used. ROM is included in the process of HR-MD and works to improve MP-ranking. In the HR-MD decision which is one of the hypothetical reasoning tasks in RAF, MP-ranking for each AGV destination in the competing hypotheses is decided through three processes, <1> giving priority rankings to group MCs, <2> giving priority rankings to parts warehouse and products warehouse and <3> giving priority rankings to the selection orders for each MC. ROM improves the MP-ranking determined through three processes by rewriting the MP-ranking so that it is likely to be a realizable true hypothesis. New procedures to do this, which are carried out after the above three procedures, are included in ROM. These three new procedures are <4> refresh MP-ranking using memory, <5> refresh the oblivion degree, and <6> add a new memory and delete an existing memory. They are incorporated in HR-MD as shown in Fig. 13.10.

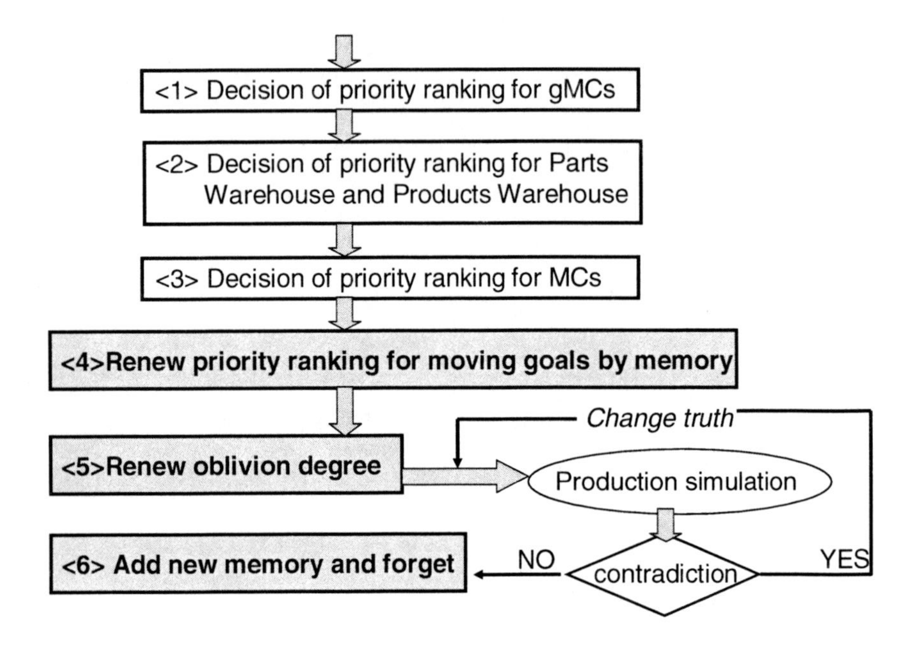

Fig. 13.10. ROM and HR-MD

Therefore, the MP-ranking decided in <1>~<3> is called a temporary MP-ranking and the ranking decided by the new three procedures, <4>~<6> of ROM, is called the final MP-ranking.

Second, this section first defines how the MP-ranking is refreshed using memory in <4>, which is one of the new three procedures of ROM.

[Definition] Refresh MP-ranking using memory: carry out the following three procedures.

[Procedure 4-1] Find the degree of similarity between the current production situation and the content of each M_{number} single memory in whole memory.

[Procedure 4-2] Find the single memory whose priority ranking is the highest and insert the determined action into the top position in the temporary MP-ranking.

[Procedure 4-3] Move the priority rankings in the second and lower positions down in the order. □

In order to carry out Procedure 4-2, it is necessary to check out the similarity between the current production situation and the content of each M_{number} single memory in the whole memory. The similarity is obtained using the following three similarity degrees.

<similarity degree ${}_{move}A$>: the similarity in the MP-ranking
<similarity degree ${}_{agv}A$>: the similarity in the current positions of each AGV
<similarity degree ${}_{parts}A$>: the similarity in the current positions of each part

The similarity degree ${}_{move}A$ is the value expressing the difference between MP-ranking (a), which is in the memory, and the current temporary MP-ranking which is acquired during HR-MD. The similarity degree ${}_{agv}A$ is the value expressing the difference between AGV positions (b) and the current positions of each AGV in AD-FMS. The similarity degree ${}_{parts}A$ is the value expressing the difference between parts positions (c) and the current positions of each part in AD-FMS. The numerical values of the similarity degrees are calculated by adding 1 if the current production situation is the same as a situation in memory, and adding 0 if the current production situation is not the same as any in memory. Specifically, the similarity is expressed numerically with the following algorithm for similarity degree, from which the degree of similarity between each single memory in whole memory and the current production situation is obtained. The symbols used in the algorithm are described in the following.

- $G_{memory}(n)$: the destination of each AGV inmemory (a) in a single memory when n=1, 2,・・・, n'. $G_{memory}(n)$ is expressed as follows.

$$G_{memory}(n) \in \{Partswarehouse, \Pr oductswarehouse, MC_{1-1}, MC_{1-2}, \cdot\cdot\cdot\}$$

- $P(n)$: the final MP-ranking of $G_{memory}(n)$. The values of $P(x)$ are 1, 2, 3, 4,...... .
- ${}_{AGV}L_{memory}(v)$: the location of each AGV in memory (b) in a single memory when v=1, 2,・・・, v'. ${}_{AGV}L_{memory}(v)$is expressed as follows.

$${}_{AGV}L_{memory}(v) \in \{Parts\_warehouse, \Pr oducts\_warehouse, MC_{1-1}, MC_{1-2}, \cdot\cdot\cdot\}$$

- ${}_{Parts}L_{memory}(w)$: the current location of each part in memory (c) in a single memory when w=1, 2,・・・, w'.
- ${}_{Parts}L_{memory}(w)$ is expressed as follows.

$${}_{Parts}L_{memory}(w) \in \{Parts\_Warehouse, \Pr oducts\_warehouse, MC_{1-1,}MC_{1-2}, \cdot\cdot\cdot, AGV_1, AGV_2, \cdot\cdot\cdot\}$$

[Algorithm for Similarity Degree]

Step 1: The initial values are set as follows: x=1, ${}_{move}A$=0, ${}_{parts}A$=0.

Step 2: Find $G(x)$ and $G_{memory}(n)$ corresponding to $P'(x)= P(n)$ and then carry out the following rule.

if $(G(x)= G_{memory}(n))$, ***then*** (establish $_{move}A= {}_{move}A +1$ and go to Step 3)
else (go to Step 3)

Step 3: Establish $x=x+1$ and carry out the following rule.
if $(x>x')$, ***then*** (go to Step 4)
else (return to Step 2)

Step 4: Establish $m=1$ and $v=1$.

Step 5: Carry out the following rule.
If $({}_{AGV}L(m)= {}_{AGV}L_{memory}(v))$, ***then*** (establish $_{agv}A= {}_{agv}A +1$and go to Step 6)
else (go to Step 6)

Step 6: Establish $m=m+1$ and $v=v+1$ and carry out the following rule.
if $(m>m')$, ***then*** (go to Step 7)
else (return to Step 5)

Step 7: Establish $s=1$ and $w=1$.

Step 8: Carry out the following rule.
if $({}_{Parts}L(s)= {}_{Parts}L_{memory}(w))$, ***then*** (establish $_{parts}A= {}_{parts}A +1$ and go to Step 9)
else (go to Step 9)

Step 9: Establish $s=s+1$ and $w=w+1$ and carry out the following rule.
if $(s>s')$, ***then*** (go to Step 10)
else (return to Step 8)

Step 10: Consider the value of $_{move}A+{}_{agv}A+{}_{parts}A$ as the similarity degree of single memories and finish the algorithm. □

In Step 2 and Step 3 of the above algorithm, similarity degree $_{move}A$ is acquired, in Step 4~Step 6, similarity degree $_{agv}A$ is acquired and, in Step 7~Step 9, similarity degree $_{parts}A$ is acquired. The processes of Step 2 and Step 3 may be explained using

Table 13.7. (a) Example in memory

n	1	2	3	4	5	6	7
$G_{memory}(n)$	$MC_{1\text{-}1}$	$MC_{1\text{-}2}$	$MC_{1\text{-}3}$	$MC_{2\text{-}1}$	$MC_{2\text{-}2}$	Parts Warehouse	Products Warehouse
$P(n)$	**4**	3	6	**2**	**5**	1	7

Table 13.8. (b) Example in memory

v	1	2	3	4
AGV_s	AGV_1	AGV_2	AGV_3	AGV_4
$_{AGV}L_{memory}(v)$	$MC_{1\text{-}2}$	$MC_{1\text{-}1}$	$MC_{2\text{-}2}$	$MC_{2\text{-}1}$

Table 13.9. (c) Example in memory

w	1	2	3	4	5
Parts	P_1	P_2	P_3	P_4	P_5
$_{Parts}L_{memory}$ *(w)*	$MC_{1\text{-}1}$	$MC_{2\text{-}1}$	$MC_{2\text{-}2}$	$MC_{1\text{-}3}$	AGV_2

Table 13.4 and Table 13.7. Table 13.4 shows the current temporary MP-ranking and Table 13.7 shows the final MP-ranking in memory. In Table 13.4, when x=1, $G(1)$ becomes $MC_{1\text{-}1}$ and the temporary MP-ranking of $MC_{1\text{-}1}$ is $P'(1)$=4. Finding $G_{memory}(n)$ in Table 13.4 that corresponds to $P'(x)=P(n)$, $G_{memory}(1)=MC_{1\text{-}1}$ is also found. That means $_{move}A$ becomes 1 when its initial value was 0 because $G(x)=G_{memory}(n)$. Also, when x=2, $G(2)$ is equal to $MC_{1\text{-}2}$ and as a result, $P'(2)$ becomes 1. Finding $G_{memory}(n)$ that corresponds to $P'(x)=P(n)$, $G_{memory}(6)$=*Parts Warehouse* is also found. That means the value of $_{move}A$ is not added because $G(x)$ is not $G_{memory}(n)$. In the same manner, the priority values 4, 2 and 5 in Table 13.8 are matched with the if condition of Step 2 and, as a result, $_{move}A$ is calculated to equal 3.

An example of the Step 4 process is shown using Table 13.5 and Table 13.8. Consider Table 13.5 to show the AGVs current destination and Table 13.8 the AGVs' destination in memory. Since m corresponding to $_{parts}L(s)={}_{parts}L_{memory}(v)$ is 3 or 4, $_{agv}A$ is calculated to equal 2.

An example of the Step 5 process is shown using Table 13.6 and Table 13.9. Consider Table 13.6 to show the current parts locations and Table 13.9 to show the parts locations in memory. Since s corresponding to $_{parts}L(s)={}_{parts}L_{memory}(w)$ is just 1, $_{parts}A$ is calculated to equal 1.

From Step 6, the similarity degree between the current production situation and memory corresponding to Table 13.4 ~ Table 13.9 is calculated as $_{move}A+{}_{agv}A+{}_{parts}A$ and its value is 3+2+1=6.

By using the algorithm for similarity degree in each single memory of M_{number} units, the similarity degree between the current production situation and the single memory for M_{number} units can be calculated. Next, [Procedure 4-2] is carried out. It selects the single memory with the highest similarity degree and inserts the determined action (d) of the single memory into the top position in the temporary MP-ranking. Third, [Procedure 4-3] moves the existing priority rankings down one place in order and, as a result, a new MP-ranking or the final MP-ranking can be decided.

Next, the followings definitions correspond to <5>, refreshing the oblivion degree.

[Definition] <5> Refresh the oblivion degree: Insert the single memory with a highest similarity degree into the lowest place, M_{number}, of the oblivion degree in the whole memory. □

Due to refreshing a single memory once selected is not easily removed from whole memory.

Finally, the function of <6>, which adds a new memory and deletes an existing memory, is defined as below.

[Definition] <6> Add a new memory and deletes an existing memory: The following two procedures are carried out.

[Procedure 6-1] The AGVs next action that is finally decided without contradictions is considered to be a new (d) determined action. The (a) MP-ranking, (b) AGV positions and (d) parts locations at this time including the new (d) determined action are memorized in the middle range of the oblivion degree among the whole memory as a new single memory.

[Procedure 6-2] Single memories whose oblivion degrees are higher than the new single memory are in turn moved up and, as a result, the single memory that has had the highest oblivion degree is deleted from the whole memory. □

Through this procedure <6>, whenever a real AGV's destination is decided, a new memory is added and at the same time a memory with a low frequency of selection as an AGV destination is deleted.

13.7 Application Examples of ROM

The ROM proposed in the last section was applied to 3 types of DA-FMS created on computer in simulation experiments. The 3 DA-FMSs differed by number of MC groups, number of MCs, and number of AGVs. Style 1 had 9 types of parts, 8 MC groups, 1 of each MC, and 5 AGVs. Style 2 had 9 types of parts, 8 MC groups, 2 of each MC, and 5 AGVs. Style 3 had 9 types of parts, 8 MC groups, 3 of each MC, and 5 AGVs. The plant layouts for Styles 1, 2, and 3 are shown in Figs. 13.11, 13.12, and 13.13, respectively.

On the assumption that unanticipated troubles occur during actual FMS operation, trouble conditions were set in which each AGV randomly broke down 3 times a day, each time for 5 minutes, and part processing time on each MC was randomly extended 10%. Ten different random number sequences were used as the random conditions for these troubles. The number of simple memories stored in the total memory of ROM was set at 20.

Operating results for Style 3 are summarized in Table 13.10. The table shows the number of hypotheses until the true hypothesis was selected (hypothesis selection number) during operation when the AD-FMS was operated for 24 hours with 10 types of trouble. For comparison, the table also shows the hypothesis selection number when using HR-MD with a temporary MP-ranking (conventional method). From the table, we see that the hypothesis selection number with ROM was reduced to less than half that with the conventional method for all of the 10 types of trouble. For example, the mean hypothesis selection number for Style 3 operation was 899,943.5 with the conventional method and 445,763.2 with ROM. Similarly, the hypothesis selection number was reduced by about half with the other 2 styles, Style 1 and 2 as well. This indicates that the MP-ranking of ROM is better than the temporary MP-ranking at ordering hypothesis in order of possibility of being selected as true, confirming that there is a reduction in the numbers of wasted true/false hypothesis replacements and production simulations that are executed.

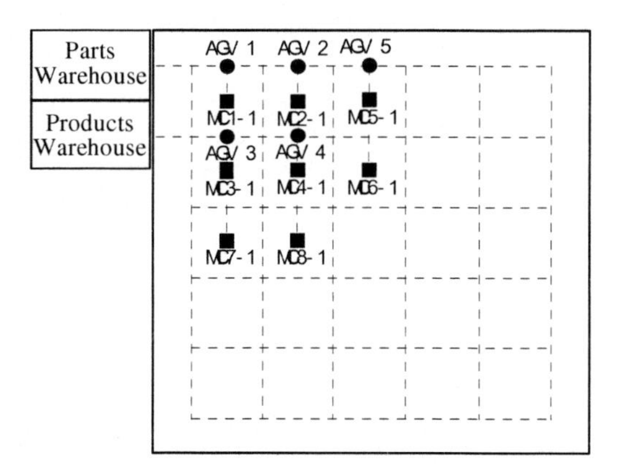

Fig. 13.11. Style 1 of AD-FMS

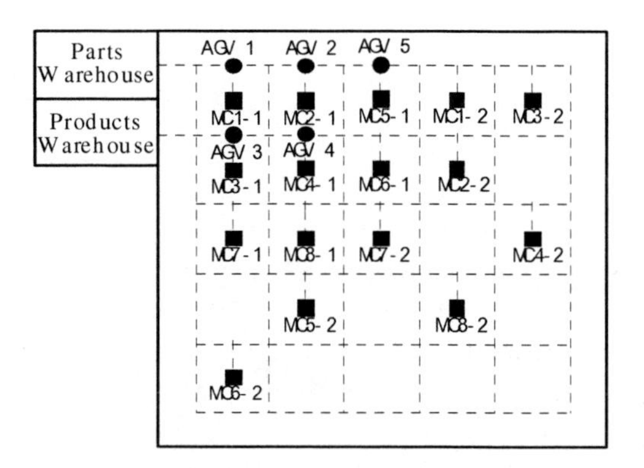

Fig. 13.12. Style 2 of AD-FMS

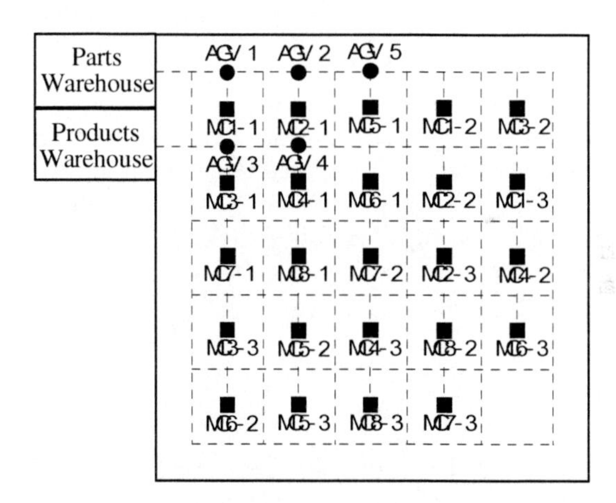

Fig. 13.13. Style 3 of AD-FMS

Table 13.10. Results of Style 3

Hypotheses Selection Number	Conventional method	**ROM**
Condition 1	893,532	452,539
Condition 2	905,398	445,126
Condition 3	921,744	441,059
Condition 4	902,827	446,284
Condition 5	886,551	446,901
Condition 6	898,857	446,815
Condition 7	899,596	444,296
Condition 8	900,342	445,106
Condition 9	891,872	444,910
Condition 10	898,716	444,599
Average	899,943.5	445,763.5

This simulation experiment compared not only hypothesis selection number but also production output. With Style 1, the mean production output was 454.5 units with the conventional method and 525.7 units with ROM. The respective figures were 460 and 698.5 with Style 2, and 476 and 754.4 with Style 3. In each case ROM gives a higher production output and improved reasoning efficiency without decreasing production efficiency.

13.8 Conclusions

This chapter described advanced technology towards a decentralized autonomous FMS. That is the intelligent method to decide an AGV action plan to operate a DA-FMS and developed ROM to anticipate the FMS operating situations happening in the near future by using information from each agent and forecasting several steps ahead of AGV's practicable next actions. ROM consists of a hypothetical reasoning that regards practicable AGV next actions as competitive hypotheses and a discrete simulator that simulates the future alternative possibilities. Because of the decision, AGV next moving destinations and which part is transferred next are decided and both a prior AGV moving plan and a prior parts input schedule are unnecessary. The numerical experiments were executed by applying the developed ROM for a DA-FMS that exists on a computer. As a result, it was ascertained that the DA-FMS can have the product ratio very close to the target production ratio even when a prior parts input schedule is not used. Compared with the result of the case that looked just one step ahead as a forecastable action range, it was also ascertained that the developed reasoning method to forecast several steps ahead could get the better outputs.

As another advanced technology towards DA-FMS, the chapter described a system called ROM that determines a hypothesis priority ranking using memory for improving the reasoning efficiency of RAF. ROM stores previous production situations and the action of the AGV decided at those times as memories. Using these memories, hypotheses are ranked according to possibility of being selected from among competing hypotheses as true, with the aim of improving the efficiency of the RAF. This ROM repeatedly adds new memories and deletes existing ones, so that the memory contents are constantly refreshed.

ROM was applied to DA-FMSs constructed on a computer and the number of hypothesis replacements until a true hypothesis was determined was compared with that from a conventional system. It was found that under all conditions ROM reduced the number of hypothesis replacements to half that with a conventional system, demonstrating the validity of this system.

The described two kinds of technology started as a basic research to decide the AGV actions plan of a DA-FMS with the idea to look several steps ahead. Although there are still some problems left, such as how many steps to forecast is optimal and how closer to the target production ratio is achieved, it was ascertained that the ideas corresponding to RAF and ROM is an efficient methods.

References

1. Arkin, R.C., Murphy, R.R.: Autonomous Navigation in a Manufacturing Environment. IEEE Transactions on Robotics and Automation 6(4), 445–454 (1990)
2. Laengle, T., et al.: A Distributed Control Architecture for Autonomous Robot Systems. In: Workshop on Environment Modeling and Motion Planning for Autonomous Robots. Series in Machine Perception and Artificial Intelligence, vol. 21, pp. 384–395. World Scientific, Singapore (1994)
3. Kouiss, K., Pierreval, H., Mebarki, N.: Using multi-agent architecture in FMS for dynamic scheduling. Journal of Intelligent Manufacturing 8(1), 41–47 (1997)
4. Moriwaki, T., Hino, R.: Decentralized Job Shop Scheduling by Recursive Propagation Method. International Journal of JSME, Series C 45(2), 551–557 (2002)
5. Yamamoto, H., Rizauddin, R.: Real-time Decision Making of Agents to Realize Decentralized Autonomous FMS by Anticipation. International Journal of Computer Science and Network Security 6(12), 7–17 (2006)

6. Dole, J.: A Truth Maintenance System. Artificial Intelligence (1979)
7. Poole, D.: A methodology for using a default and abductive reasoning system. International Journal of Intelligent Systems 5(5), 521–556 (1990)
8. Provetti, A.: Hypothetical reasoning about actions: from situation calculus to event calculus. Computational Intelligence 12(3), 478–498 (1996)
9. Baldoni, M., Giordano, L., Martelli, A.: A modal extension of logic programming: Modularity, beliefs and hypothetical reasoning. Journal of Logic and Computation 8(5), 597–635 (1998)

Author Index

Editors

Professor Lakhmi C. Jain is a Director/Founder of the Knowledge-Based Intelligent Engineering Systems (KES) Centre, located in the University of South Australia. He is a fellow of the Institution of Engineers Australia.

His interests focus on the artificial intelligence paradigms and their applications in complex systems, art-science fusion, e-education, e-healthcare, unmanned air vehicles and intelligent agents.

Ngoc Thanh Nguyen currently works as a professor of Computer Science in Wroclaw University of Technology, Poland. His scientific interests consist of knowledge integration methods, intelligent technologies for conflict resolution, multi-agent systems and E-learning methods. He serves as Editor-in-Chief of *International Journal of Intelligent Information and Database Systems*, Editor-in-Chief of two book series for IGI Global (*Advances in Applied Intelligence Technologies* and *Computational Intelligence and its Applications*), Associate Editor of *International Journal of Computer Science & Applications*; *Journal of Information Knowledge System Management* and *KES Journal* and a Member of Editorial Boards of several other international journals. He serves also as an expert for Ministry of Science and Higher Education and Ministry Regional Development of Poland in evaluating projects. He is an Associate Chair of KES International and the Chair of KES symposium series on Agent and Multi-agent Systems. He is a Senior Member of IEEE and ACM.